Phase Change Materials for Energy Management and Efficiency

This book explores the prospective applications of Phase Change Materials (PCMs) in energy storage systems, thermal system temperature control, peak shifting, and energy management. It starts with definitions and a brief history of energy storage systems, followed by an exploration of different types of PCMs, encapsulation techniques, heat transfer enhancement methods, and the applications and challenges of PCMs. The book provides a comprehensive overview of PCM applications in building envelopes, free cooling, electrical appliances, lithium-ion batteries, textiles, solar panels, vehicles and logistics, and more across 14 chapters.

- Demonstrates various techniques for the enhancement of energy storage systems.
- Offers an applied approach.
- Discusses energy management, thermal control, peak shifting, energy conservation, and energy storage.

Aimed at researchers in industrial manufacturing sectors, it can also serve as a resource for undergraduate and graduate students in energy management courses. In addition, this reference is valuable for engineers and scientists focused on energy optimization and the application of sustainable materials.

Seyed Mojtaba Sadrameli is currently a Professor of Process Engineering at German University of Technology in Muscat, Oman. He earned his bachelor's degree in Chemical Engineering from Sharif University of Technology (SUT) in Tehran, Iran, in 1980 and his master's and doctoral degrees in Chemical Engineering from the School of Chemical and Process Engineering at the University of Leeds, United Kingdom, in 1984 and 1988, respectively. He joined the Faculty of Chemical Engineering at Tarbiat Modares University (TMU) in Tehran, Iran, in 1988 and retired in 2020. His current research interests focus on the applications of phase change materials for thermal energy storage systems and the production of biofuels through the thermal and catalytic pyrolysis of biomass and plastic wastes.

Phase Change Materials for Energy Management and Efficiency

Seyed Mojtaba Sadrameli

CRC Press
Taylor & Francis Group
Boca Raton London New York

CRC Press is an imprint of the
Taylor & Francis Group, an **informa** business

Designed cover image: © 2025 Shutterstock

First edition published 2025
by CRC Press
2385 NW Executive Center Drive, Suite 320, Boca Raton FL 33431

and by CRC Press
4 Park Square, Milton Park, Abingdon, Oxon, OX14 4RN

CRC Press is an imprint of Taylor & Francis Group, LLC

ISBN: 978-1-032-85459-5 (hbk)
ISBN: 978-1-032-85460-1 (pbk)
ISBN: 978-1-003-51828-0 (ebk)

DOI: 10.1201/9781003518280

Typeset in Times
by KnowledgeWorks Global Ltd.

To my loving wife,

For her boundless patience, inspiration, and encouragement throughout this journey.

Contents

Preface

As the world faces unprecedented challenges related to climate change and the increasing demand for energy, the need for sustainable and efficient energy sources such as renewable energies and management solutions has become more pressing than ever. As the demand for energy continues to rise, so does the need for innovative and efficient methods to store, manage, and utilize it. Latent heat storage which is based on heat absorption and release during a phase transition of a storage material from solid to liquid or liquid to gas (or vice versa) is one of the most efficient ways to accumulate thermal energy. Phase change materials (PCMs) capable of storing thermal energy in the form of latent heat phase transition have emerged as one of the most promising technologies for thermal energy storage (TES), offering solutions for stabilizing temperature fluctuations, peak shifting, enhancing energy efficiency, and optimizing energy consumption across various industries. PCMs are proving their versatility and efficiency enhancement from building construction materials and HVAC systems to electronic cooling and renewable energy applications such as solar panels. They should have a suitable phase transition temperature within the practical range of applications, high thermal energy density, high thermal conductivity, complete melting/solidification cycles with minimal supercooling, and chemical stability. Additionally, they should be non-corrosive, non-toxic, non-flammable, and cost-effective. A key feature of application of PCMs in TES systems is the role of PCMs in peak shifting the strategic control of energy demand to balance supply and reduce pressure on the grid during peak times especially in the hot seasons and hot climate regions. By using excess electrical energy during off-peak periods, storing it as a cooling source into the PCMs, and releasing it when demand is high especially in the summer, PCMs contribute significantly to demand-side management and grid stability. This approach not only enhances energy security but also plays a crucial role in minimizing operational and maintenance costs and environmental impacts associated with energy production.

PCMs are commonly categorized into materials such as paraffin waxes, fatty acids, polymers as organic compounds, and hydrated salts and metallic-based PCMs as inorganic materials. Among these, polymer-based PCMs demonstrate high energy density, durability, and stability, especially for many cycling tests. Throughout the chapters of this book, readers will discover advanced methodologies and experimental studies illustrating the diverse applications of PCMs in building construction materials and building envelope for energy management, electrical appliances such as household refrigerators and freezers for stabilizing temperature fluctuations, lithium-ion batteries for surface temperature control during charging and discharging, textiles for comfort temperature achievement, solar panel efficiency enhancement, cabin temperature control of vehicles, foods transportation and logistics, free heating/cooling using day and night temperature difference and solar energy, medical applications such as pain relief and cancer therapy, glasshouses for fuel savings and temperature control and other applications for energy management and high-temperature applications for solar power generation. Additionally, the book explores the latest developments in encapsulation techniques and material innovation to increase the thermal efficiency, durability, and adaptability of PCMs in various conditions. As the energy landscape evolves, integrating PCMs into energy systems offers the potential to significantly reduce energy losses and enhance efficiency.

This book serves as a comprehensive guide for researchers in both academia and industry, engineers, policymakers, and industry professionals looking to understand and leverage the capabilities of PCMs for sustainable energy solutions. By bridging the gap between theory and application, the book aspires to provide readers with the knowledge needed to develop and implement PCM-based solutions for energy storage, peak control, and overall sustainability. It is also well-suited as a textbook for both undergraduate and graduate courses in energy management and conservation.

I am deeply grateful to the contributors and experts, particularly my graduate students at the master's and PhD levels from the year 2000 to the present, whose valuable insights and research findings have made this book a valuable resource for advancing the field of energy storage and management. Grateful acknowledgement is extended to the journal publishers for granting permission to use the research publications cited in this book.

I hope that the knowledge and innovations presented in these pages will inspire fresh ideas and strategies for achieving sustainable and efficient energy system utilization.

Editor
Seyed Mojtaba Sadrameli

Abbreviations

CC	Cooling clothing
CFA	Cold fly ash
CFD	Computational fluid dynamics
CNT	Carbon nanotube
COP	Coefficient of performance
CTES	Cooling thermal energy storage
DA	Decyl alcohol
DI	Deionized
DO	Discrete ordinance
DOE	Design of experiment
DSC	Differential scanning calorimetry
DSPCM	Double shell PCM
EDX	Energy dispersion X-ray
EG	Expanded graphite
ES	Energy storage
FAO	Food and Agriculture Organization
FBR	Fixed bed regenerator
FTIR	Fourier Transform Infrared spectroscopy
GHG	Greenhouse gas
HVAC	Heating ventilation and conditioning
LFP	Lithium iron phosphate
LIB	Lithium-ion battery
LS	Lecce stone
MIT	Massachusetts Institute of Technology
M-pack	Measurement packing
MPCM	Micro phase change material
MWCNT	Multi-walled carbon nanotube
NASA	National Aeronautics and Space Administration
NPCM	Nano phase change material
PCM	Phase change material
PDD	Predicted percentage of dissatisfied
PDMS	Polydimethylsiloxane
PEG	Polyethylene glycol
PI/BN	Polyimide/boron nitrate
PMMA	Polymethyl methacrylate
PMV	Predicted mean vote
PSD	Particle size distribution
PTT	Photothermal therapy
PV	Photovoltaic
PVC	Polyvinylchloride

S2S	Surface to surface
SEM	Scanning electron microscopy
SS	Solid-solid
TES	Thermal energy storage
TGA	Thermogravimetric
TMS	Thermal management system
XRD	X-ray diffraction

1 Introduction

The earliest application of ice as a cold storage material involved collecting ice from the lakes and rivers and storing in a well-insulated underground for using during the summertime. One of the most well-known and first technical applications of harvesting ice from the lakes has been applied in the cooling systems of Hungarian parliament building in Budapest with more than 18,000 m^2 area, including 691 rooms as illustrated in Figure 1.1. The building construction began in 1885 under the direction of the architect Imre Steindl and was completed in 1896, but it did not open until 1904 when the architect who designed it died. The building cooling system has been active until 1994 using the collected ice from Lake Balaton.

One of the first technical utilizations of phase change materials (PCMs) for heating purposes harvesting solar energy has been designed and implemented in Dover Sun House by Dr. Maria Telkes and Eleanor Raymond in the 1940s in Massachusetts, USA, as part of a long-term research project of the Massachusetts Institute of Technology (MIT). Glauber salt (sodium sulphate decahydrate) with melting point of 32°C has been used as a PCM in the system to absorb solar radiation during the day and release the heat to the building during the night by circulating the hot air around the drums filled with PCM which was melted with solar energy during the day.

Application of PCMs as a medium for thermal energy storage (TES) systems became an important issue for energy efficiency and energy management following the energy crisis of 1973–1974 with a sharp increase in the price of crude oil. Using PCMs in TES system can also minimize the mismatch between energy supply and demand, improve the efficiency and reliability of energy distribution networks, and play an important role in energy conservation and environmental protection. Different types of PCMs, including paraffin waxes, hydrated salts, fatty acids, polymers, and metallics with different melting points, have been utilized for different applications in the last half-century. They are broadly classified into three groups: organics,

FIGURE 1.1 Hungarian parliament building in Budapest [1]. (Wikipedia.org)

DOI: 10.1201/9781003518280-1

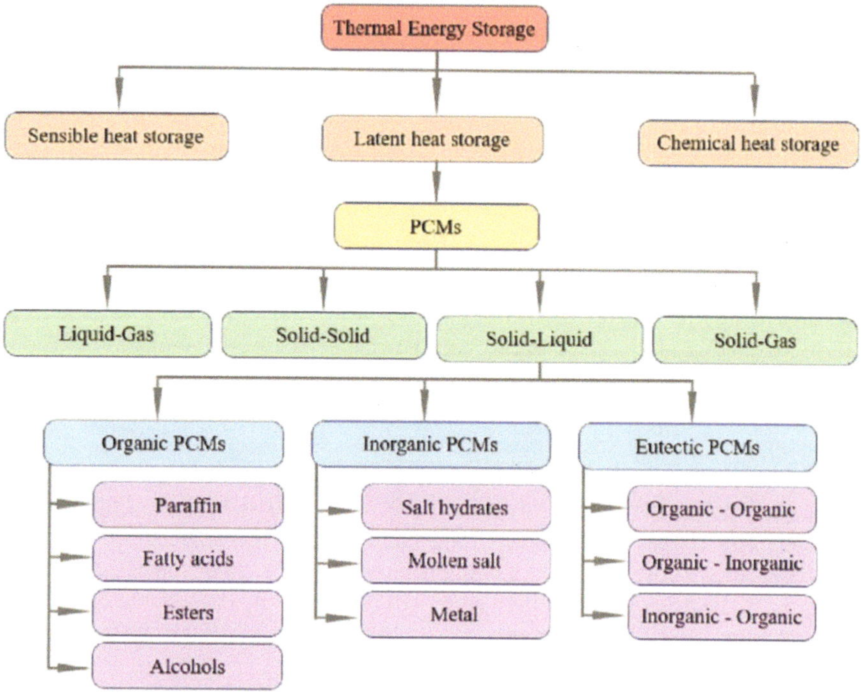

FIGURE 1.2 Phase change materials classifications [2]. (Permission from Elsevier.com.)

inorganics, and eutectics as shown in Figure 1.2 [2]. Organic materials can be divided into paraffin waxes which are open-chained saturated alkanes, and non-paraffinic compounds such as fatty acids, vegetable oils, and polymers like polyethylene glycols (PEGs). They are nontoxic, noncorrosive, and fire resistant with an acceptable latent heat density and high chemical stability but suffer from low thermal conductivity. Non-paraffinic compounds such as fatty acids and vegetable oils can be extracted from oily seeds such as canola, sunflower, corn, soyabean, and palm or from animal fats. They have an acceptable latent heat density with low CO_2 emission during the process but limited melting temperature that can be adjusted by mixing various binary mixtures of fatty acids.

Inorganic PCMs comprise salt hydrates and metallics with mostly low latent heat and being toxic to human body which exclude them from getting widely used in the thermal energy systems. One of the challenges in the application of salt hydrates is their heating density which decreases with cycling due to the congruent melting of the hydrates with the formation of the lower hydrated salt, making the process irreversible and leading to the continuous decline in their thermal storage capacity. Clogging prevention of the fluidized bed can be solved by extra water principle used in hydrated salts that reduce the storage density. However, the crystallization temperature of 30–50°C makes them very applicable in solar energy heating applications. Calcium chloride hexahydrate $CaCl_2.6H_2O$, magnesium chloride hexahydrate

$MgCl_2.6H_2O$, and magnesium nitrate hexahydrate $Mg(NO_2)_2.6H_2O$ are among the proven and useful hydrated salts in heat storage applications.

Eutectic PCMs, which are known as a composite PCM, are often created under low temperature during the crystallization process via melting and freezing of at least two PCMs with a low melting point that can be divided into three classifications based on their component PCM types, including organic eutectics, inorganic eutectics, and organic–inorganic eutectics. Varieties of eutectics from fatty acids have been created and they were shown to have appealing qualities. They melt and freeze uniformly without segregation. Most eutectics examined in the literature for use in air conditioning are composed of organic types. Some of these can't be used inside buildings due to the unpleasant smell.

Selection of PCMs for specific applications and requirements can be achieved by properly choosing characteristics of the materials such as physical, chemical, environmental, thermodynamic, kinetic, and finally the economic aspects. They can be summarized in the following properties:

- Melting or phase transition temperature ranges that match the operating temperature.
- High energy storage capacity and specific heat that can store a large amount of sensible and latent heat and decrease the size of TES systems.
- Low volume changes before and after phase change and low vapour pressure which are used to design a container for the encapsulation of PCMs.
- High thermal conductivity to enhance the heat transfer rate and uniform temperature distribution of the TES unit.
- Consistent melting that delays phase separation.
- No undercooling to ensure that the melting and freezing temperatures are the same.
- Enough crystallization rate to meet the heat transfer demand during phase transition.
- Chemical stability such as no oxidation and no chemical decomposition that extend the lifetime of PCMs.
- Compatibility with the container to prevent corrosion.
- Environmentally friendly.
- No toxicity and no possibility of fire or explosion.
- Availability and being cost-effective.

The main characteristics of PCMs for specific applications are melting point, energy storage capacity, and thermal conductivity of the materials. Their melting point must lay with the applied range of operation, melt congruently and with minimum subcooling, chemical stability, reasonable cost, non-poisonous, and corrosive resistance. Melting point less than 15°C is used for coolness storage in air conditioning and free cooling, while materials that melt at higher than 90°C are applied for absorption refrigeration and all other PCMs with melting point between these two temperatures can be utilized for solar heating, heat load levelling, and temperature control applications. Table 1.1 lists some of the measured physical properties of PCMs such as melting point, latent heat, thermal conductivity, and density [3].

TABLE 1.1
Physical Properties of Phase Change Materials [3]

Compound	Melting Temp. (°C)	Heat of Fusion (kJ/kg)	Thermal Conductivity (W/m K)	Density (kg/m³)
Inorganics				
$MgCl_2 \cdot 6H_2O$	117	168.6	0.570 (liquid, 120°C)	1450 (liquid, 120°C)
			0.694 (solid, 90°C)	1569 (solid, 20°C)
$Mg(NO_3)_2 \cdot 6H_2O$	89	162.8	0.490 (liquid, 95°C)	1550 (liquid, 94°C)
			0.611 (solid, 37°C)	1636 (solid, 25°C)
$Ba(OH)_2 \cdot 8H_2O$	48	265.7	0.653 (liquid, 85.7°C)	1937 (liquid, 84°C)
			1.225 (solid, 23°C)	2070 (solid, 24°C)
$CaCl_2 \cdot 6H_2O$	29	190.8	0.540 (liquid, 38.7°C)	1562 (liquid, 32°C)
			0.1.088 (solid, 23°C)	1802 (solid, 24°C)
Organics				
Paraffin wax	64	173.6	0.167 (liquid, 63.5°C)	790 (liquid, 65°C)
			0.346 (solid, 33.6°C)	916 (solid, 24°C)
Polyglycol E600	22	127.2	0.189 (liquid, 38.6°C)	1126 (liquid, 25°C)
			–	1232 (solid, 4°C)
Fatty acids				
Palmitic acid	64	185.4	0.162 (liquid, 68.4°C)	850 (liquid, 65°C)
			–	989 (solid, 24°C)
Capric acid	32	152.7	0.153 (liquid, 38.5°C)	878 (liquid, 45°C)
			–	1004 (solid, 24°C)
Caprylic acid	16	148.5	0.149 (liquid, 38.6°C)	901 (liquid, 30°C)
			–	981 (solid, 13°C)
Aromatics				
Naphthalene	80	147.7	0.132 (liquid, 83.8°C)	976 (liquid, 84°C)
			0.341 (solid, 49.9°C)	1145 (solid, 20°C)

As can be seen from the table almost for all types of PCMs, the thermal conductivity is less than 1 which is one of the drawbacks of the materials that will be discussed later in this chapter. Phase transition temperature also changes from 16°C in fatty acids to 117°C in inorganic materials.

One of the most applicable types of PCMs are commercial paraffin waxes with technical grades which are with reasonable price, with cyclic stability, with moderate heat of fusion of 200 J/g, and with a wide range of melting and freezing temperatures which varies with the molecular mass and chain length. Paraffins are one of the last petroleum refinery products which are in the form of greyish black sludge and are treated with bleaching agents and some other chemicals to convert into clear paraffin wax. Linear, cyclic, or branched hydrocarbons are different forms of structures for paraffins. They also exhibit minimal subcooling during melt-freeze cycles, remain stable without segregation, and are non-reactive, with only oxidation occurring, which necessitates the use of specialized containment techniques. Paraffins are noncorrosive since they have non-reactive nature but chemical similarity with some polymers such as polyolefins may affect their strength.

The disadvantages of paraffins are their flammability and low thermal conductivity which limit their applications especially in buildings. Thermal conductivity of paraffins can be improved by adding metallic powders as filling, metal matrix structure, aluminium-made finned tubes to the material before encapsulation. Thermal conductivity of paraffin wax has been enhanced using nanocomposite filling carbon-coated aluminium nanoparticles (*Al-C/PW*) by Chen et al. [4]. In this composite, a network has been performed by carbon-coated aluminium nanoparticles (*Al-C*) resulting in a remarkable improvement in the thermal conductivity of the paraffin. Their results show that paraffin thermal conductivity has been enhanced to 0.189 W/m.K with 4 wt.% of *Al-C* under 25°C which is 206.5% higher than pure paraffin wax thermal conductivity.

Latent heat density of paraffins has been enhanced by Cemil Alkan [5] using sulphonation of docosane and hexacosane with phase transition temperatures of 44.0°C and 56.3°C, respectively. The sulphonation has been performed in a standard method using sulphuric acid and acetic anhydride in 1,2-dichloroethane solution of paraffin. The measured heating value results obtained from differential scanning calorimetry (DSC) reveal that thermal properties of the sulphonated paraffins have been considerably improved.

Another drawback of commercial paraffins is their environmental impact due to the volatility of formaldehyde and vinyl chloride compounds. Therefore, long exposure to the vapours is hazardous due to the aromatic compounds which are carcinogenic and non-renewable and non-biodegradable in nature, so their disposal is a concern to the environment. PCM melting points and heating values of some of the paraffins are shown in Figure 1.3 [6].

The second group of organic PCMs are fatty acids or carboxylic acids with long linear oxygenated hydrocarbon chains, a terminal carboxylic group *($CH_3(CH_2)_{2n}$-COOH)*, especially those present as triglycerides in fats and oil. Three fatty acids are linked with a molecule of glyceride to produce molecules of triglycerides which are the components of vegetable oils. Therefore, they can be separated from glycerides

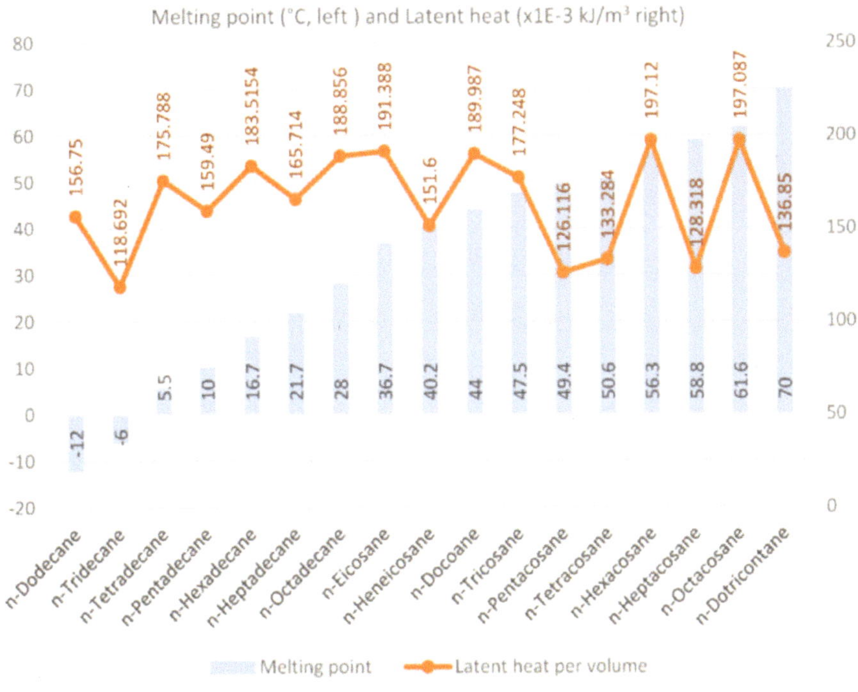

FIGURE 1.3 Melting points and heating values of some of the paraffins [6]. (Permission from Elsevier.com.)

or phospholipids and have a carbon number between 4 and 36 with saturated and unsaturated structures. Examples of saturated structures with single bonds between neighbouring carbons in the hydrocarbon chain are stearic and palmitic acids which are commonly found in meat. In the case of double bond, the fatty acid is said to be monounsaturated if there is only one double bond in the structure like oleic acid ($C_{18}H_{34}O_2$) in olive oil and if there is more than one double bond in the chain, it is called as a polyunsaturated fat such as linoleic ($C_{18}H_{32}O_2$) and linolenic ($C_{18}H_{30}O_2$) acids in canola oil. Fatty acids have been used as PCMs due to their acceptable thermodynamic and kinetic characteristics for low-temperature applications. They have higher latent heat than paraffins with reproducible melting and freezing behaviour and very little or no supercooling. The melting and freezing points, the latent heat, and the degree of crystallization gradually increase with an increasing number of carbons atoms. Fatty acids are corrosive, inflammable, and hold an undesirable odour which limits their application in buildings and textiles. In comparison to paraffin waxes, fatty acids are twice or thrice costly that makes their applications to be not economically feasible. It is also interesting to know that the thermal characteristics of carboxylic acids with an even number of carbon atoms in the structure are higher than those with an odd number of carbons. This is because the acids with an even number of carbons show a tendency for more regular alignment and denser crystalline lattice which is caused by hydrogen bonding between carboxylic acid molecules. Lauric acid with a chemical formula of $C_{12}H_{24}O_2$ is a fatty acid with melting temperature of 42°C and a very high heat of fusion in comparison to water. The stored energy in lauric acid is 70% more than water as illustrated in Figure 1.4.a, and if the operating temperature

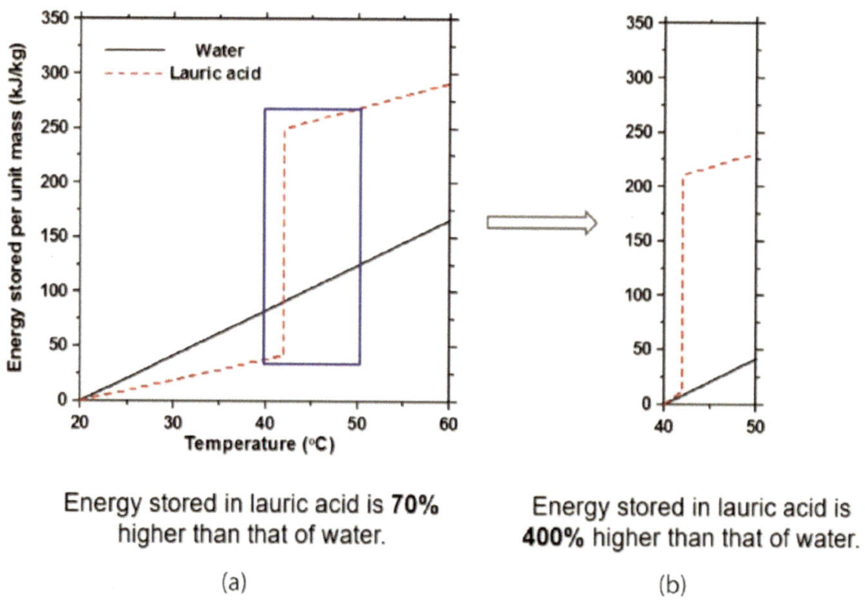

FIGURE 1.4 Energy storage in lauric acid and water (a) application in near-melting point (b) [7]. (Permission from Elsevier.com.)

is narrowed down to 10°C around its phase transition temperature, the stored energy is four times higher than that in water as shown in Figure 1.4.b [7].

Thermal properties of fatty acids such as capric, lauric, palmitic, and stearic acids and their binary mixtures with melting points of 30–65°C and latent heat of 153–182 J/g have been studied by Feldman and Shapiro [8]. Their results reveal that they are very attractive for TES in space heating applications.

A mixture of myristic acid and palmitic acid as an eutectic PCM encapsulated in polymethylmethacrylate (PMMA) shell has been synthesized by Sari et al. for solar energy applications [9]. The size distribution of prepared microcapsules measured by particle size distribution (PSD) analysis shows that the microcapsules had diameter distribution between 0.1 and 300 μm with melting point and heating density of 38.0°C and 100 J/g, respectively. The thermal properties of some of the fatty acids with carbon number of 8–22 are listed in Table 1.2 [8].

Vegetable oil-based PCMs also have been used as a medium for TES system mostly in textiles and clothes due to their safety in comparison to paraffin-based PCMs. They have melting point ranges from 90°C to 150°C and energy density of 150–220 J/g.

Inorganic PCMs can be classified into salt hydrates and metals with very high heat of fusion, non-inflammability, and sharper phase transition. Hydrated salts can be applied as a TES medium at different phase transition temperature ranges from several degrees of centigrade to more than 100°C. They are generally neutral, with fixed melting points, high dissolution heat, high heating value, low volume change during phase transition, and relatively low thermal conductivity in comparison to organic PCMs. Hydrates of alkali, alkaline earth metal halides, sulphates, phosphates, nitrates, acetates, carbonates, and chloride are among the common hydrated salt PCMs. Among the inorganic PCMs, hydrated salts such as Glauber salt (Na_2SO_4, $10H_2O$) containing 44% Na_2SO_4 and 56% H_2O with phase transition temperature of 32°C and energy storage capacity of 254 J/g (377 MJ/m^3) are very attractive due to their high volumetric heat density (350 MJ/m^3), relatively high thermal conductivity (0.5 W/m °C), and reasonable costs in comparison to paraffin waxes. Other inorganic PCMs which are more applicable as energy storage are calcium chloride hexahydrate, $CaCl_2.6H_2O$, magnesium chloride hexahydrate, $MgCl_2.6H_2O$, and magnesium nitrate hexahydrate, $Mg(NO_3)_2.6H_2O$.

TABLE 1.2
Thermal Properties for Some of Fatty Acids [8]

Fatty Acid	Number of Carbon Atoms in Molecule	Melting Temp. (°C)	Heat of Fusion (J/g)	Density (g/cm^3)
Caprylic acid	8	16.3	148	-
CA	10	31.3–31.6	163	-
LA	12	41–44	183–212	0.87
MA	14	51.5–53.6	190–204.5	0.86
PA	16	61–63	203.4–212	0.942
SA	18	70.0	222	0.94
Arachidic acid	20	74.0	227	-
Undecylenic acid	22	24.6	141	

One of the main challenges regarding the application of inorganic PCMs is their segregation and supercooling. Since most of the hydrated salts melt congruently with the formation of the lower hydrated salts, the irreversibility of the process leads to the continuous decrease of their storage efficiency and therefore the high energy storage density of the materials decreases with cycling.

Other drawbacks of hydrated salts are undercooling which is caused by the poor nucleating ability and phase segregation which is caused by their incongruent melting that limits their application for TES systems. Undercooling of inorganic PCMs can be treated by two methods:

- Addition of nucleating agents which are specified to individual hydrated salts for the nucleation ability enhancement and crystallization promotion. The thickener can enhance the viscosity and density of the solution, so the solid particles in the liquid can be distributed evenly in the solution without being soon deposited at the container bottom. Stirring and vibration can also help to make the higher density crystal nucleus suspended for a longer time to fully crystallize, decrease, or eliminate the phase separation.
- The cold finger method in which a portion of the solid PCMs called cold zone is retained gives permission to the unfused crystal to be a nucleating agent. In this method, the heat storage capacity of the PCM is reduced.

Although the first method is commonly used for the treatment of subcooling, the second method is simpler and more effective. Some thickening agent additives such as Bentonite clay with the Glauber salt can help to overcome the problem of segregation. However, this will decrease the crystallization rates and heat transfer to the salt due to the lower mixture thermal conductivity. Some nucleating agents such as borax can also decrease the subcooling, but this requires some thickening agent to prevent setting of the high-density borax. Nucleation promotion depends on many factors such as crystal structure being isomorphous and isotopic nucleating additives that fit well with the PCM structure, and solubility and stability of the hydrate.

An extensive investigation has been performed by Ryu et al. [10] on finding suitable thickening and nucleating agents such as borax, carbon, copper, aluminium, and some of the inorganic salts to compensate subcooling effects of hydrated salts. Their findings in Table 1.3 prove a significant decrease in the degree of subcooling almost to zero using suitable nucleating and thickening materials.

Solid-solid PCMs (SS-PCMs) have also been used in TES systems as they avoid the leakage problem and allow them to be mixed with other materials such as construction materials without encapsulation. The heat is absorbed and released by reversible phase change between a solid crystalline or semi-crystalline phase, and another solid amorphous, semi-crystalline, or crystalline phase. SS-PCMs can be classified into four groups such as polymeric, organic, organometallic, and inorganic. The properties of SS-PCMs like energy density, thermal conductivity, and melting points are mainly dependent on the chemistry and molecular structure of the materials. Thermal cycling stability is one of the most important parameters of SS-PCMs due to their applications that require long-term performance such as construction materials. SS-PCMs also have less phase segregation and smaller volume change as their soft segments are immobilized and therefore extend their durability by preventing degradation upon

TABLE 1.3
Subcooling Reduction by Using Different Nucleating and Thickening Materials [11]

PCM	Thickener	T_m (°C)	Nucleating Agent (Size, μm)	Subcooling (°C)	
				w/o Nucleator	w/n Nucleator
$Na_2SO_4 \cdot 10H_2O$	SAP	32	Borax (20 × 50–200 × 250)	15–18	3–4
$Na_2HPO_4 \cdot 12H_2O$	SAP	36	Borax (20 × 50–200 × 250)	20	6–9
			Carbon (1.5–6.7)		0–1
			TiO_2 (2–200)		0–1
			Copper (1.5–2.5)		0.5–1
			Aluminum (8.5–20)		3–10
$CH_3COONa \cdot 3H_2O$	CMC	46	Na_2SO_4	20	4–6
			$SrSO_4$		0–2
			Carbon (1.5–6.7)		4–7
$Na_2S_2O_3 \cdot 5H_2O$	CMC	57	K_2SO_4	30	0–3
			$Na_2P_2O_7 \cdot 10H_2O$		0–2

chemical cycling. There are differences in thermal properties of four different types of SS-PCMs as depicted in Figure 1.5. While the melting points of organic SS-PCMs are between 25 and 190°C, their energy densities change from 15 to 270 J/g. This range is even wider for polymeric and organometallic PCMs which possess melting points between 15 and 65°C and enthalpies of 10–205 J/g.

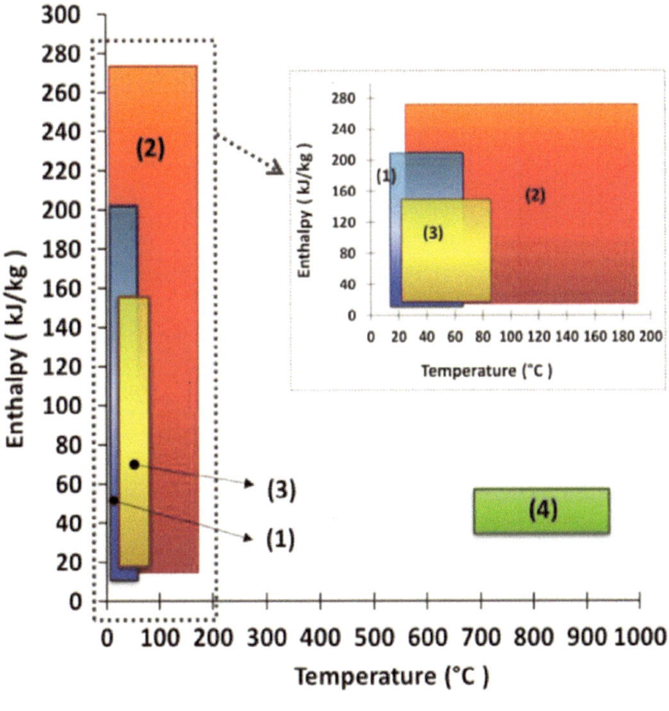

FIGURE 1.5 Heat density and temperature ranges of SL-PCMs and SS-PCMs. 1: Polymeric, 2: Organic, 3: Organometallic, 4: Inorganic SS-PCMs [12]. (Permission from Elsevier.com.)

Other important characteristics of the PCMs that limit their applications for latent heat storage system are the number of cycles that they can be stable without any change in their properties. This property can be deteriorated by the poor stability of the materials and/or corrosion between the PCM and the container used for encapsulation due to the impurities [11].

PCM containers for encapsulation can be developed based on the physical and thermal stabilities of the container material to undergo reparative heating/cooling cycles. PCM properties of some of the hydrated salts are listed in Table 1.4 [6].

Besides melting temperatures and heating values of the PCMs, there are other properties that must be considered for the selection of a PCM for specific applications. One of the most important parameters which is rarely considered during the material screening for specific applications is the health and safety measures of the PCMs. The form or condition of the material as well as its inherent properties directly affects the degree of health hazard. This is usually provided by the manufacturer in specific data sheets called material safety data sheets (MSDS) that show how hazardous the material is. This is specifically important for employees in contact with the material and for the facilities design and maintenance. Health hazard, cycling, and thermal stability of some PCMs have been studied by Miro et al. [13]. Five different PCMs from different groups with phase transition temperatures in the range of 150–200°C have been selected as salicylic acid, benzanilide, d-mannitol, hydroquinone, and potassium thiocyanate. According to their results, the suitable

TABLE 1.4
PCM Properties of Some Hydrated Salts [6]

Compound	Melting Temperature (°C)	Heat of Fusion (J/g)	Thermal Conductivity (W/mK)		Density (Solid) (10^3 kg/m^3)
			Liquid	Solid	
$LiClO_3{\cdot}3H_2O$	8	253			
$KF{\cdot}4H_2O$	18.5–19	231			1.45
$Mn(NO_3)_2{\cdot}6H_2O$	25.3	125.9			
$CaCl_2{\cdot}6H_2O$	28.0–30.0	190–200	0.540	1.088	1.80
$LiNO_3{\cdot}3H_2O$	30	256			
$Na_2SO_4{\cdot}10H_2O$	34	256			
$Na_2CO_3{\cdot}10H_2O$	33	247			
$NaCH_3COO{\cdot}3H_2O$	55.6–56.5	237–243			
$CaBr_2{\cdot}6H_2O$	34	115.5			2.19
$Na_2HPO_4{\cdot}12H_2O$	35–45	279.6	0.476	0.514	1.52
$Zn(NO_3)_2{\cdot}6H_2O$	36	146.9	0.464		
$Zn(NO_3)_2{\cdot}4H_2O$	45.5				
$Zn(NO_3)_2{\cdot}2H_2O$	54				
$Na_2S_2O_3{\cdot}5H_2O$	48–55	201			1.75
$Na(CH_3COO){\cdot}3H_2O$	58	226			1.45
$Cd(NO_3)_2{\cdot}4H_2O$	59.5				
$Na_2B_4O_7{\cdot}10H_2O$	68.1				
$Na_3PO_4{\cdot}12H_2O$	69.0				
$Na_2P_2O_7{\cdot}10H_2O$	70	184			
$Ba(OH)_2{\cdot}8H_2O$	78	266			
$(NH_4)Al(SO_4)_2{\cdot}12H_2O$	95	269			
$MgCl_2{\cdot}6H_2O$	117	169			
$Mg(NO_3)_2{\cdot}6H_2O$	89.3	150			

PCMs for the application range of 150–200°C are benzanilide and d-mannitol; however, hydroquinone can also be used in closed systems.

The molten salts, such as chlorides, nitrates, carbonates, and sulphates, are molten liquids of salts and metals that are mainly used for high-temperature applications like concentrated solar power generation and industrial heat recovery systems. Some of the drawbacks for molten salts such as low thermal conductivity and low thermal stability can also be considered when using the materials for energy storage systems.

Among the molten salts, nitrates with advantages of low cost, less corrosion, and high thermal stability have melting temperature of about 300°C and good stability below 500°C but they have a diverse energy storage capacity. The latent heat of potassium nitrate (KNO_3) is less than 100 J/g, while this value of lithium nitrate ($LiNO_3$) is higher than 200 J/g. Carbonates are superior to nitrates with higher melting point, larger energy density, with less corrosion and an attractive price but possess some challenges such as high viscosity during melting and decomposition at high temperature with high subcooling effects. Sulphates such as sodium sulphate (Na_2SO_4) with a heating value of 100 J/g can have phase transition temperature higher than 800°C.

Chloride salts have phase transition temperature around 800°C with a large latent heat slightly lower than carbonates. They are at low cost and diverse but are more corrosive. Since pure molten salts cannot meet the high-temperature PCM requirements at different fields, eutectic molten salts are produced from a mixture of two pure molten salts in a certain proportion. Details of the application of high-temperature PCMs will be presented in Chapter 12 of the book.

Eutectics, which are a mixture of two organics, two inorganics, or organic–inorganic, have a sharp phase change temperature, melt and freeze congruently without segregation, and found to have attractive properties especially in air conditioning applications. For high temperature applications eutectic molten salts with more performance stability can be used that has lower melting point compared with a single PCM.

The main challenges with the application of eutectic molten salts are low thermal conductivity which can be treated by adding high thermal conductivity materials such as graphite, CuO, or other metallic powders. Table 1.5 illustrates the properties of some eutectic PCMs [6]. Their toxicity depends on the materials whether to be organic or inorganic.

Metals and alloys also have an alternative option for high-temperature latent heat storage systems. They have very high thermal conductivity, high volumetric energy density, low pressure, low specific heat, and low latent heat per unit mass and therefore absorb heat from the source much faster than other types of PCMs.

Encapsulated copper balls coated with chromium-nickel bilayers (refractory metallic shells) with diameter in millimetres and melting point of 1,077°C and latent heat of 53.2 J/g (75% of the latent heat for the theoretical value of 71 J/g) have been used as a PCM for high-temperature applications by Zhang et al. [14]. The capsules show very good oxidation resistance and excellent stability between the core and the shell even after many heating/cooling cycles. Their results also show no leakage after 1,000 charge/discharge cycles with temperatures ranging from 1,050°C to 1,150°C and excellent thermal stability.

TABLE 1.5
Thermal Properties of Some of Eutectic PCMs [6]

Mixture	Melting Temp. (°C)	Heat of Melting (J/g)	Density (kg/m^3)
Eutectic mixtures			
45% $CaCl_2 \cdot 6H_2O$ + 55% $CaBr_2 \cdot 6H_2O$	14.7	140	
66.6% $CaCl_2 \cdot 6H_2O$ + 33.3% $MgCl_2 \cdot 6H_2O$	25	127	
50% $CaCl_2$ + 50% $MgCl_2 \cdot 6H_2O$	25	95	
48% $CaCl_2$ + 4.3% $NaCl$ + 0.4% KCl + 47.3% H_2O	27	188	
47% $Ca(NO_3)_2 \cdot 4H_2O$ + 53% $Mg(NO_3)_2 \cdot 6H_2O$	30	136	
40% $CH_3COONa \cdot 3H_2O$ + 60% NH_2CONH_2	30	200.5	
50% $Na_2SO_4 \cdot 10H_2O$ + 50% $NaCl$	18		
61.5% $Mg(NO_3)_2 \cdot 6H_2O$ + 38.5% NH_4NO_3	52	125	
58.7% $Mg(NO_3)_2 \cdot 6H_2O$ + 41.3% $MgCl_2 \cdot 6H_2O$	59	132	
53% $Mg(NO_3)_2 \cdot 6H_2O$ + 47% $Al(NO_3)_2 \cdot 9H_2O$	61	148	
59% $Mg(NO_3)_2 \cdot 6H_2O$ + 41% $MgBr_2 \cdot 6H_2O$	66	168	
14% $LiNO_3$ + 86% $Mg(NO_3)_2 \cdot 6H_2O$	72	180	
66.6% *urea* + 33.4% NH_4Br	76	161	
11.8% NaF + 54.3% KF + 26.6% LiF + 7.3% MgF_2	449		2160 (liquid)
35.1% LiF + 38.4% NaF + 26.5% CaF_2	615		2225 (liquid)
32.5% LiF + 50.5% NaF + 17.0% MgF_2	632		2105 (liquid)
51.8% NaF + 34.0% CaF_2 + 14.2% MgF_2	645		2370 (liquid)
48.1% LiF + 51.9% NaF	652		1930 (liquid)
63.8% KF + 27.9% NaF + 8.3% MgF_2	685		2090 (liquid)
45.8% LiF + 54.2% MgF_2	746		2305 (liquid)
53.6% NaF + 28.6% MgF_2 + 17.8% KF	809		2110 (liquid)
66.9% NaF + 33.1% MgF_2	832		2190 (liquid)

Polymeric PCMs such as PEG and polyols have also been used extensively in TES systems due to their flexible melting points which vary from −20°C to 70°C depending on the molecular weight of the polymer. The polymer melting point increases with the molecular weight and the heat of fusions is in the range of 110–180 J/g. The degree of crystallinity of the polymers also changes with the molecular weight which ranges from 83.8% to 96.4%. Thermal properties of some polymers are listed in Table 1.6.

TABLE 1.6
Melting Points and Heat of Fusion of Polyethylene Glycols [6]

Polymer	Melting Temperature (°C)	Heat of Fusion (J/g)
PEG 400	4.2	117.6
PEG 600	12.5	129.1
PEG 1000	40.0	168.6
PEG 3400	63.4	166.8
PEG 10000	65.9	171.6
PEG 20000	67.7	160.2
PEG 35000	68.7	166.9
PEG 100000	67.0	175.8
PEG 1000000	70.0	174.0

There are other types of polymeric-based PCMs such as glycerin, pentaerythritol, polyols, diols, pentaglycerine, neopentylglycol, and tris(hydroxymethyl)aminomethane that have an acceptable potential for TES system as they exhibit solid-state phase transformations.

Poly(styrene-co-maleic anhydride)-graft-PEG as a polymeric PCM has been synthesized by Sari et al. [15] for use in passive solar heating and cooling applications. The DSC results of the prepared copolymer samples showed the typical phase transition temperature range of 40–45°C and latent heat capacity 107 and 155 J/g. The properties of the synthesized copolymer have also been stable even after 5,000 heating/cooling treatments.

Sugar alcohols, also known as polyols, or plastic crystals with formula $HOCH_2(CHOH)_nCH_2OH$ have also received attention as biosynthetic and natural PCMs for TES applications in the range 0–250°C. Phase change temperature increases with chain length as in the polymers.

A summary of the properties and drawbacks of different types of PCMs are demonstrated in Table 1.7 [15].

Flammability of the PCMs is one of the limitations for the selection of a PCM for specific applications especially in the building construction. Selection of additives as a flame retardant must be done such that they do not affect the heat density of the material and be compatible with the PCM. Flame retardants such as volatile phosphorated compounds are normally working in the following basis [15]:

- Inhibit the PCM burning by decomposing endothermically.
- Forming inert gases to dilute the gas phase below minimum oxygen concentration (MOC).
- Release of Cl and Br as chain-terminating radicals.
- Surface and flame separation through charring.

The results show that on average, phosphorous-based radicals are the most effective combustion inhibitors and their reactivity rates are five times more effective than bromine and 10 times more reactive as the chain terminator than chlorine radicals.

TABLE 1.7
Characteristics of Different Types of PCMs [15]

PCM Type	Properties	Disadvantages
Organic paraffin wax, fatty acids, and vegetable oils	• Available in wide range of temperatures • No super-cooling and sub-cooling • No phase segregation • Stable over a number of freeze-thaw cycles • Chemically, physically, and thermally stable • Good compatibility with wide range of containers due to non-reactive nature • Non corrosive nature environmentally safe	• Poor thermal conductivity values • Large volume changes during phase transition except in case of some fatty acids. Instability at higher temperatures • No sharp phase transition • Non compatible with the plastic containers • Costly in pure form • Low phase change enthalpy • Low flash point, flammable • Varying levels of toxicity
Inorganic like salt hydrates	• High thermal energy storage capacity • Good thermal conductivity • Low cost • Easy available • Sharp melting points results in sharp phase transition • Low vapor pressure	• Show super-cooling and sub-cooling • Show phase segregation • Incompatible with metallic containers
Eutectic	• Sharp melting and boiling points • Higher volumetric storage density than the organic PCM	• Costly

TABLE 1.8
Common Flame Retardants for PCMs [16]

Common Flame Retardants	Working Principle	Material Tested
Ammonium Polyphosphate	Results in char formation which separates the flame from the material; releases vapors of water which acts as diluting agent; and sometimes used with PER	Paraffin/HDPE
Montmorillonite	As temperature rise, a rigid layer of ceramic forms on the surface acting as char	n-octadecane/HDPE-polyethylene co-vinyl acetate
Melamine	Interferes with combustion due to formation of ammonia vapors	Paraffin/HDPE
Metallic hydroxides	Break endothermically; releases water at high temperature; dilutes combustible vapors; lowers the PCM temperature	RT21/HDPE
Red phosphorous	Mechanism is not very clear. Researchers suggest it is due to char formation or formation of PO radicals which deactivate the reaction.	-
Halogenated compounds	Forms inactive halogen radicals which terminate the chain reactions, and antimony oxides acts as synergists for them	Chlorinated paraffin + 10% Antimony oxide
Expandable graphite	When exposed to heat, it expands several times acting as a charring and insulating agent due to minute air spaces in them.	-

The common and important flame retardants for PCMs are listed in Table 1.8. It is concluded from the literature that the phosphorous- and nitrogen-based flame retardants are the best solution as an inhibitor for PCM flammability in terms of their efficiency.

Most PCMs suffer from low thermal conductivity. Some of the very high thermal conductivity additives can be used to enhance the thermal conductivity of PCMs. Carbon fibre and carbon nanotubes can form a good thermal conduction network after uniform dispersion in PCMs. Other materials such as graphene, graphene oxide, and expanded graphite can also effectively improve the thermal conductivity of PCMs. The second group of materials are metal-based such as metal foams of copper, nickel, iron and aluminium, metal particles, metal nanoparticles, and metal oxides that can be added to the PCMs as thermal conductivity enhancer additives.

One of the most applicable properties of PCMs is heat storage capacity. Table 1.9 shows the comparison between the storage capacity of rock and water, with organic and inorganic PCMs. As shown in the table, the storage mass for 1,000 J in g for rock is 10 times more than that of the organic and inorganic PCMs. It is also clear from

TABLE 1.9
Energy Storage Capacities for Different Materials [17]

Property	Rock	Water	Organic PCM	Inorganic PCM
Density, kg/m^3	2240	1000	800	1600
Specific heat, kJ/kg	1.0	4.2	2.0	2.0
Latent heat, kJ/kg	–	–	190	230
Latent heat, kJ/m^3	–	–	152	368
Storage mass for 10^6 J, kg	67,000	16,000	5300	4350
Storage volume for 10^6 J, m^3	30	16	6.6	2.7
Relative storage mass	15	4	1.25	1.0
Relative storage volume	11	6	2.5	1.0

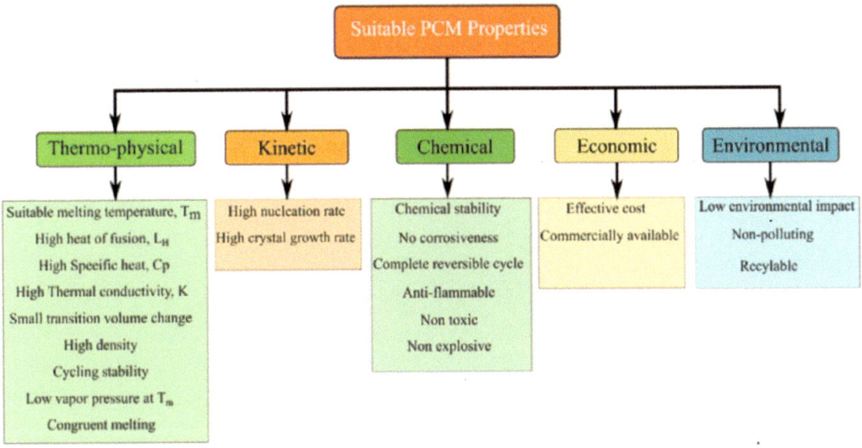

FIGURE 1.6 Critical properties of a suitable PCM [18]. (Permission from Elsevier.com.)

the data shown in the table that the volumetric thermal storage capacity for inorganic materials such as hydrated salts is almost two times higher than that for the organic materials which are caused by their higher latent heat and density [17].

For ensuring the maximum efficiency of the TES system, some critical properties of the PCM such as environmental, thermo-physical, economical, kinetic, and chemical must be considered as illustrated in Figure 1.6 [18].

The number of publications issued on the application of PCMs has increased over the last half-century (1970–2023) with a total number of 16,771 records as presented in Figure 1.7.a. The interest in PCMs for TES also began to sharply increase from the year 2011 to 2022 as shown in Figure 1.7.b. China, the USA, India, England, and

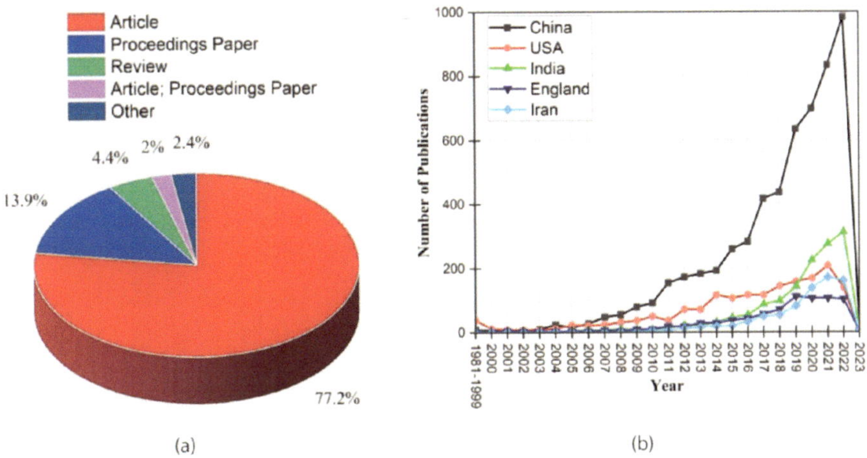

FIGURE 1.7 Overview of 16,771 publications on PCM [with the colour labels from highest to lowest appearing clockwise on the pie chart, after "Article" fills up three-quarters of the circle] (a) and number of publications on PCM per country (b) [19]. (Permission from Elsevier.com.)

Iran were among the countries in terms of total publications and citations. As can be seen from Figure 1.7.b, there is no significant difference between the countries from 1981 to 2006 since there were a few publications in this period of time. China has seen a sharp increase in the number of pulications since 2007 until 2022, which makes China a leader of scientific research in the field of PCM during 2007–2022. After China, the USA ranked second until 2019 when India surpassed the USA and took the second place in the world. The results are also based on the pupulation of the country as China and India have population of over one billion, while the USA has about 300 million.

The number of publications over the period of 1981–2010 has been presented in Figure 1.8 in six groups of five-year periods, namely group one from 1981 to 1985; group two from 1986 to 1990; group three which covers 1991–1995; group four from 1996 to 2000; group five includes the years 2001–2005, and finally group six from 2006 to 2010. From 2011 to 2022, the number of publications has increased continuously as can be seen from the figure.

The statistical analysis of the countries/regions with more than 200 publications has been listed in Table 1.10 [19] in which the abbreviations Collab, Indep, AU1, and AUC represent the number of collaborative publications, independent publications, the first author's name publications, and corresponding author name publications, respectively. TP, TC, HI, and %PR also stand for total publications, total citations, high-index factor, and the percentage of each type of publication among 16,771 records, respectively. As can be observed from the numbers in the table, China, the USA, and India have allocated 36, 11, and 9 productions, respectively, while in terms

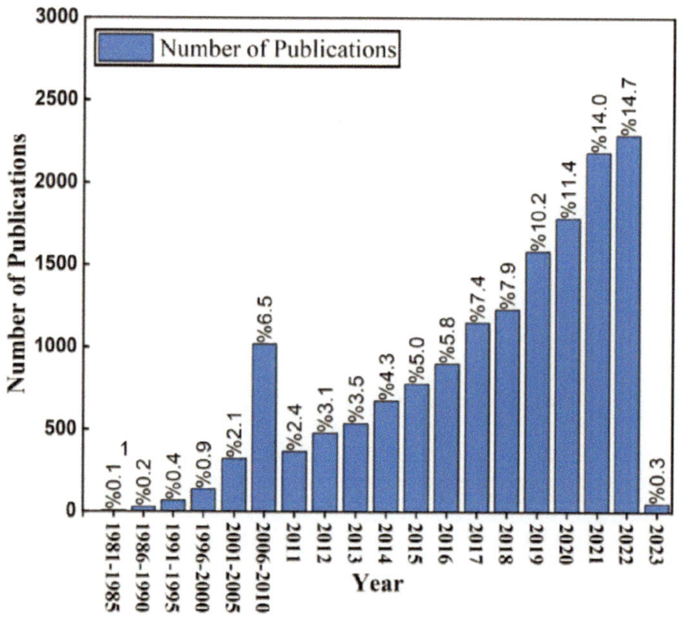

FIGURE 1.8 Number of publications on PCM over time (1981–2023) [19]. (Permission from Elsevier.com.)

TABLE 1.10
The Statistical Analysis of Countries/Regions with More than 200 Publications in PCM [19]

Country	TP	TC	TC/TP	PR (%)	Collab	Indep	Collab/TP (%)	AU1	AUC	HI
China	5663	135,911	24	36	1162	4501	21	5223	5250	133
USA	1734	51,346	29.6	11	691	1043	40	1194	1275	106
India	1390	24,940	17.9	9	323	1067	23	1225	1224	69
UK	785	25,737	32.8	5	567	218	72	382	410	78
Iran	776	19,854	25.6	5	318	458	41	646	617	70
France	676	15,745	23.3	4	417	259	62	411	397	65
Germany	651	20,511	31.5	4	349	302	54	427	426	71
Spain	633	17,946	28.4	4	310	323	49	469	469	68
South Korea	528	9999	18.9	3	140	388	27	407	449	48
Italy	525	12,311	23.4	3	261	264	50	387	395	57
Turkey	524	19,784	37.8	3	179	345	34	378	431	73
Japan	511	11,305	22.1	3	192	319	38	322	394	57
Canada	480	9715	20.2	3	235	245	49	298	317	53
Australia	434	11,199	25.8	3	279	155	64	253	247	55
Saudi Arabia	407	7298	17.9	3	370	37	91	174	145	45
Egypt	259	5876	22.7	2	168	91	65	172	133	45
Malaysia	250	7997	32	2	177	73	71	172	166	50
Singapore	220	9324	42.4	1	149	71	68	135	136	50

of total citations, China, the USA, and the UK are among the top three countries. The factor that is related to the quality of paper is TC/TP which shows the total citations to the total publications. It is interesting to note that Singapore, Turkey, and the UK are the top countries with the highest ratio of TC/TP.

Besides the increasing attention to the research, the application of these materials penetrates in our daily life as a smart material for thermal energy control, including solar panels [20], free cooling [21, 22], building construction materials [23, 24], refrigerators [25, 26], freezers [27, 28], air conditioning [29], lithium ion batteries [30, 31], and co-axial electrical cables [32], etc., which is due to their wide range of melting temperature and other properties such as high latent heat, good durability, and tunable speed of energy exchange. The final classification of PCMs which is based on their phase transition temperature and their applications are divided into four categories such as low-temperature range (–20°C to 5°C) such as ice and water gel, medium- to low-temperature range (5°C to 40°C) like hydrated salts, organic PCMs and polymers, medium-temperature range (40°C to 80°C), and high-temperature range (80°C to 200°C) as molten salts, metal alloys, and paraffins. Their specific applications for heating, cooling, and electricity generation on four different temperature ranges are illustrated in Figure 1.9. In the classification presented in Figure 1.9, low temperature between –20 and 5°C has been utilized for refrigerated products in commercial and domestic; 5–40°C has been applied for building passive cooling and heating, solar applications, air conditioning, and free cooling systems. Medium-temperature range, which is defined as 40–80°C, can be used for solar energy utilization and high-temperature range of 80–200°C can be utilized for solar absorption cooling, solar electricity production, and waste heat recovery systems. However, there are higher temperature applications of PCMs with melting points of higher than 200°C which is not included in this arrangement and can be used for solar reactors heating systems.

The aim of this book is to extend and discuss different aspects of the application of PCMs as energy storage medium for energy management and thermal energy control in our daily life, buildings, and industrial to be applicable for thermal energy management and energy conservation systems.

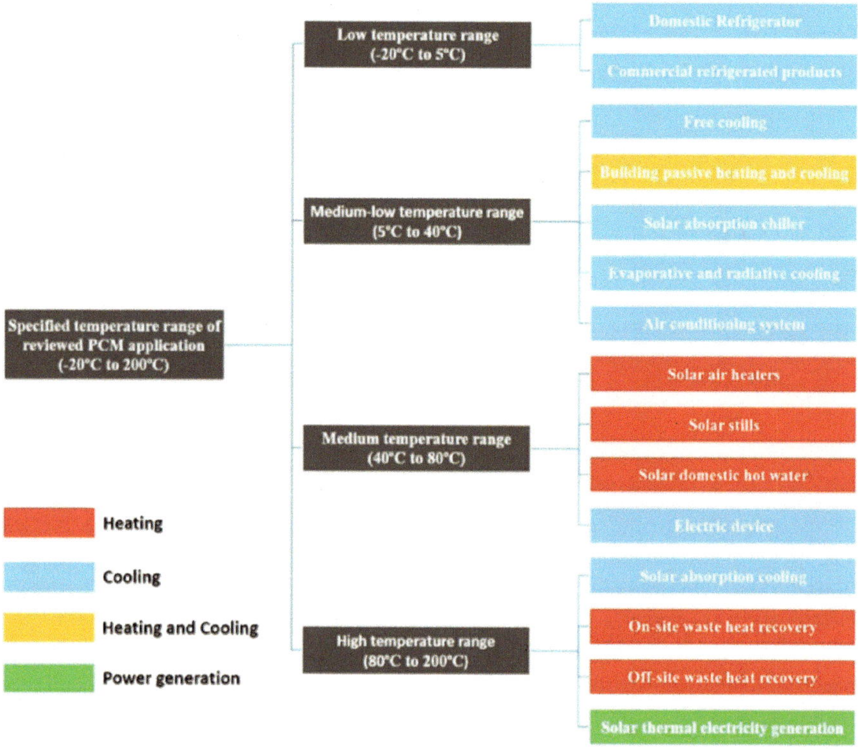

FIGURE 1.9 Applications of PCMs based on their melting point [33]. (Permission from Elsevier.com.)

1.1 SUMMARY

As discussed in this chapter, there is a wide variety of PCMs with different properties that are desirable in various applications as an energy storage material. They are basically classified as organic, inorganic, and eutectic which is a mixture of two. Among the organic PCMs, paraffin waxes are used commonly due to the lower cost and moderate latent heat. The only drawback of paraffins is their very low thermal conductivity that requires a large surface area and will be enhanced by adding metallic powders to the PCM. Fatty acids are another organic material that are very attractive PCMs in a few applications due to their safe properties but have higher costs in comparison to paraffins. Vegetable oil-based PCMs have also been used in textiles and clothes due to their more safety in comparison to paraffins. Inorganic PCMs such as hydrated salts and metals have higher energy storage capacity and higher thermal conductivity but experience phase segregation and supercooling and therefore their application is limited and requires the use of additives such as nucleating and thickening agents.

A mixture of two organic PCMs, two inorganic PCMs, or one organic with one inorganic PCM is called eutectic PCMs. They possess a sharp phase transition temperature, melt and freeze congruently without segregation, and have an attractive property especially in air conditioning applications. SS-PCMs have also been applied in TES systems as they avoid leaking problems and allow them to be mixed with other materials such as construction materials without encapsulation. Finally, the number of research publications on PCM applications has increased over the last 50 years with a total number of 16,771 records and a sharp increase in the research results is clear between 2011 and 2022.

REFERENCES

1. https://en.wikipedia.org/wiki/Hungarian_Parliament_Building#/media/File:Parliament_of_Hungary_November_2017.jpg
2. Wang Q., Yang L., Song J. (2023) Preparation, thermal conductivity, and application of nano-enhanced phase change materials in solar heat collection: a review, J. Energy Storage, 63, p. 107047.
3. Lane G.A. (1980) Low temperature heat storage with phase change materials, Int. J. Energy Res., 5, pp. 155–160.
4. Chen Y., Luo W., Wang J., Huang J. (2017) Enhanced thermal conductivity and durability of paraffin wax nanocomposite based on carbon-coated aluminum nanoparticles, J. Phys. Chem. C, 121(23), pp. 12603–12609.
5. Alkan C. (2006) Enthalpy of melting and solidification of sulfonated paraffins as PCMs for thermal energy storage, Thermochim. Acta, 451, pp. 126–130.
6. Diaconu B.M., Cruceru M., Anghelescu L. (2024) Phase change materials in space system. Fundamental applications, material, and special requirements – a review, Acta Astronautica, 216, pp. 163–213.
7. Ali H.M. (2023) Phase change materials for heat transfer, Chapter One. Netherlands: Elsevier Inc. p. 6.
8. Feldman D., Shapiro M.M. (1989) Fatty acids and their mixtures as phase-change materials for thermal energy storage, Sol. Energy Mater., 18, pp. 201–216.
9. Sari A., Bicer A., Alkan C., Ozcan A. (2019) Thermal energy storage characteristics of myristic acid-palmitic eutectic mixtures encapsulated in PMMA shell, Sol. Energy Mater. Sol. Cells, 193, pp. 1–6.
10. Ryu H.W., Woo S.W., Shin B.C., Kim S.D. (1992) Prevention of subcooling and stabilization of inorganic salt hydrates as latent heat storage materials, Sol. Energy Mater. Sol. Cells, 27, pp. 161–172.
11. Lane G.A. (1991) Phase change materials for energy storage nucleation to prevent subcooling, Sol. Energy Mater. Sol. Cells, 27, pp. 135–160.
12. Fallahi A., Guldentops G., Tao M., Granados-Focil S., Dessel S.V. (2017) Review on solid-solid PCMs thermal energy storage: molecular structure and thermal properties, Appl. Therm. Eng., 127, pp. 1407–1441.
13. Miro L., Barreneche C., Ferrer G., Sole A., Martorell I., Cabeza L.F. (2016) Health hazard, cycling and thermal stability as key parameters when selecting a suitable PCM, Thermochim. Acta, 627–629, pp. 39–47.
14. Zhang G.C., Li J.Q., Chen Y.F., Xiang H., Ma B.Q., Xu Z., Ma X.G. (2014) Encapsulation of copper-based PCMs for high temperature thermal energy storage, Sol. Energy Mater. Sol. Cells, 128, pp. 131–137.
15. Sari A., Bicer A., Alkan C. (2017) Thermal energy storage characteristics of poly (styrene-co-maleic anhydride)-graft-PEG as polymeric solid-solid PCMs, Sol. Energy Mater. Sol. Cells, 161, pp. 219–225.
16. Chandel S.S., Agarwal T. (2007) Review of current state of research on energy storage, toxicity, health hazards and commercialization of phase changing materials, Renew. Sustain. Energy Rev., 67, pp. 581–596.

17. Hasnain S.M. (1998) Review on sustainable thermal energy storage technologies, part 1: heat storage materials and techniques, Energy Convers. Manag., 39, pp. 1127–1138.
18. Faraj K., Khaled M., Faraj J., Hachem F., Castelain C. (2020) PCM thermal energy storage systems for cooling applications in buildings; A review, Renew. Sustain. Energy Rev., 119, p. 109579.
19. Qin Y., Ghalambaz M., Sheremet M., Fteiti M. (2023) A bibliometrics study on PCMs, J. Energy Storage, 73, p. 108987.
20. Mousavi Baygi A., Sadrameli S.M. (2018) Thermal management of photovoltaic solar cells using polyethylene glycol 1000 (PEG1000) as a phase change material (PCM), Therm. Sci. Eng. Progr., 5, pp. 405–411.
21. Mirahmad A., Sadrameli S.M., Seifi H. (2014) Theoretical and experimental studies on a latent heat thermal energy storage system (LHTES) containing flat slabs of phase change materials, Int. J. Smart Grid Clean Energy, 3(2), pp. 234–240.
22. Alizadeh M., Sadrameli S.M. (2018) Numerical modeling and optimization of thermal comfort in building: Central composite design and CFD Simulation, Energy Build., 164, pp. 187–202.
23. Hasanabadi S., Sadrameli S.M., Soheili H., Moharrami H., Heyhat M.M. (2019) A cost effective form stable PCM composite with modified paraffin and expanded perlite for thermal energy storage in concrete, J. Therm. Anal. Calorimetry, 136, pp. 1201–1216.
24. Jamekhorshid A., Sadrameli S.M., Barzin R., Farid M.M. (2017) Composite of wood-plastic and micro-encapsulated phase change material (MEPCM) used for thermal energy storage, Appl. Therm. Eng., 112, pp. 82–88.
25. Pirvaram A., Sadrameli S.M., Abdolmaleki L. (2019) Energy management of a household refrigerator using eutectic environmental friendly PCMs in cascade condition, Energy, 181, pp. 321–330.
26. Javeri I., Fakhroleslam M., Sadrameli S.M. (2023) Application of phase change materials for performance enhancement of open-display supermarket refrigerators: numerical simulation and parametric study, J. Energy Storage, 66, 107506, pp. 1–12.
27. Abdolmaleki L., Sadrameli S.M., Pirvaram A. (2020) Application of environmental friendly and eutectic phase change materials for the efficiency enhancement of household freezers, Renew. Energy, 145, pp. 233–241.
28. Pirvaram A., Sadrameli S.M., Abdolmaleki L. (2021) Optimization of energy consumption and temperature fluctuations for a household freezer using non-toxic and non-flammable eutectic phase change materials with a cascade arrangement, Int. J. Energy Res., 45(2), pp. 1775–1788.
29. Alizadeh M., Sadrameli S.M. (2019) Indoor thermal comfort assessment using PCM based storage system integrated with ceiling fan ventilation: experimental design and response surface approach, Energy Build., 188–189, pp. 297–313.
30. Azizi Y., Sadrameli S.M. (2016) Thermal management of a $LiFePO_4$ battery pack at high temperature environment using a composite of phase change materials and aluminum wire mesh plates, Energy Convers. Manag., 128, pp. 294–302.
31. Sadrameli S.M., Azizi Y. (2024) Thermal management of a $LiFEPO_4$ battery pack in a cold temperature environment using phase change materials (PCMs), JOM, 76(2), pp. 853–862.
32. Jafaripour M., Sadrameli S.M., Seyed Mousavi S.A.H., Soleimanpour S. (2021) Experimental investigation for the thermal management of a coaxial electrical cable system using a form-stable low temperature phase change material, J. Energy Storage, 44, p. 103450.
33. Du K., Wang J., Wu Y., Liu H. (2018) A review of the applications PCMs in cooling, heating, and power generation in different temperature ranges, Appl. Energy, 220, pp. 242–273.

2 Encapsulation of Phase Change Materials

2.1 INTRODUCTION

Encapsulation of phase change materials (EPCMs) is a practical method of providing support to the PCM, improving their heat transfer rate, preventing potential interaction and reactivity with the surrounding environment, controlling volume changes during phase transition, and avoiding PCM liquid outflow throughout the melting process [1]. Encapsulation methods must be thoroughly reviewed and improved before PCMs can be implemented practically to ensure that they are used safely and efficiently in the thermal energy storage systems. Encapsulation of PCMs can also address the challenges caused by using PCMs in thermal energy storage systems such as flammability, corrosion, low thermal conductivity, subcooling, phase change separation, etc. This chapter examines and discusses different methods of PCM encapsulation and provides helpful information for the scientists working in this field. Three common classes of encapsulation based on the size of the capsules such as macroencapsulation, microencapsulation, and nanoencapsulation are also presented with their applications in thermal energy systems.

A protective layer around the PCMs with encapsulation avoids the PCM from being contacted to the environment especially in some cases that the chemical stability of the PCM is affected by the surrounding conditions and prevents corrosion of TES system components. PCM encapsulation techniques can be classified into three categories, namely chemical, physical, and physico-chemical as illustrated in Figure 2.1 [2]. Each technique has its own advantages and disadvantages with a diverse area of applications, but chemical encapsulation methods are mainly used for the PCM microencapsulation. Physical encapsulation techniques are done with comparatively lower costs and can be easily scaled up, but the formed capsules have non-uniformity in size and with imperfect properties and not smaller than few hundred microns. However, chemical methods can provide desired properties with exact particle structure.

Three common classes of PCM encapsulations based on the size of the produced capsules are nanoencapsulation, microencapsulation, and macroencapsulation. When a substance is enclosed in a shell with size of nanoscale which is practically between 1 and 100 nm in size, the process is called nanoencapsulation which finds different applications across various fields such as drug delivery.

For the creation of microcapsules, which range in size from tens to hundreds of micrometres, a technique known as microencapsulation is applied in which individual particles or droplets are coated with a continuous film [3]. Microencapsulation

DOI: 10.1201/9781003518280-2

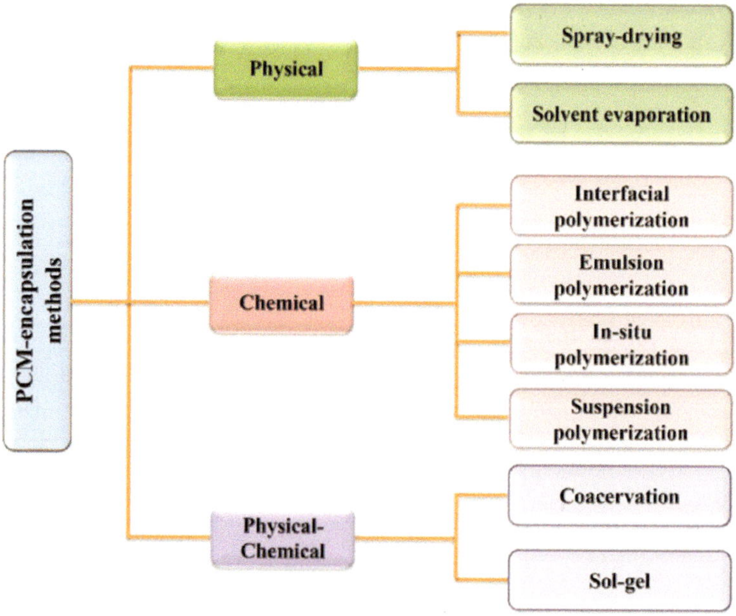

FIGURE 2.1 Different techniques for encapsulation [2]. (Permission from Elsevier.com.)

is performed in a capsule, including a core-shell composite construction in which small-sized capsules or containers are consisted of two parts: (a) core contains PCMs and (b) shell of polymer or other materials which may have core-shell, multi-shell, and polynuclear structures as shown in Figure 2.2 [4]. Two types of microcapsules can be created: an irregular shape and in a regular shape-like spherical, tubular, or oval. Applications of PCMs microcapsules in the textile filaments and mixing with paint which can then release pigments or other colourants upon application are examples of microencapsulation.

A process of confining phase change materials in observable capsules, pouches, tubes, thin plates, shells, or balls which are normally in the size range of between millimetres and centimetres in diameter is named as macroencapsulation. This technique provides easy corporation of PCM in the buildings and provides control and protection against environmental degradation. However, it suffers from leakage issues, low heat transfer characteristics, and thermal stratification. This technique also has a lower performance improvement compared to micro and nano due to the heat conduction with the capsule. Ice balls which are used in cooling energy storage systems are one of the examples of macroencapsulation. Particle size distribution, particle morphology, and shell material will affect the final PCM properties and performance in nano- and microencapsulation. Different types of corrosions such as pitting, crevice, erosion, fretting, or oxidation can be obtained by the selection of non-appropriate material for containment of PCM. Pitting corrosion is developed by direct contact of PCM with metals like aluminium and the corrosion rate is increased if used with copper due to a galvanic effect which leads to engineering failure.

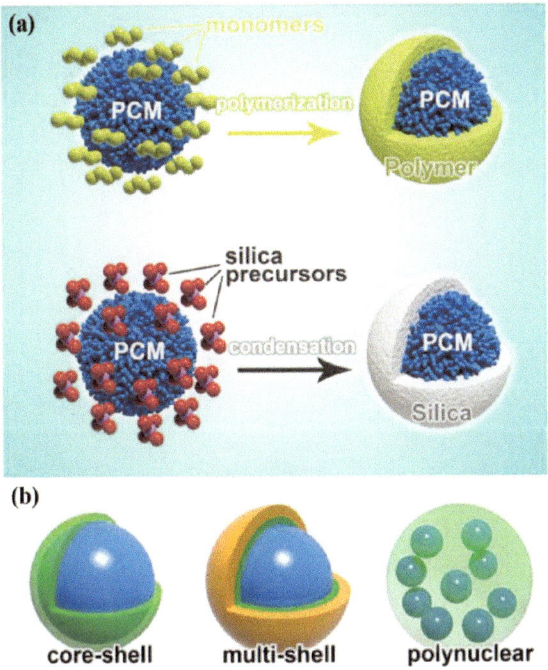

FIGURE 2.2 Stages of an encapsulation process [4]. (a) Polymerization and condensation. (b) Core shell, multi-shell, and polynuclear. (Permission from John Wiley.)

This can be prevented by coating the container with a more reactive metal which undergoes decomposition or galvanizing the container with zinc to protect the container material. In both techniques, care must be taken that no cracks or crevices are produced which causes pitting corrosion. Paraffins are not basically corrosive but if encapsulated with plastics leading to softening of the container. However, fatty acids and salt hydrates are very corrosive and care must be taken for the selection of the encapsulation materials. Stainless steel and carbon-based materials can be used in the case of corrosive PCMs [5].

There are some other characteristics requirements for shell materials in encapsulation applications such as chemical stability, being neutral to the surroundings, being thermally stable at applied temperature, with higher smooth surface morphology, minimal porosity, and not leaking at temperature higher than PCM's melting point. Aluminium, copper, stainless steel, and nickel can be used as metallic shell materials. They have a high mechanical strength, can form easily, with high thermal stability, and high potential for corrosion. However, none of them are cost-effective for shell manufacturing. Ceramic based materials such as silicon dioxide, titanium dioxide, calcium carbonate have high thermal stability and mechanical strength with less expensive than metallic shells but have disadvantages like porous shell that can lead to leakage and difficult to accommodate volume expansion. Finally, polymeric materials such as polystyrene (PS), high-density polyethylene (HDPE), and

TABLE 2.1
Suitable Container Materials for Some of the Important PCMs [5]

PCM	Suitable Container Materials
Paraffin's	All materials are suitable; Care must be taken with plastics
$CaCl_2.6H_2O$	Aluminium is not used: Stainless steel is a good choice
$Na_2SO_4.10H_2O$	Stainless steel
$Na_2S_2O_3.5H_2O$	Carbon steel
$MgSO_4.7H_2O$	Aluminium and iron

polymethylmethacrylate (PMMA) can be used for different encapsulation methods with maximum temperatures of 300–400°C. Some of the suitable container materials for PCMs are listed in Table 2.1 [5].

To considerably enhance the heat transfer efficiency of a carrier fluid, such as water, microencapsulated PCMs can be utilized in powder form or disseminated in the fluid which is known as slurry solution of microencapsulated phase change material (MPCM) [6, 7]. Zhang et al. [8] proposed adding nanoparticles to microencapsulation of PCMs to increase the material's thermal conductivity. The instability and aggregation of nanoparticles and MPCMs in carrier nano-fluid [9, 10] are the main drawbacks of the MPCM containing nanoparticle, though.

It is possible to microencapsulate PCMs with melting points between −10°C and 80°C [3]. The physical and chemical characteristics of the materials to be employed affect the microencapsulation method. Encapsulation is performed using a variety of physical, chemical, and physico-chemical processes [11, 12]. The most popular microencapsulation techniques are depicted in Figure 2.3. Pan coating, air

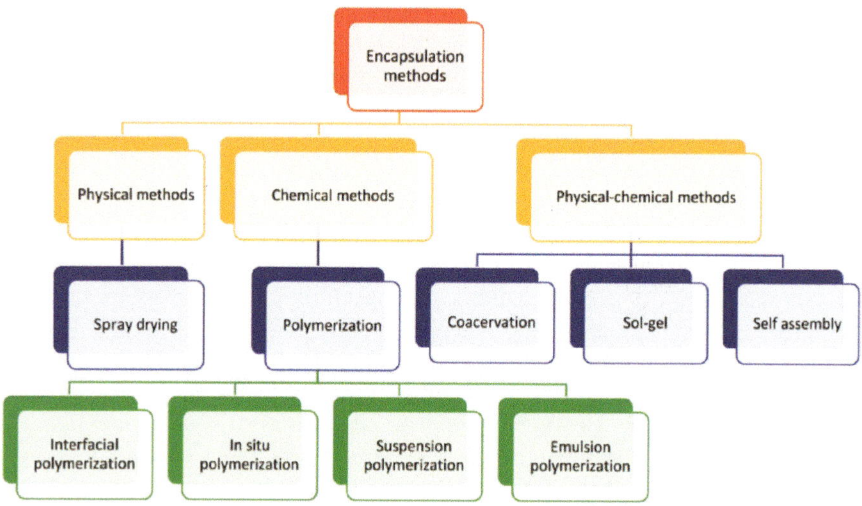

FIGURE 2.3 Encapsulation techniques [15]. (Permission from Elsevier.com.)

suspension coating (fluidized bed), centrifugal extrusion, vibrational nozzle, spray drying, and solvent evaporation are the most often examples of physical encapsulation processes. Spray drying, solvent evaporation, centrifugal extrusion, and fluidized bed processes are mainly used for the microencapsulation [13]; however, as mentioned by Jamekhorshid et al. [14], fluidized bed or suspension coating is not a suitable technique for the microencapsulation production. Chemical methods involve interfacial, suspension, and emulsion polymerization. Physico-chemical techniques include ionic gelation, coacervation, and sol-gel. Interfacial polymerization, suspension polymerization, and emulsion polymerization are three chemical processes.

2.2 PHYSICAL ENCAPSULATION METHODS

In the physical encapsulation technique, the microcapsule cores are surrounded, or coated physically or mechanically, by the microcapsule wall and the composition of the shell does not change during the process. Although an extensive application of physical techniques such as pan coating, air suspension coating and spry drying are in the pharmaceutical and food industries but only the spray drying technique is frequently applied for covering phase change materials. Centrifugal extrusion, vibrational nozzle, and solvent evaporation are among other techniques for physical encapsulation.

2.2.1 Pan Coating

Tablets and small-coated particles are made by one of the oldest industrial processes which is called pan coating technique and is widely utilized in the pharmaceutical industry. A dry coating material is combined with solid particles, the temperature is increased to cause the coating material to melt and encase the core particles, and the coated material is cooled to solidify it. In a different technique, the coating material might be sprayed or progressively applied to the core particles as they tumble in a vessel as opposed to being completely mixed with the core particles at the beginning of encapsulation.

Macroencapsulation of n-octadecane (OD) and HDPE as core materials and calcium alginate *(CaAlg)* as the shell, with wall thickness of 30–50 μm, using pan coating has been performed by Wang et al. [16]. The permeability and hydrophilicity of PCMs have been improved in addition to the roughness, with the pores of 3 μm by using chromic acid.

In this method, if the core melting temperature is higher than the shell melting temperature, PCM encapsulation is not possible. Since this method cannot be used in the micro- and nanoranges and due to the melting point restriction, pan coating technology has not been studied widely in the literature.

2.2.2 Air Suspension Coating

By altering the intervals at which the core particles pass through the coating zone, microencapsulation by air suspension offers greater control of the shell thickness and flexibility compared to pan coating technique [17]. In this method, the solid particles are coated and dried while suspended in an air stream that is travelling upwards.

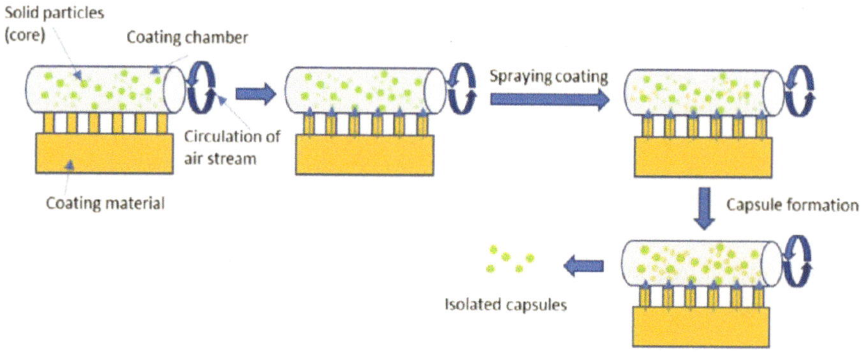

FIGURE 2.4 Encapsulation and air suspension coating [20]. (Permission from Elsevier.com.)

Coating materials are used as suspensions and solutions in both water and volatile organic solvents. The solid particles as core materials are suspended in the coating chamber using a recirculating air stream. A coating material is dissolved in water, or an organic solvent is sprayed onto the moving particles while they move through the chamber as depicted in Figure 2.4. The particles finally get dried and coated after multiple circulations based on the coating thickness.. This method, however, was created for the food sector, cosmetic products, and the pharmaceutical business [18, 19], which is not appropriate for PCM encapsulation.

2.2.3 Centrifugal Extrusion

In the centrifugal extrusion, the liquid phase of the core material will be in an inner tube in the centrifugal extrusion method which has been employed first in Southwest Research Institute (SwRI) [21], while the coating material, which should be miscible with the core material, flows through an annular tube around it.

The core and coating material emerge from the orifices at the ends of the tubes as the head rotates or vibrates, and surface tension forces cause them to disintegrate into spherical droplets. Finally, using heat or the proper chemical processes, these are consolidated in a bath as illustrated in Figure 2.5.

Centrifugal extrusion technique is a simple and feasible method of encapsulation, but the diameter range of capsules is large with a range of 1–5,000 μmm. Although it appears possible, there are no reports of PCM being encapsulated using this technique.

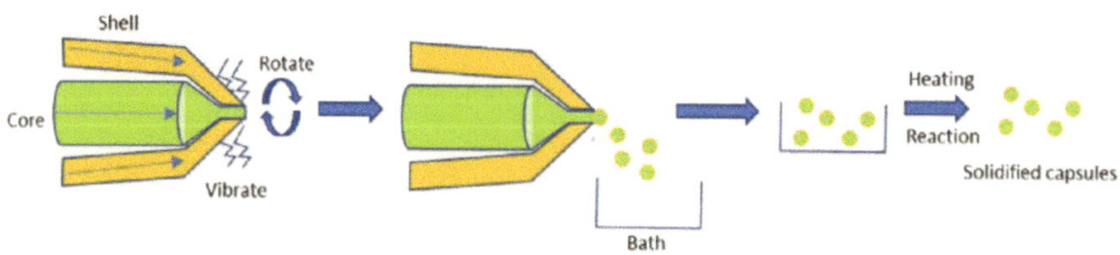

FIGURE 2.5 Centrifugal extrusion process for encapsulation [20]. (Permission from Elsevier.com.)

2.2.4 Vibrational Nozzle

In order to execute core-shell encapsulation or micro-granulation (matrix encapsulation), a laminar flow via a nozzle is typically used, together with an extra vibration of the nozzle or the liquid. Very homogeneous droplets are produced by vibration [22]. Although form-stable PCM composites were created using a similar technique but without vibration [23, 24], this approach has not been applied for PCM microencapsulation.

2.2.5 Spray Drying

Spray drying is one of the most applicable and low-cost commercial methods of PCM encapsulation in which the microcapsules are created by rapidly evaporating an intimate mixture of core and shell material in a heated chamber. The steps of the spray drying method are listed below [25], and are demonstrated in Figure 2.6.

1. Making tiny droplets using an atomizer to spray the feed solution or emulsion.
2. Direct contact of the preceding dispersion with the gas stream at the temperature required for the solvent to completely evaporate.
3. Filtration of solid particles carried by the gas phase using cyclones and/or filters as shown in Figure 2.7 [16].

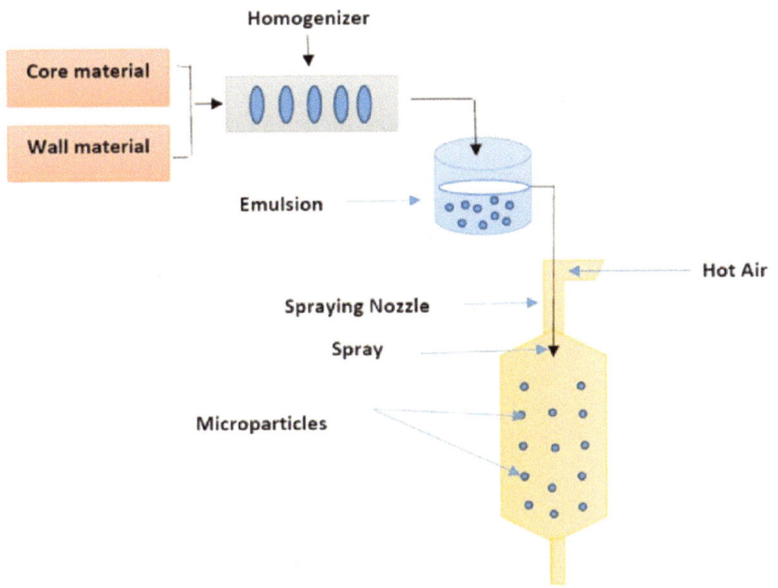

FIGURE 2.6 Flow diagram of a typical spray drying encapsulation method [25]. (Molecules, open access.)

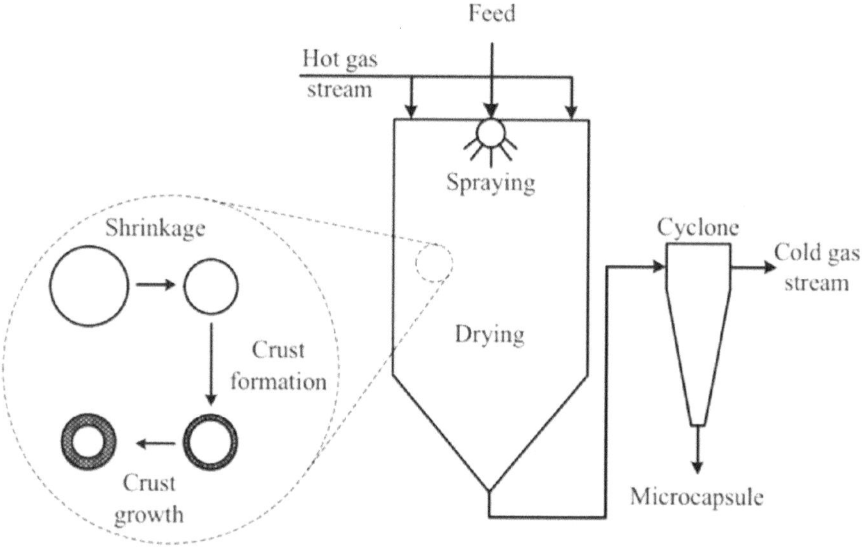

FIGURE 2.7 Spray drying equipment for encapsulation of PCM [16]. (Permission from Elsevier.com.)

Typically, spray drying approach produces polynuclear or matrix-type microcapsules. Particle agglomeration and uncoating are frequent issues with this quick microencapsulation technique [26]. However, this method is simple to scale up, and by creating the right atomizer, it is possible to produce homogeneous microcapsules with the necessary particle size [27].

Spray drying techniques have been studied by a few researchers in the past. Microencapsulation of PCMs using paraffin wax as a PCM and gelatine/acacia as a shell by coacervation method and spray drying has been performed by Hawlader et al. [28] in which different parameters on the encapsulation performance have been investigated. Their results of DSC and SEM analyses show that capsules with a uniform size distribution and high latent heat storage can be obtained by this method. N-octadecane (C18) with a high loading of 80% has been used as a PCM coated with titania shell by a rapid aerosol process with a hydrothermal post-pretreatment to produce microcapsules by Fei et al. [29]. They produced capsules of 0.1–5.0 μm in diameter with two structures of dense and hollow spheres and enthalpy of 92–97 J/g. Encapsulation of paraffin wax PCM (Rubitherm ®RT27) with and without nanofibres with a minimum average particle size of 3.9 μm has been performed by Borreguero et al. [30] using low-density polyethylene-polyvinylacetate as a polymeric shell. Their results show the encapsulation yield of 49% with an enhancement in thermal conductivity using nanofibres with PCM. Figure 2.8 illustrates the spray dryer device used for the encapsulation of PCM.

2.2.6 Solvent Evaporation

When the materials which are used as PCMs are resistant to heat, solvent evaporation is used for microencapsulation as a physical technique. In this method, a high

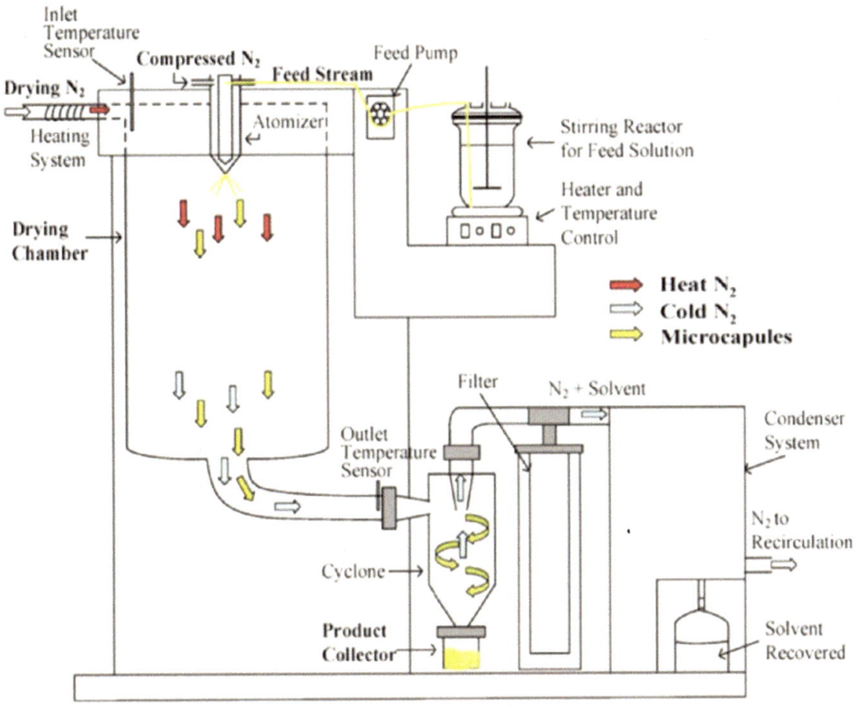

FIGURE 2.8 Fluidized bed spray dryer used for the encapsulation of paraffin wax [30]. (Permission from Elsevier.com.) [The red Heat N_2 arrows point through the Heating System and the upper area of the Drying Chamber. The blue Cold N_2 arrows point through the lower Drying Chamber, and move straight through the Cyclone to the N_2 + Solvent. The yellow Microcapsules arrows point through the upper and lower Drying Chamber, and rotate through the Cyclone.]

heat is provided to evaporate the low boiling point solvent, thus leading to deposition of shell materials on the PCMs surface. An aqueous phase with a suitable emulsifier and oil phase containing a polymer are prepared as a container material in which the core material and low boiling solvent could dissolve core materials in the water system to form microcapsules. The last step will be to wash and dry the microcapsules. The steps for solvent evaporation method are shown in Figure 2.9 [20].

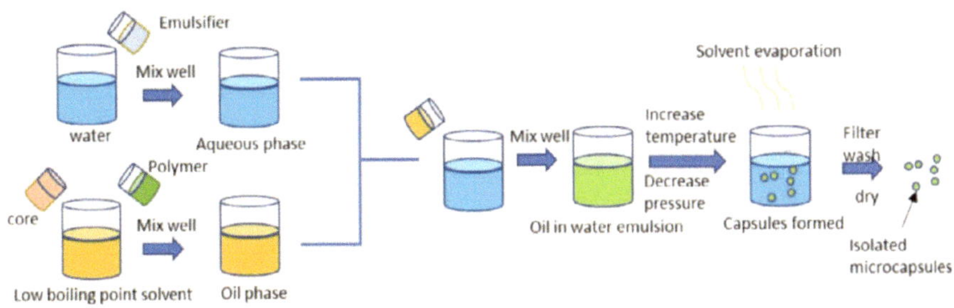

FIGURE 2.9 Steps for solvent evaporation technique of encapsulation [20]. (Permission from Elsevier.com.)

TABLE 2.2
Summary of the Advantages and Drawbacks of Physical Encapsulation Techniques [31–32]

Technique	Advantages	Disadvantages	Applicable to MEPCM Production
Pan coating	Low-cost equipment	Difficult to control High skill level required	No
Air-suspension coating	Low cost and higher production volume	High skill level required and agglomeration of particles	No
Centrifugal extrusion	Suitable for bioencapsulation	High temperature	Yes
Vibrational nozzle	High yields of production and easy to scale up	High temperature	Yes
Spray drying	Equipment and know how widely available, versatile, and easy to scale up	High temperature and agglomeration of particles Remaining uncoated particles	Yes
Solvent evaporation	Low cost	Lab scale production	Yes

Table 2.2 summarizes the advantages and drawbacks of the physical encapsulation techniques.

2.3 CHEMICAL ENCAPSULATION METHODS

Although there are different techniques for the chemical encapsulation of PCMs, in situ polymerization is the most significant chemical process which encompasses interfacial, emulsion, and suspension polymerization [33]. The various chemical encapsulation techniques utilized to create PCM microcapsules are depicted and contrasted in Figure 2.10 and further explained below.

2.3.1 Interfacial Polymerization

Rapid polymerization of hydrophilic and lipophilic monomers at the interface of an oil-in-water emulsion causes wall development in this method. The solution containing the lipophilic reactant X in the core material is emulsified in an aqueous phase that also contains an emulsifier, as shown in Figure 2.11. At that point, the aqueous phase receives the hydrophilic reactant Y, initiating the development of the shell and interfacial polymerization. The wall that forms once a reaction is started becomes a barrier to diffusion and eventually starts to restrict the interfacial behaviour rate for various applications. Interfacial polymerization of hydrophilic monomer hexamethylene-1.6-diamine (HMDA) through the polymeric shell followed by a polymerization reaction with the hydrophobic monomer hexamethylene-1,6-diisocyanate (HMDI) at the inner surface has been performed by Yadav et al. [34] as one of the earliest attempts to model microencapsulation via interfacial polymerization. Based on the results, the encapsulation time was

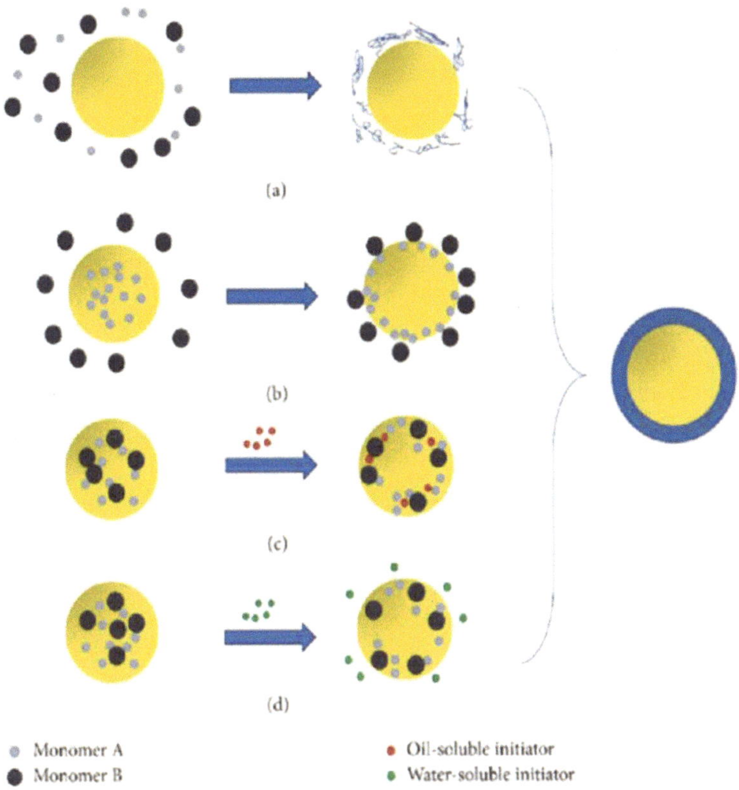

FIGURE 2.10 Schematic diagram of various chemical encapsulation techniques, (a) in situ polymerization [Monomers A and B change their shape and position], (b) interfacial polymerization [Monomers A and B change so that they line up along the circumference borders], (c) suspension polymerization [when oil-soluble initiator is added along the "change arrow," after the change these move in next to Monomer A and between Monomer B], and (d) emulsion polymerization [when water-soluble initiator is added along the "change arrow," after the change these initiators move outside the circumference border] [33]. (Wiley, Hindawi, open access.)

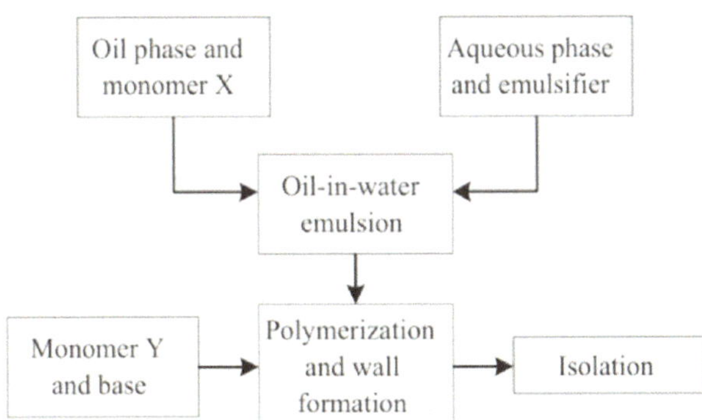

FIGURE 2.11 Flow diagram of interfacial polymerization technique [21]. (Permission from Elsevier.com.)

proportional to the size and the square of size in the kinetically controlled and diffusion-controlled regimes, respectively.

Successfully creating urea-formaldehyde (UF) microcapsules with aromatic lemon oil was accomplished by Park et al. [35]. Additionally, three PCMs (n-pentadecane, n-eicosane, and a paraffin wax) were encapsulated by Tseng et al. [36] using a UF shell.

Microencapsulation of a refined vegetable oil as a bio-based PCM with silica-coated inorganic shell for thermal energy storage application has been synthesized by Ismail et al. [37] in which chemical etching was used on the cenospheres to create holes through which melted PCM was loaded to produce microcapsule. The silica-based coating was used to cover the perforation and avoid leakage of PCM from the microcapsule as depicted in Figure 2.12. Different analysis like microstructural, chemical compatibility, thermal, and leakage properties of the produced microcapsules have been performed and the results revealed no leakage, with higher thermal stability, and higher conductivity for the microcapsules. The latent heat capacity of the cementitious composite was 12.9 J/g at 30% vol. replacement.

Microcapsules of hexadecane as a core and melamine formaldehyde as a shell has been produced by Khakzad et al. [38] in which influence of stabilizer type and amount, surfactant amount, and homogenization conditions such as speed and time on the particle size, morphology, and thermal properties of the microcapsules have been investigated. Appropriate morphology and thermal properties have been achieved by using polyvinyl acetate with higher molecular weight as a stabilizer with the amount of 1 wt.%. Other optimum parameters were 4 wt.% of surfactants (80:20 mixture of sodium dodecylsulphate and Triton X-100), homogenization speed of 6,000 rpm, and homogenization time of 20 minutes. Paraffin has been encapsulated using polyamide shell by interfacial polymerization technique by Wei et al. [39] and the produced microcapsules could disperse in water and organic solvents easily with good thermal and chemical stability.

Fabrication of paraffin wax as nano-PCMs core materials and polyurethane as a shell using three-step interfacial polymerization has been done by Nikpourian et al. [40] as illustrated in Figure 2.13.

Creation of spherical nanoparticles in the presence of sodium dodecyl sulphate (SDS) as surfactant, modification of the surface with 2,4 diisociate (TDI), and

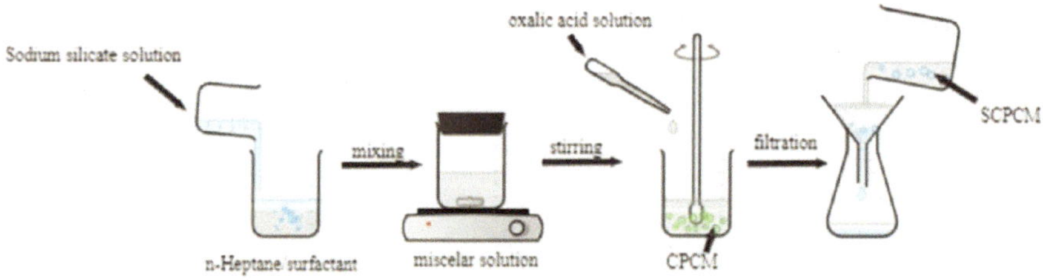

FIGURE 2.12 Schematic diagram of silica coating process for microencapsulation [37]. (Permission from Elsevier.com.)

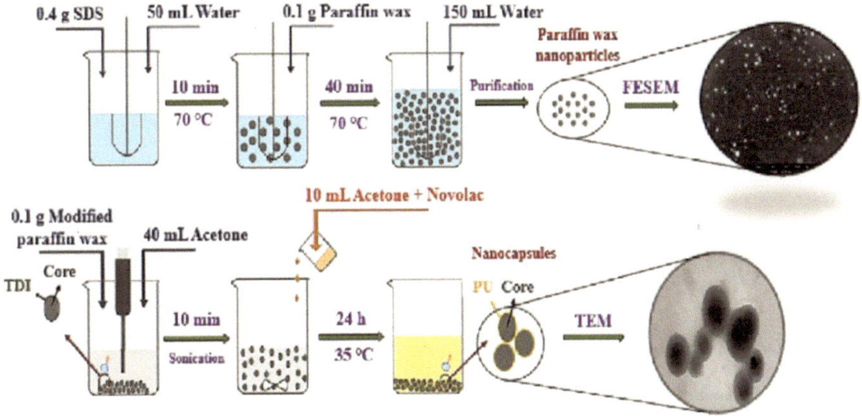

FIGURE 2.13 Three-step interfacial polymerization for nano-EPCMs [40]. (Permission from Elsevier.com.)

encapsulation of the modified nano-PCMs with a thin layer of polyurethane shell through interfacial polymerization were three steps involved in the production process. The average size of nanoparticles was between 25 and 185 nm in the presence of SDS and the mean particle diameter of 83.38 nm. The maximum heating density of the nano-EPCMs samples was 153.9 J/g and the samples showed very good thermal stability after undergoing 100 thermal cycles with melting and crystallization.

2.3.2 Suspension Polymerization

One of the most convenient, eco-friendly, cheap, and efficient methods of chemical encapsulation is suspension polymerization which includes two steps [41]. In the first step, preparation of a suspension containing monomers, PCMs, and initiators are done in an aqueous continuous liquid phase using mechanical stirring and surfactants. Initiation of polymerization at the desired thermal condition until completion will be the second step of encapsulation. The size of the prepared capsules is affected by the surfactant concentration, monomer phase, stirring rate, and droplets and suspension medium viscosities. These droplets develop by stirring the liquid vigorously and dissolving the stabilizers in the aqueous phase. Then, polymerization starts and continues at the chosen temperature. This technique is highly applied for manufacturing polymers such as polyvinyl chloride (PVC), polymethyl methacrylate (PMM), and PS.

Microencapsulation of n-octadecane with different copolymer shells using suspension polymerization has been performed by Li et al. [42] in which the influence of initiator type has been investigated and oil-soluble initiator has been selected as a suitable initiator for microencapsulation of n-octadecane. They repeated this study by adding homo-dispersed polypyrrole (PPy) particles to n-octadecane and encapsulated with PMMA/PAMA copolymer shell to eliminate the supercooling by adding

4–14 wt.% PPy in the core, while PPy had no effect on the morphology and particle distribution of the microcapsule [43]. Although microencapsulation of hydrated salt PCMs is difficult due to its water-solubility but this has been achieved by Huang et al. [44] using $Na_2HPO_4.7H_2O$ with modified PMMA as coating polymer via suspension polymerization solvent technique. Microcapsules of 6.8 μm average diameter with heating density of 150 J/g have been produced and the TGA results showed a weight loss of 10% in the temperature range of 30–84°C which was due to the water loss. In another investigation, they used MMA crosslinked with ethyl acrylate for the microencapsulation of $Na_2HPO_4.12H_2O$ as a core and UF resin as a shell [45]. They found that the weight loss of microcapsules with UF is lower than those with PMMA shell.

Microencapsulation of paraffin wax as a core PCM and PS as a shell has been produced by Jamekhorshid et al. [46] using suspension polymerization in which influence of four experimental parameters such as initiator/styrene mass ratio, paraffin wax/styrene mass ratio, stabilizer/styrene mass ratio, and water/styrene mass ratio on the microcapsule's properties has been investigated. Their results prove that while PCM/styrene mass ratio was the most important parameter affecting latent heat of the PCM, the particle size is affected by the percentage of stabilizer/styrene and water/styrene mass ratios. The optimum parameters for the maximum heating density of 148.5 J/g were 2.18 wt.% of initiator/styrene, 1.94 wt.% of PCM/styrene ratio, 8.84 wt.% stabilizer/styrene, and 11.67 wt.% for the ratio of water/styrene with an encapsulation ratio of 78.5%.

2.3.3 Emulsion Polymerization

A wide range of polymers like adhesives, binders, impact modifiers, drug delivery systems, paper and construction materials have been manufactured using an emulsion polymerization technique.

In contrast to suspension polymerization, emulsion polymerization uses a surfactant to emulsify the monomer in the polymerization medium, while the initiator is soluble in the aqueous phase. As shown in Figure 2.14, the monomer is distributed between droplet emulsion, surfactant micelles, and bare minimum water phase. The polymerization reaction begins in the aqueous phase (i.e., outside the droplets and

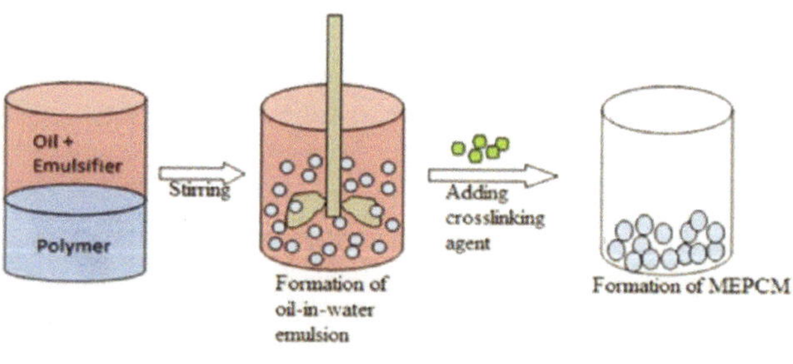

FIGURE 2.14 Schematic representation of emulsion polymerization [47]. (Permission from Elsevier.com.)

micelles) since the initiator is only present in this phase, and it then progresses to the micelles. The percentage of the monomer molecularly dissolved in the aqueous phase mostly influences the average diameter of the resultant particles. Additionally, it is affected by the polymerization temperature, emulsifier concentration, and initiator concentration [47].

Emulsion polymerization can be classified into conventional emulsion polymerization, inverse emulsion polymerization, dispersion emulsion polymerization, micro-emulsion polymerization, and mini-emulsion polymerization.

Microencapsulation of paraffin wax with heat density of 174.1 J/g as a core PCM and melamine-urea-formaldehyde (MUF) as a shell material using in situ polymerization has been performed by Han et al. [49] in which the influence of different parameters like core/shell ratio, reaction temperature, and reaction time on the phase change behaviour has been investigated. They found that by increasing the core/shell ratio, the size of the droplets and microcapsules also increases from average size of 3.02 μm at a weight ratio of 1.0 to the size of 3.79 μm at the core/shell ratio of 2.5. The optimum temperature and time with the minimum impurity has been achieved at 80°C and 2 hours, respectively, with higher quality of the microcapsules. The produced microcapsule had latent heat of crystallization and melting of 133.1 J/g and 134.3 J/g, respectively, with an encapsulation ratio of 77.1%.

Thermal energy storage characteristics of eutectic mixture of myristic and palmitic acids (MA-PA) encapsulated in PMMAs shell using emulsion polymerization have been synthesized by Sari et al. [50] and the prepared microcapsules have been characterized using Fourier transform infrared (FTIR) spectroscopy for confirmation of success encapsulation of the PCM in PMMA shell. The results clearly show that the prepared capsules of PMMA/(MA-PA) with a ratio of 1:2 are an entirely spherical profile which have diameter distributions between 0.1 and 300 μm. The phase transition temperature and heating density of the samples have been measured by the differential scanning calorimetry (DSC) to be 38.0°C and 100.5 J/g, respectively. The thermogravimetry (TGA) data also reveals that the microcapsules are durable thermally up to 250°C. Finally, the produced capsules tested for chemical stability and the results show excellent thermal and chemical stability after 5,000 heating/cooling thermal cycles test.

A modified emulsion polymerization technique has been used by Doguscu et al. [51] for the microencapsulation of eutectic n-alkanes in PS for solar thermal application in which the melting points and combination ratios of four eutectic mixtures of n-alkanes (C_{17}-C_{18}, C_{20}-C_{17}, C_{20}-C_{17}, and C_{20}-C_{24}) have been calculated theoretically and proven experimentally before encapsulating. In the modified emulsion polymerization method, the monomer was previously mixed with core material and homogeneously distributed in water phase using a suitable surfactant and then the system was maintained at polymerization temperature for several hours until the microcapsules have been performed. DSC results for the prepared microcapsules show the melting point and heating values of the samples to be 21–35.9°C and 61.2–146.1 J/g, respectively.

Figure 2.15 lists the advantages and disadvantages of the chemical techniques for the encapsulation of PCMs [48].

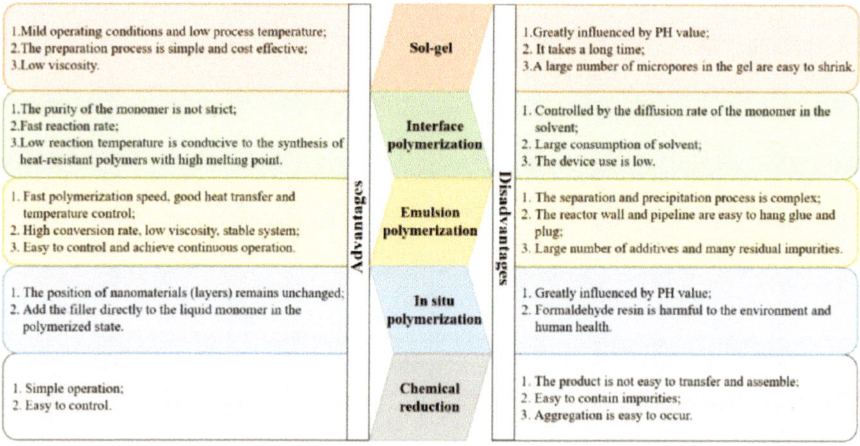

FIGURE 2.15 Advantages and disadvantages of the chemical methods [48]. (Permission from Elsevier.com.)

2.4 PHYSICO-CHEMICAL ENCAPSULATION METHODS

In the physico-chemical encapsulation process, gelation or coacervation makes a solid and stable particle.

2.4.1 Ionic Gelation

The ionic gelation process, which is frequently employed in the pharmaceutical business, has not yet assessed for the encapsulation of PCMs. This technique is based on polyelectrolytes' capacity to crosslink and produce hydrogels when exposed to multivalent counterions like Ca^{2+}, Ba^{2+}, and Al^{3+} [52]. Calcium alginate *(CaAlg)* microcapsules, for instance, would result from the ionic gelation of alginate with calcium ions [53].

2.4.2 Coacervation

The Latin word "acervus," which means "heap," is where the word "coacervation" came [54]. Coacervation is the phase separation of one or more hydrocolloids from an initial solution and the further deposition of the newly formed coacervate phase around an active ingredient suspended or emulsified PCM. Coacervation can be performed by either a simple or complex methods. The phase separation is induced by adding an alcohol or a salt, varying the temperature or pH in the simple technique, while an oppositely charged polymer is used in the complex coacervation. The solution includes three immiscible phases: the core material, the coating material, and the solvent are formed in the simple coacervation, while in the complex coacervation, a liquid-liquid phase separation process of two oppositely charged polymers through an electrostatic interaction between a dense coacervate phase and a dilute one in equilibrium is involved. The production of the emulsion through the dispersion of

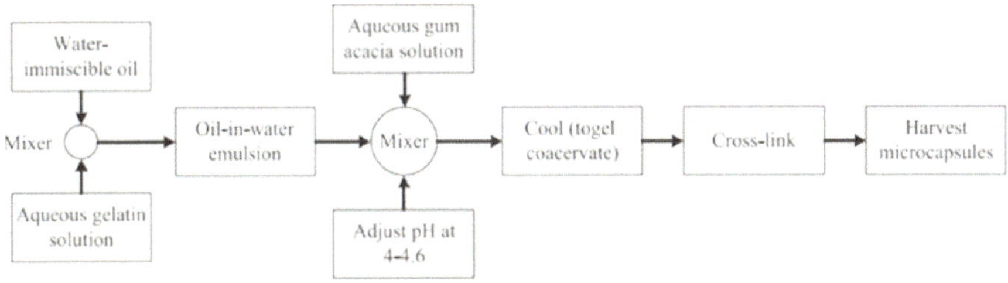

FIGURE 2.16 Block diagram of a complex coacervation encapsulation method [23]. (Permission from Elsevier.com.)

the core material (often oil) into an aqueous polymer solution is the first stage in the complicated coacervation technique. The subsequent addition of the second aqueous polymer solution, followed by the addition of salt, a change in pH, temperature, or medium dilution, causes the deposition of shell material onto the core particles. The stability of the microcapsules through crosslinking, desolvation, or thermal treatment is the last stage [6].

The flow diagram of a typical complex coacervation process, which is frequently used to encapsulate PCMs, is shown in Figure 2.16. Gum acacia, a polyanion or negatively charged polymer, is then added to the gelatine solution after the core material has been dispersed in an aqueous gelatine solution at a temperature range where the wall material solution is liquid. After that, the pH and polymer concentration are adjusted to create a liquid complex coacervate. The system is cooled to gel coat once the liquid coacervate has formed. The final process involves employing formaldehyde or glutaraldehyde to solidify the microcapsules [55].

2.4.3 Sol-Gel

The combination of solution and gelation involves the sol-gel method which can also be named as polycondenstion reaction. For the sol-gel process, in a liquid phase, a molecular precursor undergoes polycondensation processes to create a colloidal solution (sol), which is then transformed into an oxide network (gel) as illustrated in Figure 2.17.

Wang et al. [56] conducted the first study on the inorganic encapsulation of PCM (n-pentadecane) with silica shell using a combination of the O/W emulsion approach and the sol-gel method. The generation of spherical microcapsules in the 4–8 m size

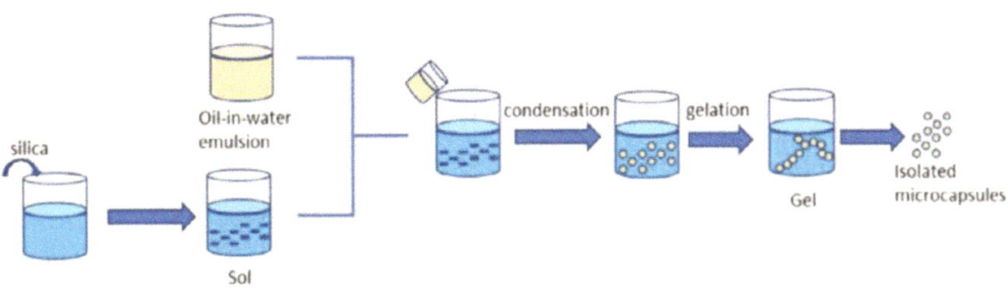

FIGURE 2.17 Schematic diagram of sol-gel method [20]. (Permission from Elsevier.com.)

TABLE 2.3
Comparison of the Physico-Chemical Encapsulation Methods [34]

Technique	Advantages	Disadvantages	Applicable to MEPCM Production
Ionic gelation	Low cost and temperature	High wall permeability	Yes
Coacervation	Versatile	Aldehyde as hardener	Yes
	Efficient control of the particle size	Difficult to scale-up	
Sol-gel	Inorganic shell with high thermal conductivity	Agglomeration	Yes
		Still under research	

range was confirmed by SEM data. The microencapsulated paraffin with SiO_2 shell was effectively produced and described by Fang et al. [1] and Zhang et al. [57] using the sol-gel process. Both of them succeeded in encasing more than 85% of the paraffin. Chen et al. [58] produced microencapsulated stearic acid, one of the saturated fatty acids with melting point of 56.1°C with SiO_2 shell material using the sol-gel technique. A superb encapsulation ratio of 90.7% has been reported. In a different work, in an alternative study, Chen et al. [59] used a SiO_2 shell material to successfully microencapsulate and analyse paraffin wax. The microcapsules melted at 58°C with a latent heat of 156.9 J/g, according to their DSC data.

The benefits and drawbacks of physico-chemical techniques are enumerated in Table 2.3.

Table 2.4 lists a summary of the relative particle size range, encapsulation ratio, and materials employed for the core and shell in these microencapsulation processes.

TABLE 2.4
Relevance of PCMs Microencapsulation Techniques [57]

Microencapsulation Process	Particle Size Range (μm)	Encapsulation Ratio (%)	Common Shell Material	Common PCM Type
Spray drying	0.1–5000	38–63	LPDE/EVA	Paraffin wax
			Gelatin/acacia gum	
			Titania	
Coacervation	2–1200	6–68	Gelatin/acacia gum	Paraffin wax
			SF/CHI	Fatty acid
Sol-gel	0.2–20	30–87	Silica	Paraffin wax
Interfacial polymerization	0.5–1000	15–88	PU	Paraffin wax
			Urea/formaldehyde	
			Melamine/formaldehyde	
Suspension polymerization	2–4000	7–75	Polystyrene	Paraffin wax
	0.05–5		PMMA	
			MMA/St	
Emulsion polymerization		14–67	Polystyrene	Paraffin wax
			PMMA	

2.5 NANOENCAPSULATION OF PCM

Today's technological advancements have made it possible to nanoencapsulate PCM. The PCM capsule's size is crucial and may open up new possibilities for encapsulated PCM applications. Sukhorukov et al.'s research [60] demonstrated that nanocapsules are structurally more stable than microcapsules. When the same force was given to capsules 10 nm in size compared to those 10 µm in size, they showed much less deformation. MEPCMs are not suitable for long-term circulation in several applications, particularly MPCM, because they are prone to fracture during flow and pumping. Additionally, they raise the viscosity of the fluid [61–63]. So it is inevitable that nanoencapsulated PCM (NEPCM) will evolve.

Preparation of nanocapsules of $NiNO_3$/NaCl as phase change materials have been synthesized by Mo et al. [64] using the sol-gel method. FTIR, EDS, SEM, and XRD analysis have been performed to characterize the nanocapsules and their results indicated that the encapsulation has been successfully done in spherical shapes and the encapsulation did not change the crystal structure of the PCM. DSC results showed that the phase transition temperature and heating density of the nanocapsules have been 222.4°C and 228.7 J/g, respectively, with high reliability of 99.7% after 50 thermal cycles. Measurement of thermal conductivity by the Hot Disk method also revealed that an approximate 31.7% increase in thermal conductivity can be achieved by adding 2% *MgO* to the naocapsule suspension.

An eutectice mixture of n-dodecanol and cetostearyl alcohol as an eutectic PCM (e-PCM) has been synthesized with PMMA by Yadav et al. [65] as a novel approach to control the temperature of chocolate in a packaging box using ultrasound-assisted nanocapsules of PCM. DSC results for the eutectic nanocapsules revealed that the melting point and latent heat are 18.8°C and 192.6 J/g, respectively. The prepared nanocapsules can be used in packaging, buildings, and slurry applications when the application temperature range is between 18°C and 22°C.

2.6 MATERIALS FOR ENCAPSULATION OF PCMs

Encapsulation with different sizes from nano to macro involves encapsulation of the PCM in a specific selected material container with certain geometry and size based on the TES system. PCMs thermal properties such as thermal conductivity, phase change temperature, and the heat associated with the phase transitions are the key points for the selection of encapsulation materials; however, other properties like stability, density, volatility of organic compound emission, flammability, permeability, and mechanical properties are among the factors that play a main role for specific applications of capsules. Morphology and size distribution are also added to the above parameters in the case of micro- and nanoencapsulation. Shell materials can be classified in the three main catergories of metallic, polymeric, and ceramic which define the mechanical and thermal properties of capsules with some advantages and disadvantages as illustrated in Figure 2.18.

Despite possible corrosion problems, due to the high mechanical strength and high thermal stability, matallic shells are mainly used for the fabrication of high-temperature macroencapsulations. Based on the literature, among the materials used

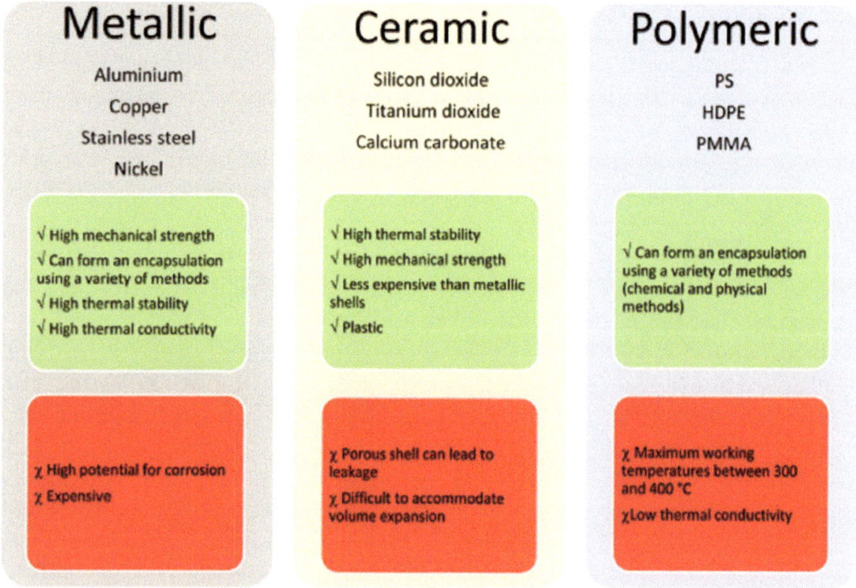

FIGURE 2.18 Materials for encapsulation with advantages and disadvantages [66]. (Permission from Elsevier.com.)

for the shell carbon steel and stainless steel, nickel and its alloys, sodium silicate, silicon dioxide, calcium carbonate, and titanium dioxide were recommended for further research [66]. The second group of shell materials are ceramics with low cost that possess outstanding thermochemical stability at high-temperature conditions under molten salt compared to metallic ones; however, the microporous structure of ceramics can impact the strength, thermal conductivity, and leakage of PCM capsules [67].

Organic shell materials such as polymer have also been used extensively as wall materials of microcapsules despite low thermal conductivity, slow thermal responses, some chemical incompatibilities, and inability to work at high temperatures [68].

A great interest has been risen for the encapsulation of bio-based and biodegradable PCMs that can be used in the food industry, textile, bio-catalysts, pharmaceuticals, and medical care and therefore more research can be done and the field is still trying to tackle the technological challenges.

2.7 SUMMARY

Thermal performance of the thermal energy storage system can be improved by encapsulation of PCMs in which the PCM is protected by a layer that prevents the PCM to be in contact with the environment and to avoid corrosion of TES system components. Encapsulation of PCMs can be classified in three sizes of macroencapsulation, microencapsulation, and nanoencapsulation.

Three categories of preparation techniques for encapsulation of PCMs were identified in this chapter as physical, chemical, and physico-chemical approaches.

The requirements of the microcapsules, such as the needed capsule size, the materials of the core/shell, the thickness of the microcapsule shell, the thermal and mechanical properties of the capsule, etc., have a significant impact on the choice of various microencapsulation procedures. As a result, the procedure needs to be specifically adjusted in order to provide a suitable result. The most often mentioned processes in the literature for creating MEPCM include spray drying, coacervation, emulsion polymerization, suspension polymerization, and interfacial polymerization. NEPCMs and multiphase core materials might be able to combat MEPCM instability and supercooling, which still require further research. Although encapsulation of PCMs stands as a feasible solution to tackle some of the challenges with PCMs applications, the technology is still at the development stage and is likely to branch to other sectors.

REFERENCES

1. Fang G., Chen Z., Li H. (2010) Synthesis and properties of microencapsulated paraffin composites with SiO2 shell as thermal energy storage materials, Chem. Eng. J., 163, pp. 154–159.
2. Albdour S., Haddad Z., Sharaf O., Alazzam A., Abu-Nada E. (2022) Micro/nano-encapsulated phase change materials (ePCMs) for solar photochemical absorption and storage: fundamentals, recent advances, and future directions, Progr. Energy Combust. Sci., 93, p. 101037.
3. Tyagi V.V., Kaushik S.C., Tyagi S.K., Akiyama T. (2011) Development of phase change materials based microencapsulated technology for buildings: a review, Renew. Sustain. Energy Rev., 15, pp. 1373–1391.
4. Arshad A., Jabbal M., Yan Y.Y. (2019) The micro/nano-PCMs for thermal energy storage systems: a state-of-the-art review, Int. J. Energy Res., 43, pp. 5572–5620.
5. Chandel S.S., Agarwal T. (2017) Review of current state of research on energy storage, toxicity, health hazards and commercialization of PCMs, Renew. Sustain. Energy Rev., 68, pp. 581–596.
6. Chen Z., Fang G. (2011) Preparation and heat transfer characteristics of microencapsulated phase change material slurry: a review, Renew. Sustain. Energy Rev., 15, pp. 4624–4632.
7. Sabbah R., Farid M.M., Al-Hallaj S. (2009) Micro-channel heat sink with slurry of water with micro-encapsulated phase change material: 3D-numerical study, Appl. Therm. Eng., 29, pp. 445–454.
8. Zhang P., Ma Z.W., Wang R.Z. (2010) An overview of phase change material slurries: MPCS and CHS, Renew. Sustain. Energy Rev., 14, pp. 598–614.
9. Salunkhe P.B., Shembekar P.S. (2012) A review on effect of phase change material encapsulation on the thermal performance of a system, Renew. Sustain. Energy Rev., 16, pp. 5603–5616.
10. Kwak K., Kim C. (2005) Viscosity and thermal conductivity of copper oxide nanofluid dispersed in ethylene glycol, Korea–Australia Rheol. J., 17, pp. 35–40.
11. Azari A., Kalbasi M., Rahimi M. (2013) Numerical study on the laminar convective heat transfer of alumina/water nanofluids, J. Thermophys. Heat Transf., 27, pp. 170–173.
12. Boh B., Šumiga B. (2008) Microencapsulation technology and its applications in building construction materials, Tehnologija mikrokapsuliranja in njena uporaba v gradbenih materialih, RMZ–Mater. Geoenviron., 55, pp. 329–344.
13. Zhao C.Y., Zhang G.H. (2011) Review on microencapsulated phase change materials (MEPCMs): fabrication, characterization and applications, Renew. Sustain. Energy Rev., 15, pp. 3813–3832.

14. Jamekhorshid A., Sadrameli S.M., Farid M. (2014) A review of microencapsulation methods of PCMs as a thermal energy storage (TES) medium, Renew. Sustain. Energy Rev., 31, pp. 531–542.
15. Sheikh Y., Hamdan M.O., Sakhi S. (2023) A review of micro-encapsulated PCMs used for thermal management and energy storage systems: fundamentals, materials, synthesis, and applications, J. Energy Storage, 72, p. 108472.
16. Wang J.P., Zhang X.X., Wang X.C. (2011) Preparation, characterization and permeation kinetics description of calcium alginate macro-capsules containing shape-stabilized PCMs, Renew. Energy, 30(11), pp. 2984–2991.
17. Poncelet D. (2006) Microencapsulation: fundamentals, methods, and applications. In: Blitz J., Gunoko V., editors. Surface chemistry in biomedical and environmental science. Netherlands: Springer, pp. 23–34.
18. Venkatesan P., Manavalan R., Valliappan K. (2009) Microencapsulation: a vital technique in novel drug delivery system, J. Pharm. Sci., 1, pp. 26–35.
19. Wurster D.E. (1959) Air-suspension technique of coating drug particles: a preliminary report, J. Am, Pharm. Assoc., 48, pp. 451–454.
20. Huang Y., Stonehouse A., Abeykoon C. (2023) Encapsulation methods for PCMs - a critical review, Int. J. Heat Mass Transf., 200, p. 123458.
21. Venkatesan P., Manavalan R., Vallippan K. (2009) Microencapsulation: a vital technique in novel drug delivery system, J. Pharm. Sci. Res., 1(4), pp. 26–35.
22. Heinzen C., Berger A., Marison I. (2002) Use of vibration technology for jet breakup for encapsulation of cells, microbes and liquids in monodisperse microcapsules, Landbauforsch Volkenrode, 241, p. 19.
23. Wang L., Meng D. (2010) Fatty acid eutectic/polymethyl methacrylate composite as form-stable phase change material for thermal energy storage, Appl. Energy, 87, pp. 2660–2665.
24. Wang W., Yang X., Fang Y., Ding J. (2009) Preparation and performance of form-stable polyethylene glycol/silicon dioxide composites as solid–liquid phase change materials, Appl. Energy, 86, pp. 170–174.
25. Mohammed N.K., Tan C.P., Manap Y.A., Muhialdin B.J., Hussin A.S.M. (2020) Spray drying for encapsulation of oils: a review, Molecules, 25(17), p. 3873.
26. Gravalos Moreno J., Calvo Herrera I., Mieres Royo J., Cubillo Capuz J., Simon B., Maria A. (2009) Procedure for microencapsulation of phase change materials by spray drying, EP Patent 2, 119, p. 498.
27. Borreguero A.M., Valverde J.L., Rodríguez J.F., Barber A.H., Cubillo J.J., Carmona M. (2011) Synthesis and characterization of microcapsules containing Rubitherm ®RT27 obtained by spray drying, Chem. Eng. J., 166, pp. 384–390.
28. Hawlader M.N.A., Uddin M.S., Khin M.M. (2003) Microencapsulated PCM thermal energy storage system, Appl. Energy, 74, pp. 195–202.
29. Fei B., Lu H., Qi K., Shi H., Liu T., Li X. (2008) Multi-functional microcapsules produced by aerosol reaction, J. Aerosol. Sci., 39, pp. 1089–1098.
30. Borroeguero A.M., Valverde J.L., Rodriguez J.F., Barber A.H., Cubillo J.J., Carmona M. (2011) Synthesis and characterization of microcapsules containing Rubitherm ®RT27 obtained by spray drying, Chem. Eng., J., 166, pp. 389–390.
31. Ghosh S.K. (2006) Functional coatings: by polymer microencapsulation. Germany: Wiley-vch.
32. Dorati R., Genta I., Modena T., Conti B. (2013) Microencapsulation of a hydrophilic model molecule through vibration nozzle and emulsion phase inversion technologies, J. Microencapsul., 30, pp. 559–570.
33. Peng G., Dou G., Hu Y., Chen Z. (2020) PCM microcapsules for thermal energy storage, Adv. Polym. Technol., 2020, p. 9490873.

34. Yadav S.K., Khilar K.C., Suresh A.K. (1997) Release rates from semi-crystalline polymer formed by interfacial polymerization, J. Memb. Sci., 125, pp. 213–218.
35. Park S.-J., Shin Y.-S., Lee J.-R. (2001) Preparation and characterization of microcapsules containing lemon oil, J. Colloid. Interf. Sci., 241, pp. 502–508.
36. Tseng Y.-H., Fang M.-H., Tsai P.-S., Yang Y.-M. (2005) Preparation of microencapsulated phase-change materials (MCPCMs) by means of interfacial polycondensation, J. Microencapsul., 22, pp. 37–46.
37. Ismail A., Zhou J., Aday A., Daidoff I., Odukomaiya A., Wang J. (2023) Microencapsulation of biobased PCMs with silica coated inorganic shell for thermal energy storage, J. Build. Eng., 67, p. 105981.
38. Khakzad F., Alinejad Z., Shirin Abadi A., Ghasemi M., Mahdavian A. (2013) Optimization of parameters in preparation of PCM microcapsules based on melamine formaldehyde through dispersion polymerization, Colloid Polym. Sci., 292(2), pp. 1–14.
39. Wei J., Li Z., Liu L., Liu X. (2013) Preparation and characterization of novel polyamide paraffin microencapsulated PCM by interfacial polymerization technique, J. Appl. Polym. Sci., 127, pp. 4588–4593.
40. Nikpourian H., Bahramian A.R., Abdolahi M. (2020) On the thermal performance of novel PCM nanocapsule: the effect of core/shell, Renew. Energy, 151, pp. 322–331.
41. Li M., Mu B., Altabey W.A. (2018) A review study on preparation and application of microencapsulated PCMs, Int. Sustain. Mater. Struct. Syst., 3(2), pp. 151–170.
42. Li W., Song G., Tang G., Chu X., Ma S., Liu C. (2011) Morphology, structure, and thermal stability of microencapsulated PCM with copolymer shell, Energy, 36, pp. 785–791.
43. Wang H., Wang J.P., Wang X., Li W., Zhang X. (2013) Preparation of microencapsulated PCMs containing two-phase core materials, Ind. Eng. Chem. Res., 52, pp. 14706–14712.
44. Huang J., Wang T., Zhu P., Xiao J. (2013) Preparation, characterization, and thermal properties of the microencapsulation of a hydrated salt as PCM storage materials, Thermochim. Acta, 557, pp. 1–6.
45. Wang T., Huang J., Zhu P., Xiao J. (2013) Fabrication and characterization of microencapsulated sodium phosphate dodecahydrate with different crosslinked polymer shells, Colloid Polym. Sci., 291, pp. 2463–2468.
46. Patil J.R., Mahanwar P.A., Sundaramoorty E., Mundhe S. (2023) A review of the thermal storage of PCM, morphology, synthesis methods, characterization, and application of microencapsulated PCM, J. Polym. Eng., 43(4), pp. 354–375.
47. Wang Q., Yang L., J. Song J. (2023) Preparation, thermal conductivity, and application of nano-enhanced phase change materials in solar heat collection: a review, J. Energy Storage, 63, p. 107047.
48. Han S., Chen Y., Lyu S., Chen Z., Wang S., Fu F. (2020) Effects of processing conditions on the properties of paraffin/melamine-urea-formaldehyde microcapsules prepared by in situ polymerization, Colloids Surf. A: Physicochem. Eng. Asp., 585, p. 124046.
49. Yeo Y., Baek N., Park K. (2001) Microencapsulation methods for delivery of protein drugs, Biotechnol. Bioprocess. Eng., 6, pp. 213–230.
50. Sari A., Bicer A., Alkan C., Ozcan A. (2019) Thermal energy storage characteristics of myristic acid-palmitic eutectic mixtures encapsulated in PMMA shell, Sol. Energy Mater. Sol. Cells, 193, pp. 1–6.
51. Doguscu D., Kizil C., Bicer A., Sari A., Alkan C. (2018) Microencapsulated n-alkane eutectics in polystyrene for solar thermal applications, Sol. Energy, 160, pp. 32–42.
52. Huang H.-J., Yuan W.-K., Chen X.D. (2006) Microencapsulation based on emulsification for producing pharmaceutical products: a literature review, Dev. Chem. Eng. Mineral Process, 14, pp. 515–544.

53. Jyothi N.V.N., Prasanna P.M., Sakarkar S.N., Prabha K.S., Ramaiah P.S., Srawan G. (2010) Microencapsulation techniques, factors influencing encapsulation efficiency, J. Microencapsul., 27, pp. 187–197.
54. Özonur Y., Mazman M., Paksoy H., Evliya H. (2006) Microencapsulation of coco fatty acid mixture for thermal energy storage with phase change material, Int. J. Energy Res., 30, pp. 741–749.
55. Maleki M., Sadrameli S.M., Dorkoosh F., Sharifi H. (2014) Synthetic and physical characterization of phase change materials microencapsulated by complex coacervation for thermal energy storage applications, Int. J. Energy Res., 38(11), pp. 1492–1500.
56. Wang L.-Y., Tsai P.-S., Yang Y.-M. (2006) Preparation of silica microspheres encapsulating phase-change material by sol-gel method in O/W emulsion, J. Microencapsul., 23, pp. 3–14.
57. Zhang H., Wang X., Wu D. (2010) Silica encapsulation of n-octadecane via sol-gel process: a novel microencapsulated phase-change material with enhanced thermal conductivity and performance, J. Colloid. Interf. Sci., 343, pp. 246–255.
58. Chen Z., Cao L., Shan F., Fang G. (2013) Preparation, and characteristics of microencapsulated stearic acid as composite thermal energy storage material in buildings, Energy Build., 62, pp. 469–474.
59. Chen Z., Cao L., Fang G., Shan F. (2013) Synthesis and characterization of microencapsulated paraffin microcapsules as shape-stabilized thermal energy storage materials, Nanoscale Microscale Thermophys. Eng., 17, pp. 112–123.
60. Sukhorukov G., Fery A., Möhwald H. (2005) Intelligent micro- and nanocapsules, Progr. Polym. Sci., 30, pp. 885–897.
61. Fang Y., Yu H., Wan W., Gao X., Zhang Z. (2013) Preparation and thermal performance of polystyrene/n-tetradecane composite nanoencapsulated cold energy storage phase change materials, Energy Convers. Manag., 76, pp. 430–436.
62. Tahan Latibari S., Mehrali M., Mehrali M. (2013), Indra Mahlia TM, Cornelis Metselaar HS. Synthesis, characterization, and thermal properties of nanoencapsulated PCMs via sol-gel method, Energy, 61, pp. 664–672.
63. Fang Y., Kuang S., Gao X., Zhang Z. (2008) Preparation, and characterization of novel nanoencapsulated phase change materials, Energy Convers. Manag., 49, pp. 3704–3707.
64. Mo S., Xiao B., Li J., Jia L., Chen Y. (2024) $LiNO_3$/NaCl nanocapsules with high thermal properties for medium-temperature thermal energy storage, J. Energy Storage, 83, p. 110672.
65. Yadav A., Singh J., Reddy V.J., Chattopadhyay S. (2024) A novel approach to control the temperature of chocolate in a packaging box using ultrasound-assisted nanoencapsulated eutectic PCMs powder, Powder Technol., 435, p. 119375.
66. Palacios A., Navarro-Rivero M.E., Zou B., Jiang Z., Harrison M.T., Ding Y. (2023) A perspective on PCM encapsulation: guidance for encapsulation design methodology from low to high-temperature thermal energy storage applications, J. Energy Storage, 72, p. 108597.
67. Dele-Afolabi T.T., Hanim M.A.A., Norkhairunnisa M., Sobri S., Calin R. (2017) Investigating the effect of porosity level and pore former type on the mechanical and corrosion resistance properties of agro-waste shaped porous alumina ceramics, Ceram. Int., 43(12), pp. 8743–8754.
68. Liu H., Wang X., Wu D. (2019) Innovative design of microencapsulated PCMs for thermal energy storage and versatile applications: a review, Sustain. Energy Fuels, 3, pp. 1091–1149.

3 Thermal Energy Storage Systems

3.1 INTRODUCTION

Phase change materials (PCMs) release and absorb a significant quantity of heat at constant temperature during the phase transition. This characteristic makes them very attractive for thermal energy storage (TES) systems following the 1973–1974 energy crisis. Today, effective utilization of energy is essential due to the limited fossil fuel reserves and greenhouse gas emission concerns. Using PCMs in TES provides an elegant and realistic solution to increase the efficiency of the storage and consumption of energy in many domestic and industrial sectors. This chapter explains two different types of TES systems, namely sensible and latent heat storage. Different applications of TES in buildings, solar panels, textiles, electrical appliances, lithium-ion batteries, etc., for energy management and temperature control are presented and discussed.

The use of PCMs for energy storage minimizes the imbalance between supply and demand, enhances the efficiency and dependability of energy distribution networks, and contributes significantly to the overall energy conservation. Solid-liquid, liquid-gas, and solid-gas phase transitions are possible for the energy storage system; however, PCMs with solid-liquid phase transitions are the most popular due to their wide temperature range, high storage capacity, and minimal volume changes during the phase transition. High energy storage capacity is also present in PCMs with liquid-gas transition, although their usage is constrained by significant volume fluctuations [1]. Solid-solid PCMs are particularly appealing since they don't need any external energy sources to store energy because the material's crystalline state changes. However, their energy capacity is much lower than the other two types of phase transitions [2–6]. As explained in Chapter 1, the most commonly applicable PCMs are paraffin waxes, salt hydrates, fatty acids, and their eutectics.

3.2 ENERGY STORAGE TECHNIQUES

An available and not used thermal energy either cooling or heating can be stored in an appropriate medium for further application. The process involves charging, storing, and discharging. Energy storage systems can be categorized into three forms, sensible, latent, and chemical reaction energy storage, as illustrated in Figure 3.1. If the packing material in the medium is with no phase change, then the system is sensible energy storage, while for the systems filled with PCMs, the process is called latent heat storage. The energy storage in the sensible energy storage

DOI: 10.1201/9781003518280-3

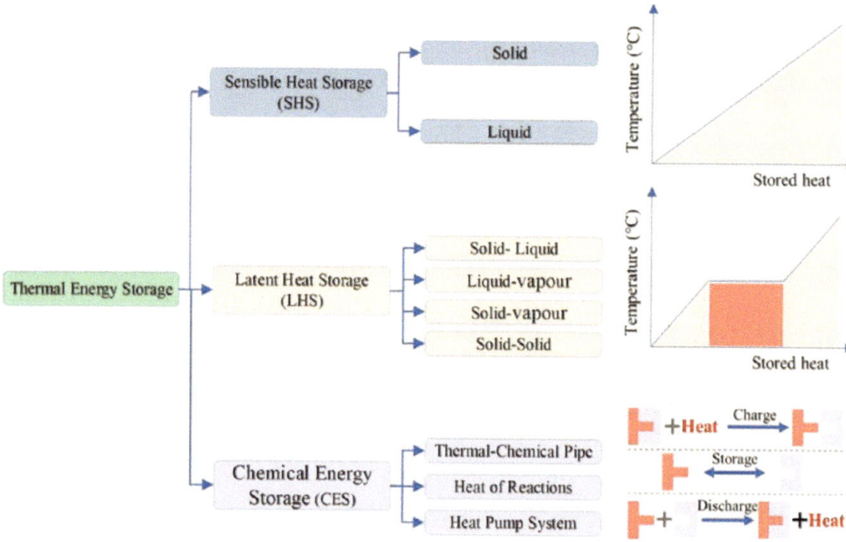

FIGURE 3.1 Classification of energy storage systems [7]. (Permission from Elsevier.com.)

depends on temperature swings, mass, and heat capacity of the materials, while for the latent heat storage, the storage is a function of latent heat of the materials. As shown in Figure 3.1, there are four types of PCMs based on the phase change process: solid-liquid, liquid-vapour, solid-vapour, and solid-solid. In the chemical energy storage systems, which are used as an inter-seasonal energy storage, the energy storage capacity relies on the chemical reaction with exothermic and endothermic processes. These systems have not been practically used for large-scale development due to their complex reaction processes, high initial investment costs, and low reliability [7].

3.2.1 Sensible Energy Storage

In sensible energy storage systems, materials with no phase transition are used for the delivery of heating/cooling energy from one stream to another. Metals, ceramics, pebbles, rocks, concrete, and sand are some of the materials that are frequently employed as packings inside the system. Their energy density is low and requires a significant increase in unit size due to the enormous volume of storage they need, making them ideal for modest applications where cost is a top priority. Thermal regenerators are among the most common sensible energy storage systems.

3.2.1.1 Thermal Regenerators

Thermal regenerators are packed bed heat exchangers that use materials with high heat capacities to absorb and release heat during hot and cold periods. For more than two centuries, they have been employed in the industrial and building sector for energy

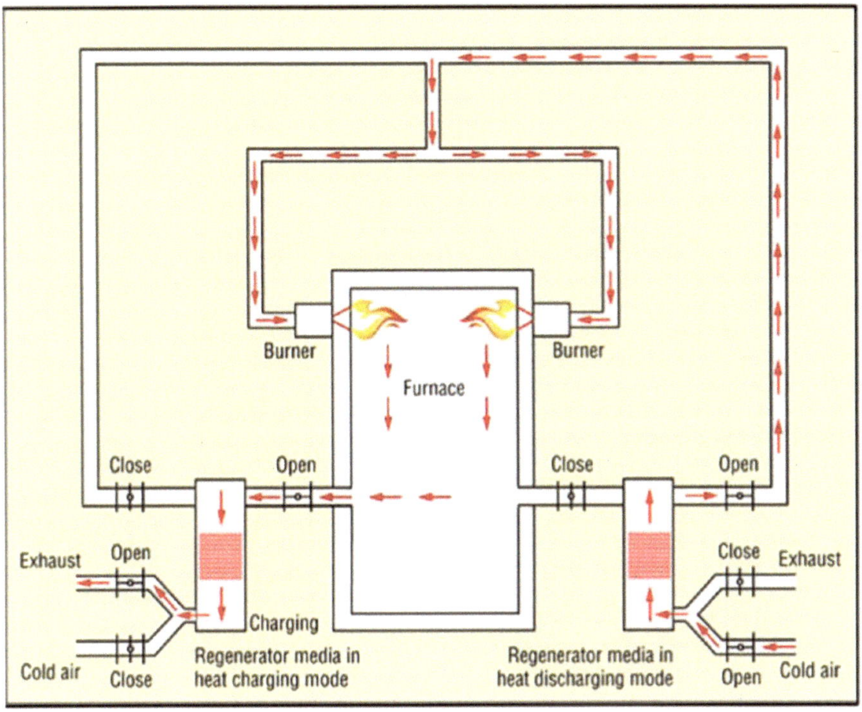

FIGURE 3.2 Schematic diagram of a fixed-bed regenerator [8]. (Permission from Elsevier.com.)

efficiency enhancement. The regenerators are used as heat recovery systems in two different types: fixed-bed and rotary [8]. In fixed-bed regenerators (FBRs), the storage bed is stationary, and the hot and cold streams alternately pass through the bed using solenoid valves for releasing and absorbing the heat alternately as shown in Figure 3.2. They are used in aluminium, steel, and glass furnaces for the recovery of waste flue gas heat from the stack for air preheating to the furnace. Due to the periodic nature of the system, at least two distinct matrix beds must always be used for the continuous operation of the system so that one bed is being heated (hot period) and the other is being cooled (cold period) (Figure 3.2).

Rotary regenerators, which are also called air preheaters, heat wheel, or Ljungstrom, are mainly used in the power plants, process industries, and air conditioning systems from low- to medium-temperature ranges. In such systems, the rotor is a sizable porous disk fabricated from some materials having high heat capacity which rotates between two side-by-side ducts: one facing the waste hot gas and the remaining facing the cold gas normal air. The fraction of the rotor surface area facing the hot gas normally with 50% of the total rotor surface absorbs the heat from the hot gas (waste gas) and releases it to the cold gas (air) during the rotation of the rotor. The sketch of the rotor is illustrated in Figure 3.3.

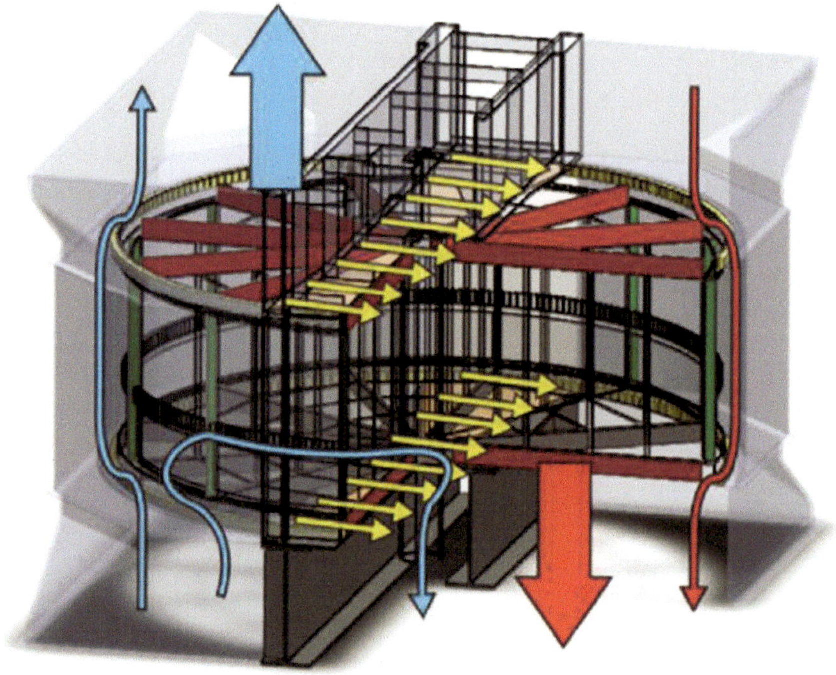

FIGURE 3.3 Sketch of the rotor in a rotary regenerator [8]. (Permission from Elsevier.com.)

3.2.2 Latent Heat Energy Storage

This type of energy storage relies solely on the material's phase change enthalpy, which doesn't involve a change in temperature. Phase changes in the material can occur between solid and liquid, liquid and gas, or solid and gas. In the latent heat energy storage system, high energy storage as much as 50–100 times more than sensible heat can be achieved with reduced temperature fluctuations resulting in minimized heat loss and container size [9]. Due to its greater energy storage capacity per unit volume and practically constant temperature during the heat storage, latent heat storage technique is preferable to sensible heat storage methods. Additionally, these come in a wide range of temperatures to accommodate various applications. Solid-liquid phase transition materials are the focus of the current work since they are frequently employed in a wide range of applications for thermal management of heating/cooling systems, including free cooling in conditioning [10, 11], electrical appliances [12, 13], solar panels [14], lithium-ion batteries [15], and building materials for building energy management [16]. PCM must have the following properties to be used as a heat storage medium: workable melting temperature that is within the system's mean temperature range; high latent heat capacity; compatibility with the container; minimal volume changes; stability of properties during melt-freeze cycles; nontoxicity; environmental safety; non-flammability; cost-effectiveness; and ease of availability. Table 3.1 provides a summary of the necessary characteristics for a material to be used as PCM in the TES system.

TABLE 3.1
Characteristics of the PCMs Used in the TES Systems

Thermal	Kinetic	Chemical	Physical	Others
Melting point in the desired operating temperature range	No super-cooling	Chemical stable over a no of freeze-thaw cycles	Small vapour pressure	Low cost
High latent heat capacity	No sub-cooling	Noncorrosive	Small phase transition volume change	Easily available
High thermal conductivity	No phase segregation	Nontoxic	High density	
High specific heat	Good nucleating properties			

3.3 APPLICATIONS OF TES

3.3.1 Building Energy Management

Building energy consumption has increased sharply in recent years and the amount of CO_2 emissions has tripled between 1990 and 2022 as illustrated in Figure 3.4 [17].

Almost 30–40% of the total energy has been utilized in building applications in most of the countries in the world such as the USA and the UK although in some countries such as China, this has been decreased to 21% [18–20]. Most of this energy is utilized in the air conditioning and heating equipment. Therefore, to save building energy, an innovative system must be designed and applied for energy efficiency enhancement in the building.

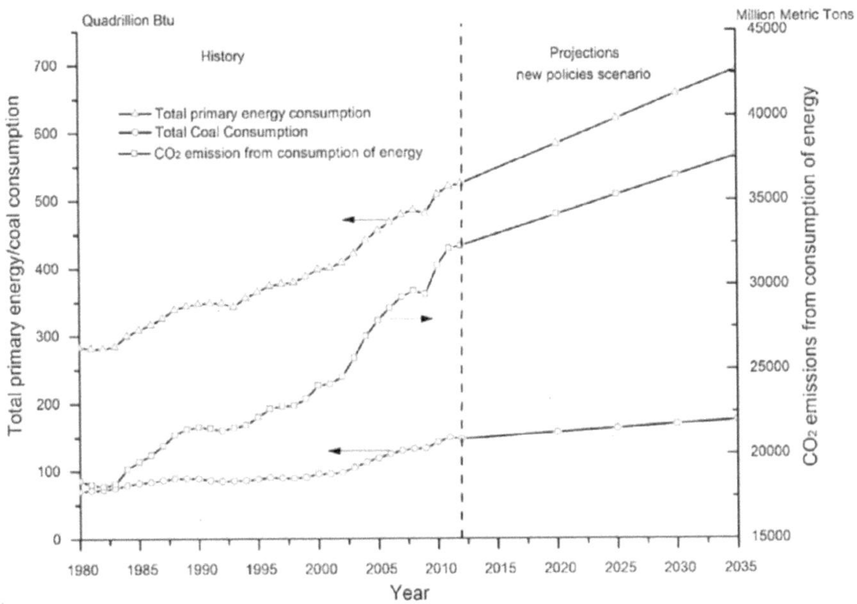

FIGURE 3.4 World energy consumption and CO_2 emissions [17]. (Permission from Elsevier.com.)

PCMs as a medium for TES systems in the buildings can have an attractive role for decreasing cooling/heating energy consumption in the residential and commercial buildings [21]. Application of PCMs in the buildings can be classified to the main building envelope such as walls, windows, floors, and roofs and the auxiliary equipment of the building such as electrical appliances, window blades, and shutters. A detailed study on the application of PCMs in building energy management will be presented in the following chapters.

3.3.2 Electrical Appliances

Almost all electrical appliances consume electrical energy which is the backbone of modernization in our societies, and this is going to be increased due to the rapid industrialization and progression of living standards. Refrigerators and freezers are the most domestic household appliances with a total world number of more than one billion, due to their continuous operation. Efficiency enhancement of the household refrigerators and freezers, which mainly depends on the compressor efficiency and on/off periods, thermal load, ambient temperature, and door openings, plays a crucial role in the electrical energy consumption in the country [22]. Efficiency enhancement of the household refrigerators can be implemented by the following modifications [23]:

- Using highly efficient compressors
- Advanced control system optimization
- Insulation upgrading
- Condenser and evaporator heat transfer improvements

In all electrical refrigerators and freezers, the compressors run in on/off mode. During the compressor work, the refrigerant in the evaporator starts absorbing cabin heat, while during the off time, there is a sharp increase in the evaporator cabinet temperature which is due to food heat release inside the fridge cabin and ambient conditions. At the same time, the compressor must work to release the heat from the condenser surface by convection. Generally, performance enhancement of the refrigeration systems can be performed by several ideas from which the main method is to enhance heat transfer from the heat exchangers such as condenser and evaporator [23]. If the condenser surface temperature can be decreased by any external agency either active or passive, the temperature inside the cabinet is maintained which in return increases the compressor off-cycle time [24]. The role of PCMs is to alleviate the condenser heat transfer by absorbing extra heat during the compressor on time and altering its phase from solid phase to liquid phase at constant temperature. This, in turn, extends the off-time cycle of the compressor and directly decreases the electrical consumption. PCMs can also play an extensive role in the electrical consumption of refrigerators and freezers when used in the evaporator and compartment of the systems to absorb extra heat from the system. A detailed study of applications of PCMs in refrigerators and freezer is presented in Chapter 5.

3.3.3 Lithium-Ion Batteries

Lithium-ion batteries can be utilized as one of the most acceptable and cost-effective alternatives to fossil fuels in the present century which can be used in different areas such as electric vehicles and motorcycles. However, they face some challenges such as lithium resources storage, life cycle ability, thermal degradation, and overheating during charging and discharging. Batteries overheating may lead to explosions and fire in critical cases [25]. Among cooling techniques that have been studied by scientists using water, air, and refrigeration, PCMs have shown superior cooling features due to their advantages in comparison with traditional methods. They work as a passive thermal management system in electric and hybrid vehicles and are cheaper than refrigeration and cooling methods that require external agencies such as electrical pumps and fans [26]. Application of PCMs in lithium-ion batteries does not require any stimulant components and the generated heat from the batteries can be absorbed by PCM during the life span of the batteries [27]. One of the main drawbacks of PCMs is their low thermal conductivity that can reduce the rate of heat transfer from the battery surface to the environment. This can be solved using additives or fillers that can be divided into carbon-based, metal-based, or alternative materials [28]. Metal foams, extended graphite, nano-graphite sheets, and carbon fibres are among the most thermal conductivity enhancements for PCMs which enhance the cooling efficiency and operating performance of the batteries as illustrated in Figure 3.5.

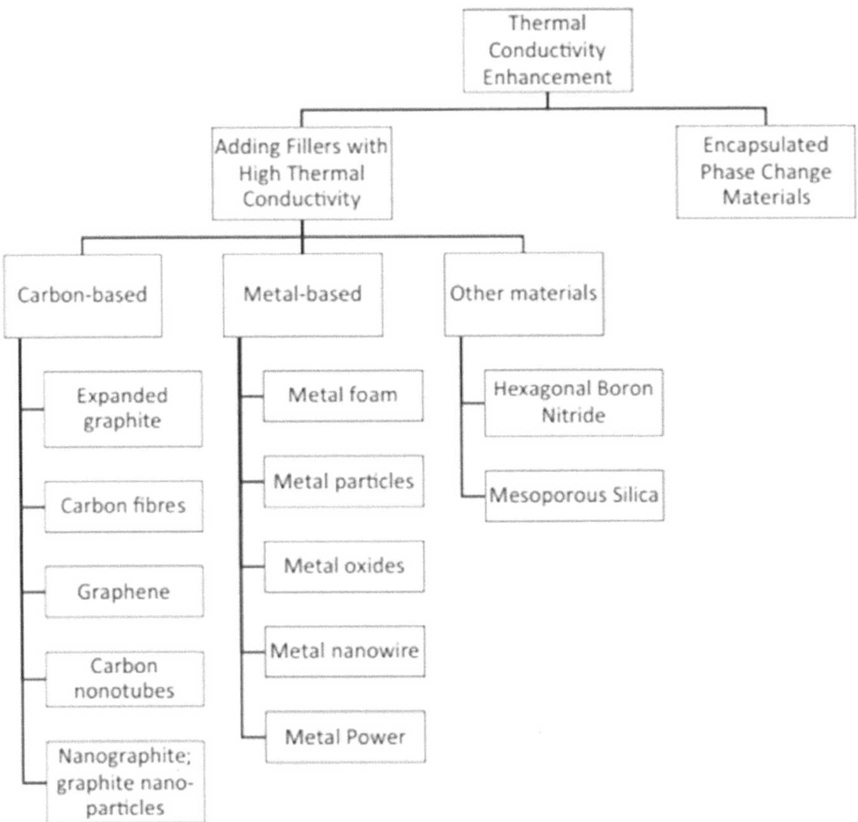

FIGURE 3.5 Thermal conductivity enhancement methods for PCMs [30]. (Permission from Elsevier.com.)

A detailed review on the application of PCMs in lithium-ion batteries is presented by Najafpour Esfahani et al. [29]. Applications of PCMs in lithium-ion batteries will be discussed in detail in Chapter 6.

3.3.4 Solar Panels

Solar energy is one of the main sources of renewable energy with the contribution of 27% of global electricity generation by 2050, using photovoltaic (PV) and concentration power systems accounting for 60% and 40%, respectively [31]. Devices that convert radiant energy from the sun into heat energy, including PV/T system, solar water heater, and solar water desalination, are called solar heat collection systems. The main application of solar energy is illustrated in Figure 3.6 [32].

One of the challenges in the operation of solar panels is their panel surface temperature control. The efficiency of the solar panels decreases considerably with increasing panel surface temperature. The power efficiency is decreased 0.5% for every 1°C increase in the solar panel surface temperature [33]. Therefore, thermal management of solar panels is invariably cardinal to the overall performance of either solar PV cell that utilizes photoconductive properties of the cell or thermal collector using concentrated radiative heat from the sun [34]. The application of PCMs in solar systems is promising and PCMs can fill the gap of intermittent solar energy supply as a representative material for the energy storage system. The first study on the application of PCMs in a solar-driven system in which PCMs can absorb and release heat during phase transition at constant temperature has been published by Telkes in 1953 [35]. The application of PCM in water heaters has been initiated since the 1980s in which PCMs

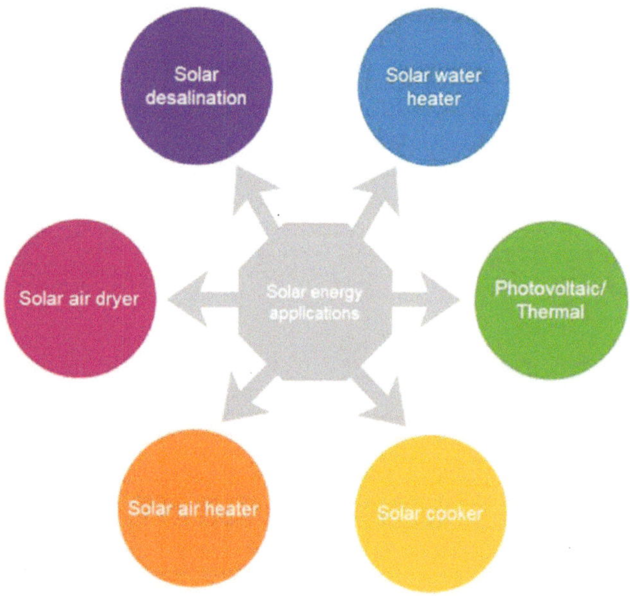

FIGURE 3.6 Main application of solar energy [32]. (Permission from Elsevier.com.)

were located at the bottom of the heater for storing and releasing solar energy. Another economical application of solar energy in households is solar cookers especially used in Third World countries such as India and Pakistan. In this device which consumes no fuel and has no operating costs with no pollution, solar energy from the sunlight is used to heat food to cook or sterilize it. PCMs can also be used in such systems to store the solar energy during the day and release it during the nighttime. The sunlight in the cooker is focused into a narrow cooking area and converts it into heat that must be trapped in the cooker to prevent losing by convection and radiation. The feasibility of utilizing PCMs as a heat storage in solar cookers has been investigated since 1995 [36]. Among the PCMs used in the solar cooker are acetamide with melting point of 82°C, acetanilide, 118°C, erythritol, 118°C, and magnesium nitrate hexahydrate with melting point of 90°C. Application of PCMs in solar panels will be discussed in Chapter 8.

3.3.5 Free Cooling

The cooling potential and performance of the AC system can be enhanced by the integration of a PCM-based TES in which the coldness of the ambient during the night can be stored in the system and releases to the input air to the AC during the day. This maintains excellent indoor air quality, with reduction of greenhouse gas emissions and maintenance costs in the building. This is applied best in the areas with the diurnal temperature difference of greater than 15°C. In this case, the melting temperature of the PCMs is at the middle of the diurnal temperatures and therefore equal temperature difference is available for charging and discharging periods. The first experimental study on free cooling has been conducted by Turnpenny et al. [37] in which the coldness of the night air was stored in the PCM and released during the daytime using heat pipes embedded in PCM to enhance heat transfer between the air and PCM as shown in Figure 3.7.

Zelba et al. did the first feasibility study on the application of PCMs for free cooling in which flat plates of PCMs with melting temperature of 20–25°C have been installed

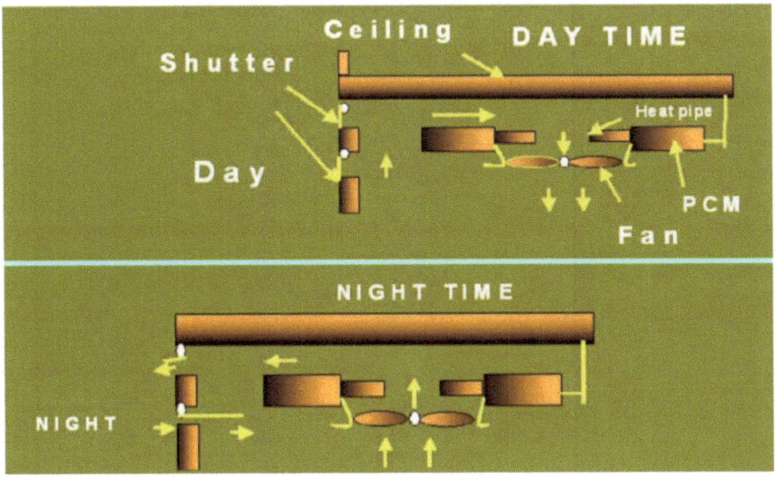

FIGURE 3.7 Free cooling system proposed by Turnpenny [37]. (Permission from Elsevier.com.)

in the roof of the building rooms [38, 39]. The melting temperature of PCM during discharge was measured to be 28°C and 30°C and during freezing or charging were 16°C and 18°C. Symmetric surface for the melting and freezing of PCM, controllable heat transfer with PCM thickness, and high area to volume ratio were some of the advantages of the flat-plate encapsulated PCM used in this study. Due to the free convection in the liquid phase, the solidification time is lesser, and the melting time is more. However, when the thickness of the encapsulated PCM was lower, the temperature difference between air and melting temperatures of the PCM was higher, and with higher air flow rate, the solidification process was improved. The performance of the PCM-based free cooling system is based on the suitability of PCM for the specific climate conditions with high latent heat, high thermal conductivity, small volume change, and less sub-cooling during the phase change and calculation of optimal mass of the PCM for the selected geometry and performance parameters of the TES system. The conventional melting temperature of PCMs for free cooling is 15–20°C. Detailed discussion regarding the application of PCM for free cooling will be presented in Chapter 10.

3.3.6 Textiles

Human environmental protection to achieve a comfortable condition is very important in our daily lives. This protection by clothing includes water protection, extreme cold and hot conditions, open fire, high voltage, propelled bullets, toxic chemicals, nuclear radiation, and biological toxins. The clothing materials can be selected for the textile products to have more comfortable conditions for the human being with all conditions in the environment. PCMs are one of the intelligent materials that can fulfil the comfortable environmental condition for us. In smart textiles, including PCMs, the extra heat can be absorbed and stored in the materials and discharges as they oscillate between solid and liquid forms based on the various changes in the environmental temperatures.

Microcapsules of PCM in the garment layers can create small, transitory heating and cooling effects by change of phase when the layers' temperature reaches the PCM melting temperature. The performance of such a system can be maximized when the wearer of the cloth is frequently going through the temperature transitions, i.e., going back and forth between cold and hot conditions.

Microencapsulation of PCMs ranges from a few microns to a few millimetres and can be achieved using several chemical techniques such as interfacial polymerization, sol-gel, coacervation, phase separation, and in situ polymerization. They can be categorized into three types, namely Monocore, Polycore, and Matrix, as illustrated in Figure 3.8.

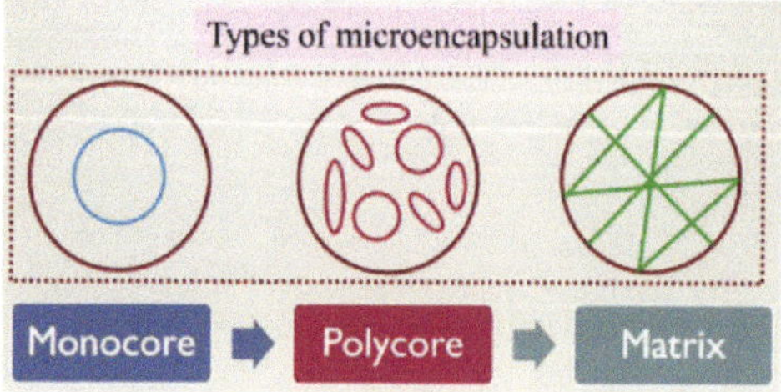

FIGURE 3.8 Different categories of PCM encapsulation [40]. (Open access journal.)

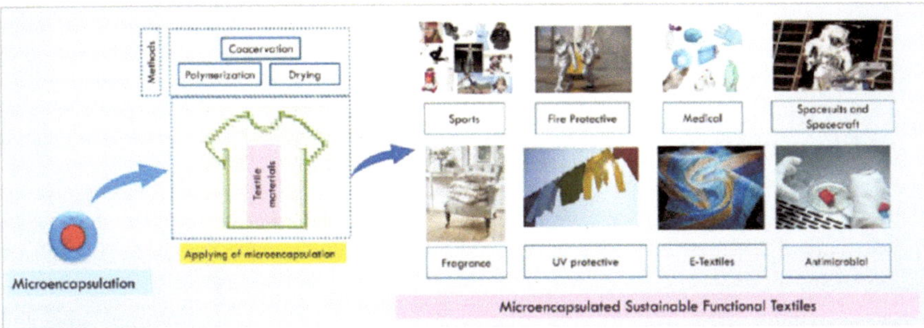

FIGURE 3.9 Different applications of microencapsulated PCMs in textiles [40]. (Open access journal.)

Spraying, coating, padding, impregnation, exhaustion dyeing, and stamping techniques are among the accomplishment processes of microcapsules in textiles which utilize microencapsulated liquids in the solid powder form deposit onto the textile fibres. They can also be easily absorbed into the fibre without any feel or colour change using binders such as starch, acrylic, polyurethane, or silicone which secures the capsules onto the garment to even after washing. Microencapsulated PCM can be incorporated into silk, cotton, and synthetic fibres to make them for the variety of applications such as sports, medicine, fragrance, E-textiles, antimicrobials, thermochromic, UV protection, and dyes [41] as shown in Figure 3.9.

Although the application of PCMs in textiles has some beneficial advantages but presents some technological challenges that need to be studied in future research which include fixing agents, short washing life span and physical property alteration. Therefore, microencapsulation of PCMs and its application in textiles requires innovative engineering and novel process technologies. The detailed description of this application is explained in Chapter 7.

3.4 COOLING THERMAL ENERGY STORAGE (CTES)

One of the highest energy consumptions in buildings, specifically in the hot climate region, is cooling energy using air conditioning. The application of cooling thermal energy systems has an important role for electrical energy consumption and reduction of environmental emissions. The system can store cooling energy during the off-peak period and release it to the building during the hot hours and fill the gap between energy demand and supply. Integration of CTES with air conditioning plays a vital role in different systems like air conditioning, space cooling, milk storage, district cooling, and several other cooling applications [42]. The advantages of a CTES system can be classified as follows:

- Energy consumption and cost reduction
- Air conditioning size reduction
- Energy and financial savings

- Low initial and maintenance costs
- Higher operating flexibility
- Peak time shifting
- Air quality improvements
- Gas emissions reduction

Cooling thermal energy systems are widely used among other types based on its quick charging, higher energy storage capacity in small volume, and free convection effect in the storage system during the charging period. The system can be categorized into many types based on the materials, methods, and applications. CTES can be incorporated into many applications such as air conditioning, cool storage, and several process cooling installations. The latent heat cooling thermal energy system can store cooling energy at low temperatures by using a suitable medium, usually deionized (DI) water at low electricity cost and retrieve cooling energy during phase transition as shown in Figure 3.10.

There are various small-scale applications of cooling energy storage systems such as vaccines preservation (2°C to 8°C), milk cooling (2°C to 4°C), cut flowers (2°C to 8°C), and small-scale air conditioning at 10°C.

3.4.1 Miscellaneous Applications of PCMs

3.4.1.1 Co-axial Electrical Cables

There are other applications of PCMs as energy storage for storage and release of energy and thermal management of some processes. One of the recent applications

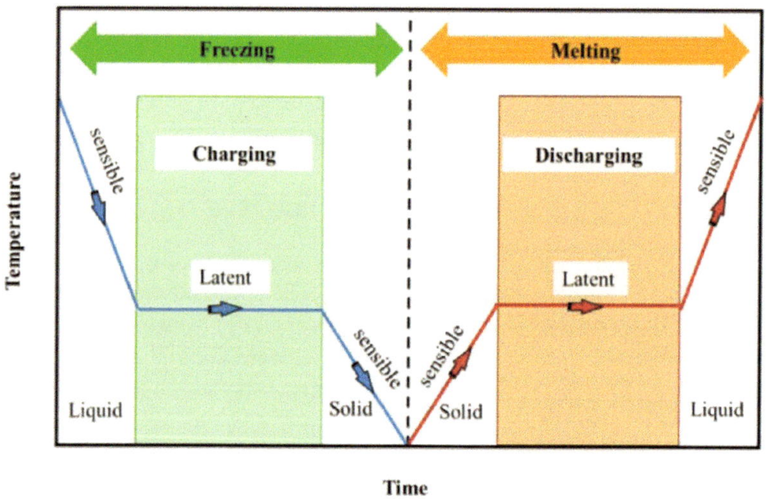

FIGURE 3.10 Schematic of PCM in cooling energy systems [43]. (Permission from Elsevier.com.)

of PCMs is in co-axial electrical cables for the stability of the cables in extreme environmental conditions [44].

Electrical cables include a central metallic tool made from copper or aluminium alloy covered by a dielectric, usually polymeric materials or air, which is used for electricity transmission in our societies. The parameters which are affected by the maximum current of electricity transmission by electrical cables are conductor cross-sectional area, cable resistance, and the attenuation which emerges as heat. Increasing the conductor temperature potentially decreases the cable performance. This can be improved by the cross-sectional area enhancement, but this increases the cost and limitation of cable application. Therefore, an effective and passive cooling technique can be a promising method in increasing the current capacity of the electrical cables [45]. One of the most applied cables in electrical applications is co-axial electrical cables which were first used in 1880 by Oliver Heaviside. They consist of two conductors divided by a dielectric, and the whole configuration is wrapped by a polymeric plastic usually made of PVC which are utilized in a variety of applications from antennas to medical devices. Thermal management of co-axial electrical cables using PCMs has been investigated worldwide in recent decades and an extensive review of the previous studies in this field has been conducted by Farid et al. [46].

3.4.1.2 Asphalt Pavement

A primary form of pavement which is widely used in urban roads and highways is asphalt pavement due to its excellent performance. One of the main challenges of asphalt pavements is their sensitivity to temperature and therefore their mechanical behaviour and performance are significantly affected by the ambient temperature due to the direct exposure to the external environment causes distresses such as rutting, bleeding, and skidding as well as low-temperature cracking after continuous high or low temperature and rapid cooling [47]. The life span of the asphalt pavement and performance of the asphalt pavement structures are mainly affected by large daily temperature difference, frequent freeze-thaw cycles, road surface frost, and snow accumulation in the cold climate regions of the world. Solving these problems was the topic of research works in the last decades and researchers have studied the material modification, structural optimization, construction technology, and quality control of the pavements and have achieved many theoretical and practical results [48]. Application of PCMs in asphalt pavement initiated in 2010 and the relevant research increased yearly. The active temperature regulation of asphalt pavement can be observed from Figure 3.11 when PCM is added to the asphalt pavement, which slows the temperature fluctuation rate and maintains the ideal working temperature more consistently.

Presently, there are many PCMs which are suitable for asphalt pavement temperature control to prevent cracking, freezing, and frosting at low temperatures but due to the non-encapsulation are easy to volatilize, leak, and cause damage to the asphalt performance when they mix with the asphalt components. This problem is solved by an effective encapsulation or shape stabilization of solid-liquid PCMs when utilized

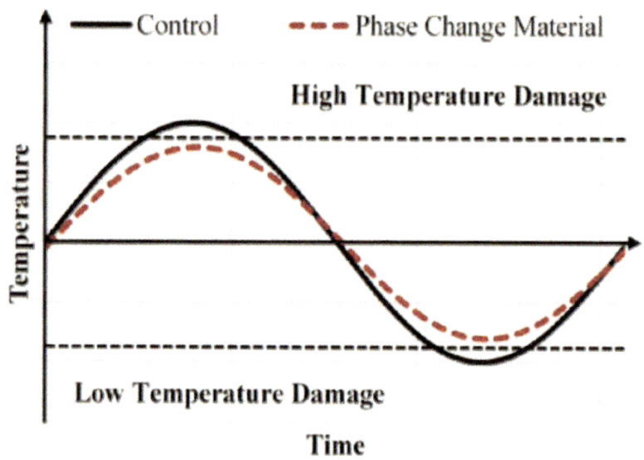

FIGURE 3.11 Temperature variation of asphalt-PCM pavement (with the Phase Change Material variation shown on the dotted line wave graph, directly beside the solid line Control wave graph) [48]. (Permission from Elsevier.com.)

in asphalt. Figure 3.12 illustrates the summary of the research performed on the application of PCMs in asphalt pavement.

3.4.1.3 Medical Applications

A broad application prospect of PCMs with excellent properties can enhance the quality of a variety of biomedical applications with some advantages over existing applications which provide novel treatment methods of disease. They can be utilized in several applications such as cold storage for vaccines and medicines, drug delivery systems, thermotherapy/cold compress therapy, and medical dressings. Some types of PCMs and their eutectics such as natural fatty acids or fatty alcohols have been utilized as thermo-response materials for cancer therapy in

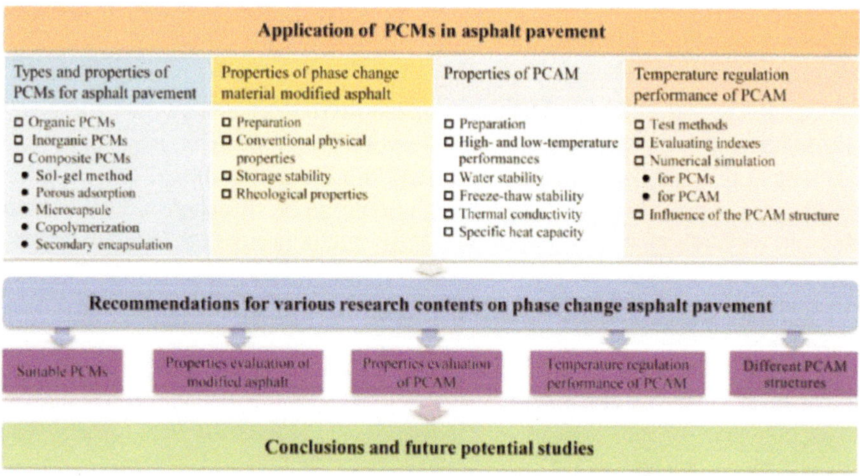

FIGURE 3.12 Application of PCMs in asphalt [49]. (Open access journal.)

TABLE 3.2
Some of the Natural PCMs Used in Drugs [51]

Species	Compound	MP (°C)
Fatty acid	Dodecanoic acid	44
	Tetradecanoic acid	54.4
	Hexadecanoic acid	63
	Octadecanoic acid	69
	11-octadecanol acid	44
	Elaidic acid	45
Fatty alcohol	1-tetradecyl alcohol	38
	Pentadecyl alcohol	41–44
	Hexadecyl alcohol	49
	1-octadecanol	59–60

recent years due to their flexible melting point (>37°C), reasonable cost, chemical stability, and excellent biocompatibility [50]. Table 3.2 lists some of the natural PCMs which have been widely utilized in packaging drugs as a control release system.

One of the novel cancer treatment strategies is photothermal therapy (PTT) in which a large amount of heat is released during the treatment process with temperature reaching the PCM phase change temperature. Application of PCMs in such therapies, including heat generation, can not only be used as gating materials to inhibit pre-release of encapsulated payloads, but also act as stimuli-responsive drug carriers for the rapid release of payloads.

An alternative promising application of PCMs is for designing novel anticancer drug system in which PCM is often utilized to encapsulate drugs into matrix material for control release. In this case, the drug and PCM are mixed in a solid matrix and with a temperature change, the PCMs undergo a discontinuous phase change or morphological change to perform control release. The role of PCMs is to protect drugs from leaking permanently during transport and increase cellular uptake of drugs by stimulating release after the drug enters the target cells. An extensive review regarding the applications of PCMs in cancer therapy has been presented by Cao et al. [52]. This application is described in detail in Chapter 11 of the book.

3.4.1.4 Applications in Greenhouses

One of the largest energy demand sectors that require energy mostly from fossil fuels for indoor environment control and plant growth and crop yield are greenhouses in which the microclimate can be controlled by proper heating, cooling, lighting, and ventilation techniques. Application of TES systems using PCMs is one of the solutions to achieve considerable energy savings in greenhouse heating and cooling processes. The role of PCMs is to store excess heat from active or passive heating sources in greenhouse such as heaters, heat pumps, solar systems, or sunlight and release it to the greenhouse when required. They can also maintain passive cooling inside the greenhouse by the absorption of excess solar radiation diurnally and passive nocturnal heating by releasing the stored heat. Various types of TES systems are applied in the thermal management of greenhouses such as sensible TES using pebble and rock beds, water reservoirs, or underground pipes and latent heat TES using PCMs [53, 54]. There are two techniques of active and passive for PCMs adaptation in heating/cooling systems of greenhouses to reduce heating/cooling loads. In active mode, air or water transfers thermal energy to the greenhouse through a duct or pipe from the heating source, while in a passive system, PCM with appropriate melting point can control the temperature by absorbing excessive thermal energy during melting and release them when required during freezing periods. Some of the applications of PCMs in greenhouse microclimates are active systems, PCM integrated with heaters and heat pump, PCM integrated with solar thermal collectors, passive systems, and PCM storage inside containers. A comfortable indoor thermal environment and efficient energy conservation can be achieved by using PCMs with appropriate thermophysical properties. A detailed description of active and passive systems utilized in the greenhouses has been studied by Nishad and Krupa [55]. Other applications of TES in vehicles, high-temperature systems, electronics, and glasshouses are presented in Chapters 9, 12, 13, and 14, respectively.

3.5 SUMMARY

TES systems, including sensible and latent, can utilize excessive energy when available and release it when required. Sensible energy storage systems are normally packed with pebbles, glass, metals, and ceramics, while in latent heat, TES systems' PCMs are used. Latent heat storage materials have much higher storage capacity in comparison to sensible TES with a narrow temperature variation during the phase transition.

Electricity consumption fluctuates during the day and night according to the demand by different sectors consumptions, causing differential tariff in the peak and off-peak periods. One of the main applications of PCMs as an energy storage system in electronics and electrical appliances is to shift the electricity usage from peak to off-peak periods that provide significant economic benefits.

The applications of TES systems are not limited to storage and release of heating/cooling energy. They can also be utilized for thermal management, efficiency enhancement, and temperature monitoring of thermal systems.

The main application of TES in electrical appliances, lithium-ion batteries, solar panels, free cooling, and textiles has been briefly described. Some other applications such as co-axial electrical cables, asphalt, medical, and greenhouse applications have also been defined in this chapter. Figure 3.13 demonstrates different types of TES systems, their applications, and PCMs properties. Detailed descriptions of the applications of PCMs in building construction materials, appliances, lithium-ion batteries, textiles, solar panels, vehicles, and medical applications will be discussed in the preceding chapters of the book.

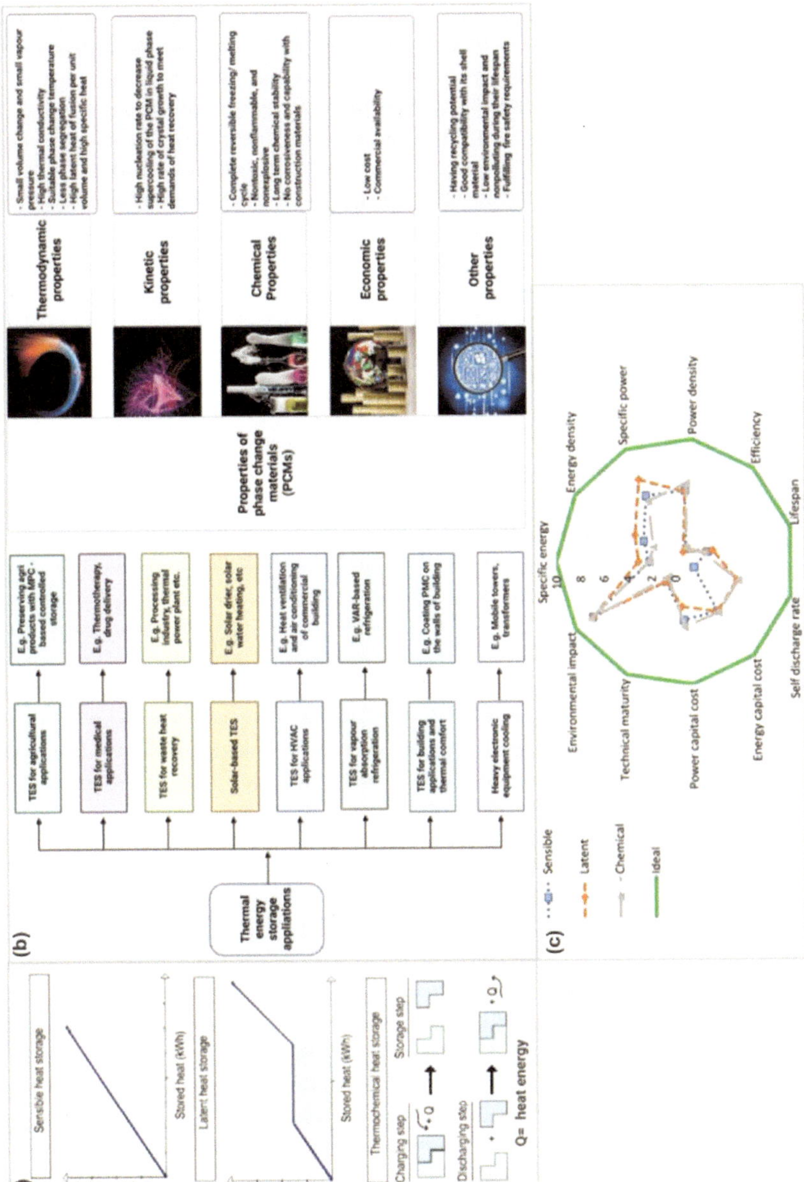

FIGURE 3.13 Summary of energy storage types (a), applications (b), and properties of PCMs (c) [56]. (Open access journal.)

REFERENCES

1. Kenisarin M., Mahkamov K. (2007) Solar energy storage using phase change materials, Renew. Sustain. Energy Rev., 11, pp. 1913–1965.
2. Garg H.P., Mullick S.C., Bhargava A.K. (1985) Solar thermal energy storage. Dordrecht, Holland: Riedel Publishing Company.
3. Lane G.A. (1983) Solar heat storage: latent heat materials. Background and scientific principles, vol. 1. Boca Raton, USA: CRC Press.
4. Lane G.A. (1985) Solar heat storage: latent heat materials. Technology, vol. 2. Boca Raton, USA: CRC Press.
5. Dincer I., Rosen M.A. (2002) Thermal energy storage: systems and applications. Chichester, England: Wiley.
6. Farid M.M., Khudhair A.M., Razack S.A.K., Al-Hallaj S. (2004) A review on phase change energy storage: materials and applications, Energy Convers. Manag., 45(9–10), pp. 1597–1615.
7. Li C., Wen X., Cai W., Yu H., Liu D. (2023) Phase change material for passive cooling in building envelopes: a comprehensive review, J. Build. Eng., 65, p. 105763.
8. Sadrameli S.M. (2016) Mathematical models for the simulation of thermal regenerators: an-state-of-the-art-review, Renew. Sustain. Energy Rev., 58, pp. 462–476.
9. Yang Z., Walvekar R., Wong W., Sharma R., Dharaskar S., Khalid M. (2024) Advances in PCMs, heat transfer enhancement techniques, and their applications in TES: a comprehensive review, J. Energy Storage, 87, p. 111329.
10. Alizadeh M., Sadrameli S.M. (2019) Indoor thermal comfort assessment using PCM based storage system integrated with ceiling fan ventilation: experimental design and response surface approach, Energy Build., 188–189, pp. 297–313.
11. Mirahmad A., Sadrameli S.M., Seifi H. (2014) Theoretical and experimental studies on a latent heat thermal energy storage system (LHTES) containing flat slabs of phase change materials, Int. J. Smart Grid Clean Energy, 3(2), pp. 234–240.
12. Pirvaram A., Sadrameli S.M., Abdolmaleki L. (2019) Energy management of a household refrigerator using eutectic environmental friendly PCMs in cascade condition energy, Energy, 181, pp. 321–330.
13. Abdolmaleki L., Sadrameli S.M., Pirvaram A. (2020) Application of environmental friendly and eutectic phase change materials for the efficiency enhancement of household freezers, Renew. Energy, 145, pp. 233–241.
14. Mousavi Baygi A., Sadrameli S.M. (2018) Thermal management of photovoltaic solar cells using polyethylene glycol 1000 (PEG1000) as a phase change material (PCM), Therm. Sci. Eng. Progr., 5, pp. 405–411.
15. Azizi Y., Sadrameli S.M. (2016) Thermal management of a LiFePO4 battery pack at high temperature environment using a composite of phase change materials and aluminum wire mesh plates, Energy Convers. Manag., 128, pp. 294–302.
16. Hasanabadi S., Sadrameli S.M., Soheili H. (2019) A cost effective form stable PCM composite with modified paraffin and expanded perlite for thermal energy storage in concrete, J. Therm. Anal. Calorimetry, 136, pp. 1201–1216.
17. Cao X., Dai X., Liu J. (2016) Building energy-consumption status worldwide and the state-of-the-art technologies for zero-energy buildings during the past decade, Energy, 128, pp. 198–213.
18. Kalnaes S.E., Jelle B.P. (2015) Phase change materials and products for building applications: a state-of-the-art review and future research opportunities, Energy Build., 94, pp. 150–176.
19. Xie Y., Gilmour M.S., Yuan Y., Jin H., Wu H. (2017) A review on house design with energy saving system in the U.K., Renew. Sustain. Energy Rev., 71, pp. 29–52.

20. Wang X., Feng W., Cai W., Ren H., Ding C., Zhou N. (2019) Do residential building energy efficiency standards reduce energy consumption in China? - A data driven method to validate the actual performance of building energy efficiency standards, Energy Policy, 131, pp. 82–98.
21. Hamada M.A., Khalil H., Abou Al-Sood M.M., Sharshir S.W. (2023), An experimental investigation of nanofluid, nanocoating, and energy storage materials on the performance of parabolic trough collector, Appl. Therm. Eng., 219, p. 119450.
22. Kumar V., Shrivastava R., Nandan G. (2016), Energy saving using phase change material in refrigerating system, 3rd international conference on manufacturing excellence – MANFEX. Noida, Uttar Pradesh: Amity University.
23. Joybari M.M., Haghighat F., Moffat J., Sra P. (2015) Heat and cold storage using phase change materials in domestic refrigeration systems: the state-of-the-art review, Energy Build., 106, pp. 111–124.
24. Khan M.I.H. (2016) Conventional refrigeration systems using phase change materials: a review, Int. J. Air-Cond. Refrig., 24, p. 1630007.
25. Liu H., Wei Z., He W., Zhao J. (2017) Thermal issues about li-ion batteries and recent progress in battery thermal management systems, Energy Convers. Manag., 150, pp. 304–330.
26. Bashirpour-Bonab H. (2020) Thermal behavior of lithium batteries used in electric vehicles using phase change materials, Int. J. Energy Res., 44, pp. 12583–12591.
27. Lu Y., Situ W., Yang X., Zhang G., Wang Z. (2018) A novel nano silica-enhanced phase change material with anti-leakage and anti-volume-changes properties for battery thermal management, Energy Convers. Manag., 163, pp. 250–259.
28. Lin Y., Jia Y., Alva G., Fang G. (2018) Review on thermal conductivity enhancement, thermal properties and applications of phase change materials in thermal energy storage, Renew. Sustain. Energy Rev., 82, pp. 2730–2742.
29. Najafpour Esfahani N., Garmestani H., Rozati M.R., Fadhil Smaisim G. (2023) The role of phase change materials in lithium-ion batteries: a brief review on current materials, thermal management systems, numerical methods, and experimental models, J. Energy Storage, 63, p. 107061.
30. Faraj K., Khaled M., Faraj J., Hachem F., Castelain C. (2023) Phase change materials in buildings. In: Pielichowska K., Pielichowski K., editors. Multicomponent phase change materials, Fundamentals, properties, and applications. Woodhead Publishing. Series in composites science and engineering. Sawston, Cambridge, U.K.
31. Hoeven, M.V.D. (2015) in: Solar Thermal Electricity-technology Roadmap, OECD Pub., 75015, pp. 1–52.
32. Shoeibi S., Kargarsharifabad H., Mirjalily S.A.A., Sadi M. (2022) A comprehensive review of nano-enhanced phase change materials on solar energy applications, J. Energy Storage, 50, p. 104262.
33. Sikiru S., Lekan O.T., Sanusi Y.K., Adewale A.A., A.A.J.C.N. OJ (2022) Advances in graphene-based materials for dye-sensitized solar cell component, electronic devices and prospective applications: a critical review, Curr. Nanosci., 18(3), pp. 336–346.
34. Sahay A., Sethi V., Tiwari A., Pandey M. (2015) A review of solar photovoltaic panel cooling systems with special reference to ground coupled central panel cooling systems (GC-CPCS, Renew. Sustain. Energy Rev., 42, pp. 306–312.
35. Sadeghi G., Mehrali M., Shahi M., Brem G., A. Mahmoudi A. (2022) Progress of experimental studies on compact integrated solar collector-storage retrofits adopting phase change materials, Sol. Energy, 237, pp. 62–95.
36. Buddhi D., Sahoo L.K. (1997) Solar cooker with latent heat storage, design and experimental testing, Energy Convers. Manag., 38(5), pp. 493–498.

37. Turnpenny J.R., Etheridge D.W., Ray D.A. (2000) Novel ventilation cooling system for reducing air conditioning in buildings. Part I: testing and theoretical modeling, Appl. Therm. Eng., 20(11), pp. 1019–1037.
38. Zalba B., Marin J.M., Cabeza L.F., Mehling H. (2004) Free cooling of buildings with phase change materials, Int. J. Ref., 27, pp. 839–849.
39. Zalba B., Marin J.M., Sanchez-Vlaverde B., L.F. Cabeze (2002) Free cooling; an application of PCMs in TES. In: IEA, ECES IA, annex 17, 3rd workshop.
40. Peng X., Umer M., Nahid Pervez M., Hasan K.M., Habib M., Islam M., Lin L., Xiong X., Naddeo V., Cai Y. (2023) Biopolymer based microencapsulation technology for sustainable textiles: a short review, Case Stud. Chem. Environ. Eng., 7, p. 100349.
41. Ozservinc A., Alkan C. (2022) Ethylene glycol based polyurethane shell microcapsules for textile applications releasing medicinal lavender and responding to mechanical stimuli, Colloid. Surf. A: Physicochem. Eng. Asp., 652, p. 129888.
42. Palanichamy S., Athiimoulam K. (2022) Influence of various additives on stability and phase change characteristics of DI water-GnP-based NFPCM for cold thermal energy storage systems, Environ. Sci. Pollut. Res., 29, pp. 66935–66949.
43. Sathishkumar A., Sundaram P., Cherlathan M., Ganesh Kumar P. (2024) Effect of nano enhanced PCMs on performance of cool thermal energy storage system: a review, J. Energy Storage, 78, p. 110079.
44. Jafaripour M., Sadrameli S.M., Seyed Mousavi S.A.H., Suleimanpour S. (2021) Experimental investigation for the thermal management of a coaxial electrical cable system using a form-stable low temperature phase change material, J. Energy Storage, 44, p. 103450.
45. Farid A.M., Farid M.M. (2009) Review of recent patents related to electrical cable cooling, Recent Patents Electr. Electron. Eng., 2, pp. 207–214.
46. Alimohammadi H., Zheng J., Buss A. (2020) Field and simulated rutting behavior of hot mix and warm mix asphalt overlays, Constr. Build. Mater., 265, p. 120366.
47. Cui X., Huang D., Liu L. (2016) A review of mechanics of asphalt pavement disease, J. Shandong Univ. (Eng. Sci.), 46(5), pp. 68–87.
48. Manning B.J., Bender P.R., Cote S.A. (2015) Assessing the feasibility of incorporating phase change material in hot and mix asphalt, Sustain. Cities Soc., 19, pp. 11–16.
49. Wang X., Ma B., Li S., Si W., Wei K., Zhang H., Zhou X., Fana Y., Kana X., Shi W. (2023) Review on application of phase change materials in asphalt pavement, J. Traffic Transport. Eng., 10(2), pp. 185–229.
50. Zhang Y., Umair M.M., Zhang S.F., Tang B.T. (2019) Recent advances in PCM based nanoplatforms for cancer therapy, J. Mater. Chem. A, 7, pp. 22218–22228.
51. Bao J., Tu H., Li J., Yu S., Gao J., Lei K., Zhang F., Li J. (2022) Applications of phase change materials in smart drug delivery for cancer treatment, Front. Bioeng. Biotechnol., 10, p. 991005.
52. Cao C., Yang N., Dai H., Huang H., Song X., Zhang Q., Dong X. (2021) Recent advances in phase change material based nanoplatforms for cancer therapy, Nanoscale Adv. (RSC), 3, p. 106.
53. Vadiee A., Martin V. (2012) Thermal energy storage strategies for effective closed greenhouse design, Appl. Energy, 109, pp. 337–343.
54. Huang B.K., Toksoy M., Cengel Y.A. (1986) Transient response of latent heat storage in greenhouse solar system, Sol. Energy, 37, pp. 279–292.
55. Nishad S., Krupa I. (2022) Phase change materials for thermal energy storage applications in greenhouses: a review, Sustain. Energy Technol. Assess., 52, p. 102241.
56. Zare M., Mikkonen K. (2023) Phase change materials for life science applications, Adv. Funct. Mater., 33, p. 2213455.

4 Building Energy Management Using Phase Change Materials

4.1 INTRODUCTION

Energy consumptions in residential and commercial buildings have sharply increased in the last decades and reached to 36% of global final energy consumption as illustrated in Figure 4.1. Nearly 40% of the energy-based CO_2 emissions in 2018 are due to the life quality enhancement, population growth, and technological development [1] and an annual increase of 1.8% is projected for energy consumption and CO_2 emissions by 2050 [2]. Application of PCMs in the building construction materials and building envelopes can store and release large amount of latent heat within a narrow range of temperature, with heat flow regulation between indoor and outdoor environments and as a passive energy strategy which is promising for building thermal comfort achievement and energy load reduction. This chapter explains and discusses different incorporation techniques of PCMs in the building construction materials and building envelopes for energy efficiency enhancement and energy consumption reduction.

The role of PCMs as a thermal energy storage in the building materials to increase the thermal performance has been used since 1989 when Hawes et al. [3] added PCMs to concrete as a main building material. Four types of aggregate have been

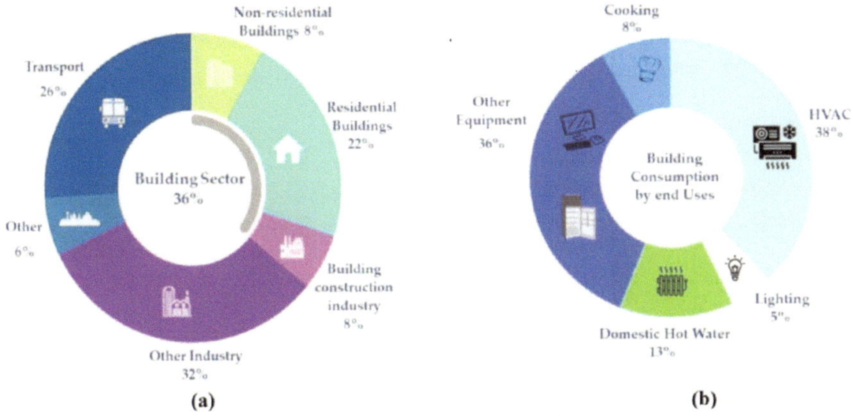

FIGURE 4.1 (a) Building's global final energy consumption in 2020; (b) world building consumption by end uses [1].

DOI: 10.1201/9781003518280-4

studied in which modification has been performed to enhance their compatibility with the PCMs. The new concretes, including organic PCMs such as fatty esters, fatty alcohols, and C_{22}-C_{24} paraffins, have prepared with lightweight aggregates like pumice, expanded shale, and expanded slag as PCM carriers for direct incorporation in the concrete. They investigated some properties of PCM such as absorption in concrete, concrete density, and hydrogen bonding on PCM penetration and stated that the stored heat dissipated by 100–130%.

A comparison between conventional concrete walls and the walls incorporated with encapsulated PCM with melting point of 26°C in two real-size concrete cubicles in Lleida, Spain, to achieve energy savings in buildings has been studied by Cabeza et al. [4]. They concluded that significant energy savings can be obtained by using encapsulated PCM in the concrete walls.

4.2 PCM INCORPORATION TECHNIQUES

PCM must be incorporated in the construction materials of buildings suitably to enhance the heat storage capacity of the building envelope. There are different techniques to incorporate phase change materials into building materials. One of the main challenges of application of PCM in building materials is the leakage and therefore, an appropriate way must be selected to incorporate it into building materials and building envelope due to the alkaline nature of the materials. Incorporation of PCM in buildings are classified into direct incorporation such as wet mixing and immersion techniques and indirect methods like different-sized encapsulation (micro and macro) and shape-stabilized PCM as depicted in Figure 4.2 [5].

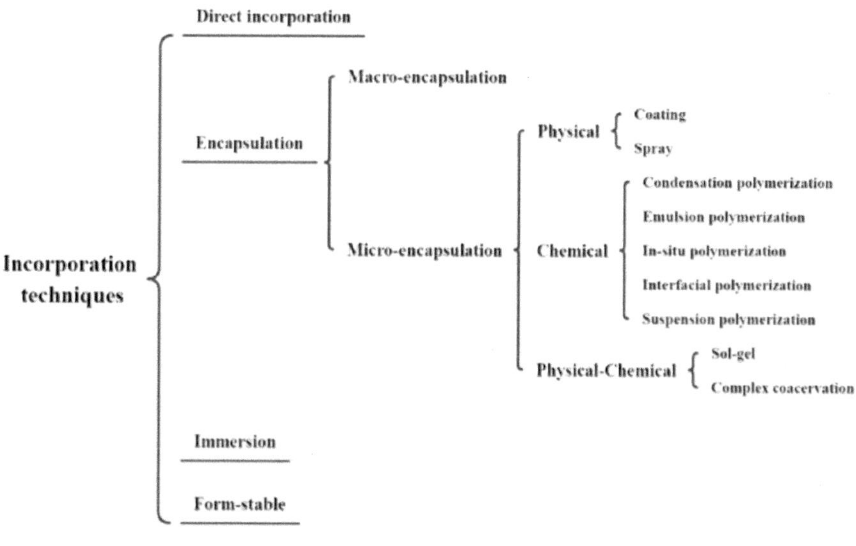

FIGURE 4.2 PCM incorporation techniques in building envelope [5]. (Permission from Elsevier.com.)

4.2.1 Direct Incorporation of PCMs

The direct incorporation method is the easiest and most economical method since it is directly mixed with construction materials and there is no need for labour and production facilities divided into wet mixing and immersion. In this method, the PCM is directly mixed with construction materials such as cement, concrete, and gypsum during production to increase the thermal energy storage capacity of the materials [6].

4.2.1.1 Wet Mixing

In the wet mixing method, the PCM is mixed with the construction materials and the successful application depends on two factors: no reaction between PCM and other components and no interference with hydration process in the material [3].

Organic PCMs have been directly incorporated into gypsum at a rate of 21% and 22% by Feldman et al. [6] to strengthen and facilitate the merging processes. The comparison between the results showed that the stored energy on the novel material is 10 times stronger than the energy stored on conventional gypsum wallboard.

4.2.1.2 Immersion

In this technique, a container filled with liquid PCM is used in which the porous construction elements such as concrete block, gypsum wallboard, and porous aggregate are dipped in and the construction element absorbs the PCM through capillary action. Absorption capacity of the concrete, temperature, and the type of PCM employed affects the efficiency and the time required to fully soak into the porous construction of the material [7].

Both methods of direct incorporation of PCM in the building materials have some various critical drawbacks which are listed as follows [8]:

- Hydration process is affected by leakage of PCM
- Loss of PCM due to the leakage after many cycles
- Longer process (4–8 hours) time
- Mechanical strength is affected by wet mixing

4.2.2 Indirect Incorporation of PCM

In this method, the PCM is first encapsulated before being incorporated into the building materials. Using encapsulated PCM with a compatible coating or shell material, the following benefits can be achieved [9]:

- Leakage and PCM loss reduction
- Enhancement of heat transfer surface area
- Structural stability increases and ease of handling
- PCM protection from environment
- Strength and durability enhancement
- Thermal reliability increase

Shape-stabilized PCM and macroencapsulation and microencapsulations are among the indirect technologies for incorporating PCM into building construction materials.

4.2.2.1 Form-Stable PCMs

In this technique, PCM reacts with a high-density material such as high-density polyethylene (HDPE), styrene, and butadiene at a high temperature using supporting materials like diatomite, expanded perlite, expanded graphite, silica fume, vermiculite, kaolin, granulated blast furnace slag, and waste glass and then cooled to form-stable and solid shape-stabilized materials with the following characteristics [3]:

- High specific temperature
- Sufficient thermal conductivity capacity
- Resistance to shape during phase transition
- Stability after many cycles
- No need for the packaging
- Can be used as internal lining
- The PCM to material ratio can reach to 80%
- No leakage
- An increase in the retention capacity of porous building materials with vacuum impregnation

One of the common PCMs which can be utilized by these techniques is polyethylene glycol (PEG) which can be easily incorporated into building materials due to its high thermal capacity, suitable melting point, low volume change during phase transition, being nontoxic and chemically stable with no super cooling [10]. Thermal properties of naturel clay and gypsum incorporated with 22% and 18% of PEG via vacuum impregnation have been measured by Sari et al. [11] using a DSC analysis. The melting point results were 10.85°C and 10.55°C, while the heat of fusions were 28.79 J/g and 24.18 J/g, respectively. The temperature reductions of gypsum with PCM during one and two hot hours were 2.08°C and 1.47°C, respectively.

One of the basic challenges in the application of PCMs in building materials is their low thermal conductivity. This can be solved by adding high thermal conductivity additives such as metallic powders to the PCM. In the study performed by Hasanabadi et al. [12, 13], graphite powder and multiwall carbon nanotubes (CNTs) have been added to paraffin with a melting point of 57°C as thermal conductivity enhancement additives. Impregnation techniques have been utilized to incorporate the modified PCM into the expanded perlite particles with the particle size of 2.38–4.78 mm (mesh #8) and thermal conductivity of 0.05 W/m.K, and a lightweight concrete was finally constructed by these form-stable perlite/paraffin composites as shown in Figure 4.3.

A DSC result of the expanded perlite, paraffin, and two kinds of prepared perlite/paraffin composites with different paraffin impregnation is illustrated in Figure 4.4. As can be observed from the results, the melting temperature of both prepared form-stable PCM composites decreased very little for form stable 1 and much more for sample 2 in comparison to pure paraffin which is due to the weak attractive

FIGURE 4.3 Form-stable composite PCM samples [12]. (Permission from Springer.)

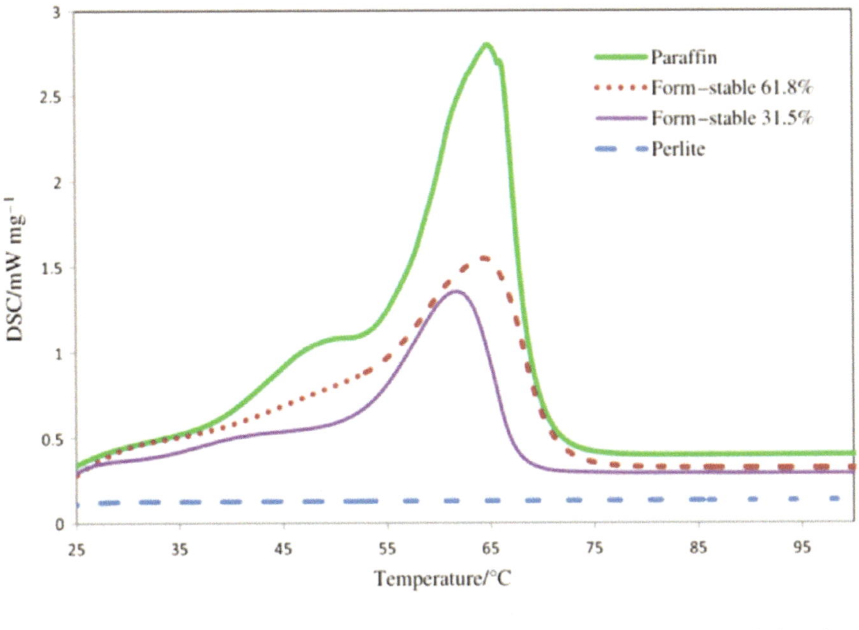

FIGURE 4.4 DSC curves of paraffin, perlite, and two samples [12]. (Permission from Elsevier.com.)

integration between paraffin and internal surfaces of the perlite porous media that affect the crystallization process inside perlite.

The heat of fusion for samples 1 and 2 decreased by 37% and 68%, respectively, in comparison to pure paraffin due to the reduction in the mass fraction of paraffin.

Thermal conductivity of paraffin has been enhanced by adding carbon additives and the following correlation has been obtained from experimental data:

$$k = 0.032x + 0.2622 \tag{4.1}$$

in which k is thermal conductivity in W/m.K and x is the graphite powder mass fraction.

In conclusion, adding thermal conductivity enhancers to PCM could improve the thermal storage behaviour of the final concrete but in a form-stable technique the stability of the additives is not significantly effective.

Two form-stable PEG-based (PEG800 and PEG1000) form-stable PCMs have been synthesized and incorporated into mortars by Sarcinella et al. [14] for building energy efficiency enhancement in different climates of continental and Mediterranean regions.

Lecce Stone (LS) has been used as a porous media for the inclusion of two PEGs to form two form-stable PCMs, namely: LS/PEG800 and LS/PEG1000 (50/50 wt.%). Thermal properties of the produced modified mortars have been measured and the materials have been assessed in a climatic chamber to be able to simulate the variation of environmental temperatures on the system at two climatic zones under consideration. Their results reveal that the modified mortars containing mixed PCM reduced the cooling and heating requirements by 8% and 13%, respectively. The reduction of 8% in the cooling costs during the summer and 12% during the spring and autumn seasons for internal heating of the building located in the Mediterranean region have been achieved.

TiO_2 nanofibres have been deposited on the drawing CNTs membrane as a supporting material of quinary fatty acid eutectics by Zhang et al. [15] for the preparation of multiscale TiO_2/CNTs nanofibrous mats for thermal storage or release. SEM, DSC, and a device made in the lab have been used for the characterization of the morphology, PCM property, and temperature regulation performance of the samples. Their results prove that channel-like structured TiO_2 distributed randomly on every layer of the drawing-aligned CNTs fibre membrane to form a 3-D structure which is beneficial for the absorption of PCMs. Heating density of the samples was about 100 J/g and there is no change after 100 cycles.

4.2.2.2 PCM Encapsulation

Encapsulation of PCM is necessary due to the leakage of materials during the melting process but the capsule materials must be corrosion-resistant, flexible, cheap, and strong at the same time [5]. The objectives of encapsulation can be summarized as follows:

- Leakage prevention
- Heat transfer area enhancement in relation to the unit volume
- Easy to handle

There are three techniques for encapsulation of PCMs: microencapsulation, macroencapsulation, and nanoencapsulation. Encapsulated PCMs can be used in wallboards, concrete, gypsum, floors, ceiling, bricks, and shuttering with high performance and efficiency as they store and release energy in the building compartment. The encapsulation methods, including physical, chemical, and physico-chemical have been discussed in Chapter 2.

4.3 PCMs IN CONSTRUCTION MATERIALS

4.3.1 PCMs in Gypsum Wallboards

One of the auxiliary materials in building construction that can be installed on the roof, wall, and floor of buildings is a gypsum board that can be utilized for sound insulation and heat insulation and fire prevention. Applications of microencapsulated PCMs in gypsum wallboards are very applicable due to the wide range of applications in buildings.

Microencapsulated PCMs with melting points of 25–28°C have been utilized in gypsum wallboards by Voelker et al. [16] and observed temperature drop of 3°C and heating power of 2k/min. In an earlier study by Hawes et al. [17], direct incorporation of 25–30% PCM to the gypsum showed temperature drop of 4°C with suitability in flammability and moisture resistance. In other investigation in northeast China, Shilei et al. [18] incorporated 26% of PCM into gypsum wallboards and collected data over three successive days in the winter season and found that PCM can keep the indoor temperature higher than outdoor as illustrated in Figure 4.5.

A simulation study performed by Darkwa et al. [19] using three types of PCMs in gypsum reveals that laminated PCM-wallboard had the best performance with an increase of 17% on night-time temperatures. Paraffin microencapsulation has been incorporated into gypsum by Kuznik et al. [20] and resulted that 5 mm thickness of gypsum wallboard had 200% energy stocked more than conventional models with higher convective heat transfer coefficient. Microencapsulated PCM has been added

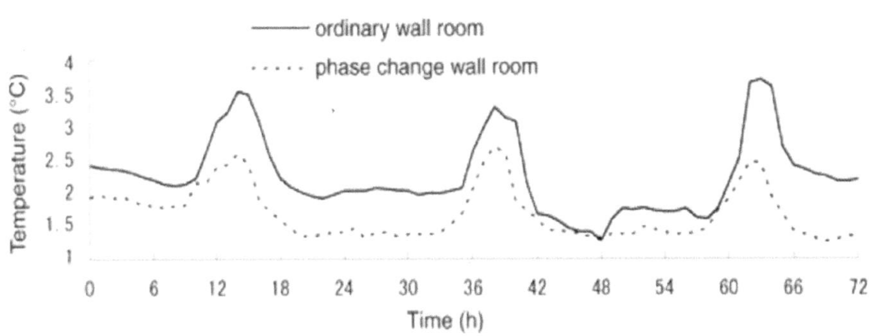

FIGURE 4.5 Indoor and outdoor temperature differences with and without PCM [18]. (Permission from Elsevier.com.)

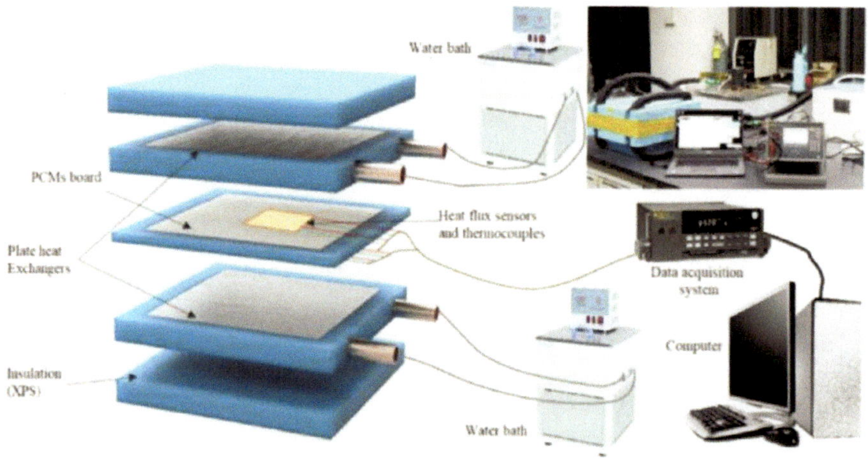

FIGURE 4.6 Application of PCM in wallbord [21]. (Permission from Elsevier.com.)

to the wallbords by Li et al. [21] as illustrated in Figure 4.6 in which they had 2.71% higher heat capacity than conventional wallboards.

In general, applications of PCMs in gypsum wallboards can improve the heat storage performance of phase change gypsum board in comparison to ordinary gypsum which affect the energy consumption in the buildings.

4.3.2 PCM in Concrete

One of the basic construction materials in buildings which is made from a mixture of variety of mineral materials is cement mortar and concrete which are cement-based composites. Using PCMs with cement-based composites works as a heat storage in a building skeleton to control the temperature in the building envelope by storing and releasing extra heat during the day and night. Hydrated calcium chloride as an inorganic PCM has been impregnated into diatomite and coated with paraffin by Li et al. [22] to have a stable shape. The composite PCM then has been mixed with magnesium oxychloride cement to form a composite cement with heat storage capacity. Their experimental results show that the temperature difference of the surface, including PCM, is much slower than that without PCM with a controlled temperature fluctuation. They have also studied the effects of PCM amount on the composite mechanical properties [23] and concluded that mechanical strength of the composite concrete decreased sharply with an increase in the percentage of hydrated salt in the composite, although this addition significantly improves the thermal storage and release performance of cement mortar.

A eutectic PCM consists of bottom ash/capric-stearic acid which has been implemented into cement using a vacuum technique at 30 wt.% by Gencel et al. [24] and measured the temperature and heat of fusion during the melting process to be

21.42°C and 13.62 kJ/kg, respectively. Indoor reduction of 2.80°C and 1.95°C has been achieved during the phase transition hours. A compressive strength of 35 Mpa has been measured based on a 28-day mechanical characterization test. An alternative eutectic PCM composed of blast furnace slag/capric acid shape-stabilized form has been tested by Gencel et al. [25] into a composite cement by a vacuum method at 45 wt.%. The melting temperature and enthalpy have been measured to be 28.52°C and 13.8 J/g, respectively, with a compressive strength of 8.16 Mpa during a 28-day test. Mechanical strength of an innovative composite of concrete with PCM has been investigated by Cabeza et al. [4] on thermal aspects of the concrete wall. Two real-size concrete cubicles have been built as an experimental setup to show the effect of encapsulated PCM in concrete. The results reveal that the concrete compressive strength and tensile splitting strength have reached 25 Mpa and 6 Mpa, respectively, and no difference occurred in the effects of PCM after six months of operation. It can be concluded that although composite of concrete/PCM can affect building energy consumption by storing and releasing some amounts of energy in the building envelope during day and night, their influence on the mechanical properties of the concrete in construction must also be taken into consideration during the application.

Microencapsulated PCM with a poly(methyl methacrylate) as a shell and 1-tetradecanol as core with a melting point of 33°C and heating density of 108.4 J/g has been produced with the Pickering emulsion technique by Alkan et al. [26] and were mixed with mortars at 5%, 7.5%, and 10% by weight of cement to investigate the thermal management of the buildings.

In the Pickering emulsions, the interfacial energy is reduced by emulsions created with matrix materials such as inorganic particles, polymer particles, and food grade particles instead of surfactants which is the key component in the formation of Pickering emulsions of colloidal particles. The comparison of the samples has been performed inside an insulated cabin and the thermal performance results revealed that the indoor temperatures were regulated at 4.7°C and 3.9°C for the heating and cooling periods with the temperature differences of 3.1°C and 2.4°C in the inner surface of the sample cementitious mortar cabin, respectively.

4.3.3 PCM in Tiles

Encapsulated PCMs as latent heat storage can be incorporated into building materials such as floor tiles for building energy savings purposes. Adding PCMs in the floor tile sharply increases its ability to store thermal energy. One application is the use of the tile in sunroom floors where it can absorb thermal energy during the day from the sun entering the space and subsequently release the energy to aid in the reduction of heating load in the winter. The reduction in heating bills will be enough to compensate for the costs of PCMs incorporated in the floor tiles.

Application of PCMs in floor tiles of the building has been investigated by Ceron et al. [27]. Prototype tiles with dimensions of 660 × 660 × 52 mm^3 in total consist of four pieces of pure clay stoneware 20 mm thick, top metal sheet 3 mm, a metal

FIGURE 4.7 Schematic diagrams of prototype and the room with tiles [27]. (Permission from Elsevier.com.)

container 32 mm thick, which contains 4.81 of paraffinic mixture; and finally coated with a layer of thermal insulation 22 mm thick, which works as acoustic insulation for the framework as depicted in Figure 4.7. Of the 30 models that have been produced, three samples were without PCM and utilized for comparison purposes.

The temperature measurement during a typical winter day with open blinds, and direct solar radiation from the window with solar radiation of 1,026 W/m^2, is shown in Figure 4.8.

The curves illustrated in the figure from the top to the bottom are temperature measurements for the tile without PCM which has the highest temperatures, tiles with PCM closed to the window, room temperature, including tiles with PCM, side room temperature without PCM, and finally the outdoor temperature during the same time. It is observed from the temperature distributions in the figure that during the day when the outdoor temperature is minimum, the PCM tiles supply

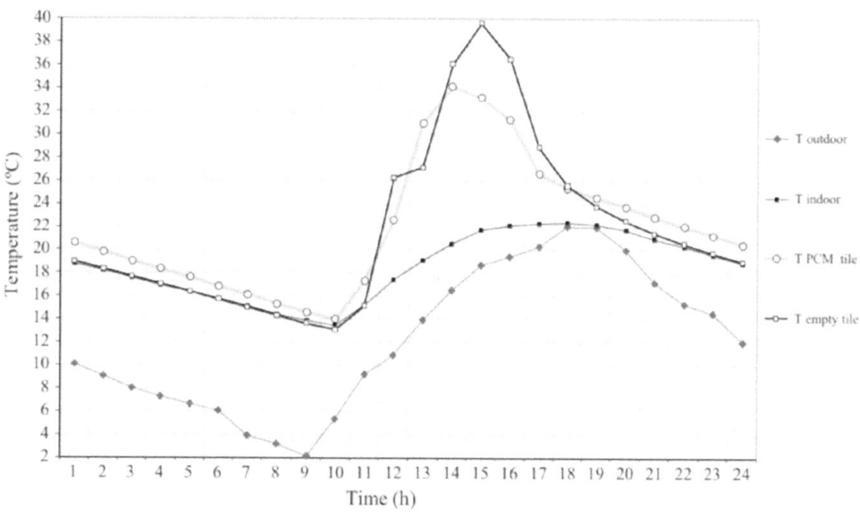

FIGURE 4.8 Temperature measurement during a typical winter day [27]. (Permission from Elsevier.com.)

the stored energy to the room and therefore the temperature increase at the tiles surfaces reaches values between 1.5°C and 2.0°C over surface temperature without PCM which has occurred until the tile start charging which is usually between 8.00 and 10.00 am.

They finally concluded that during the wintertime, the PCM-incorporated tiles on the floor that receive direct solar energy from the sun within an enclosure can work as an energy storage system to absorb solar energy during the day and subsequently release it to the room to decrease the energy consumption during the night. The tiles can also be beneficial during the summertime when the melting points of the PCMs are increased to make the role of the system as a heat sink. The results also reveal that the sunny tiles have better performance, and the contribution of the tiles removed from the front, away from the window, is very small.

Paraffin wax has been incorporated into a porous ceramic carrier fabricated from industrial waste-iron tailing with a foam-gel casting method by Li et al. [28] and the tile mechanical and thermal properties have been adjusted with fabrication parameters such as sintering temperature (Figure 4.9). Their results reveal the porosity range of 70–90% and compressive mechanical strength of 0.4–6 Mpa with good wettability for PCM.

Cycle test has been performed on the prepared samples after multiple melting/solidification periods to evaluate the paraffin leakage and composite ceramic stability and the weight loss at different porosities after 1, 5, 10, 15, 20, and 25 cycles are illustrated in Figure 4.9, which shows that all samples have a low weight loss below 10% increasing with a number of cycles. They concluded that sintering temperature has significant influence on the ceramic's properties and with an increase of temperature from 1,070°C to 1,120°C, the porosity decreases from 89% to 71%; bulk density increases from 0.29 to 0.76 g/cm^3; and compress mechanical strength increases from 0.46 Mpa to 6.19 Mpa. The tailing ceramic energy storage density is

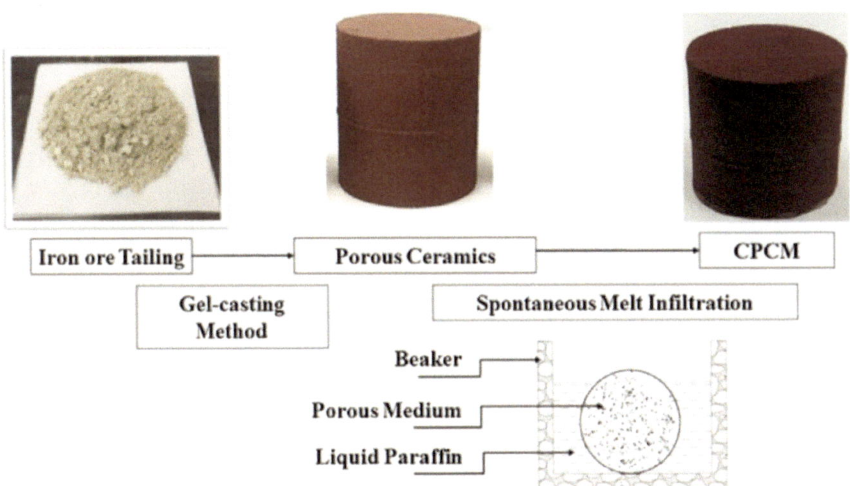

FIGURE 4.9 The ceramic/PCM preparation steps [28]. (Permission from Elsevier.com.)

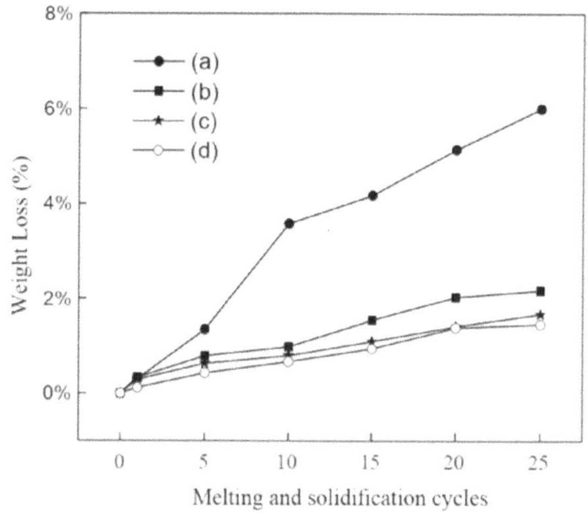

FIGURE 4.10 PCM weight loss with different porosities of 90% (a), 85% (b), 80% (c), and 70% (d) [28]. (Permission from Elsevier.com.)

also a factor of porosity and reaches 123 kJ/kg at 89% porosity with constant melting point (Figure 4.10).

It is determined from the above studies on the application of PCM in building tiles that the new composite PCM is a viable solution and promising for TES applications in building energy management. The addition of PCM in the tile can alter the thermal and properties of the tile but can be varied to determine the best mixture for the preparation of significant thermal storage in tile and meet the industrial requirements for floor tile.

4.3.4 PCM in Glass Windows

Definite opportunities and challenges can be achieved by building envelopes, including transparent and non-transparent components. The main components which are mainly accountable for the building lighting and ventilation are glazing envelopes which suffer from the defects of high solar radiation, weak insulation, and low thermal inertia as illustrated in Figure 4.11. The major heat loss contribution in buildings is considered for windows that can lose energy 10 times more than other building elements. Although in a typical building approximately 10% of the heat is lost through the windows, this can be minimized by using thermal shutters that improve the energy efficiency of the buildings. Different energy saving techniques have been developed to improve the optical and thermal performances of glazing windows.

Peak load transfer can be achieved by using glass windows filled with PCM that can effectively enhance its thermal inertia. One of the main challenges of applying PCM in windows is low energy efficiency and overheating of PCM layer. To overcome this challenge, Zhang et al. [27] proposed a new type of dynamic rotating PCM window consisting of PCM layer and a vacuum glass layer to adjust the relative

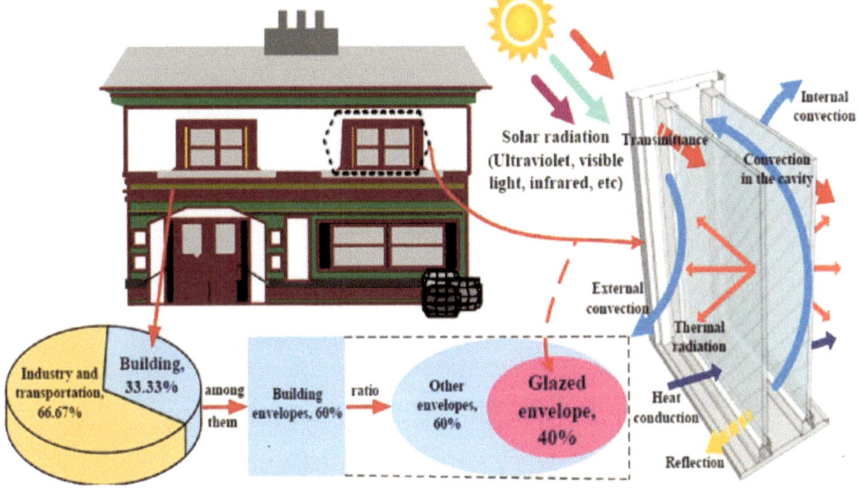

FIGURE 4.11 Energy consumption proportion in buildings [29]. (Permission from Elsevier.com.)

position of PCM and vacuum glass layers with time and therefore, heat flux changes to reduce the building load as shown in Figure 4.12.

Their results indicate that the performance of the new dynamic system is higher than the conventional static PCM/window systems; the heating load in the building has been reduced by 12.40% in winter and reduction of cooling load of 93.97% in summer. The annual building load has also been decreased by 28.70% by utilizing the new dynamic windows.

A comprehensive parametric study on the application of solid-solid PCM incorporated in windows has been conducted by Gao et al. [31] using Energy Plus software by developing a comparable model of a novel PCM window with Energy Plus's modelling capabilities.

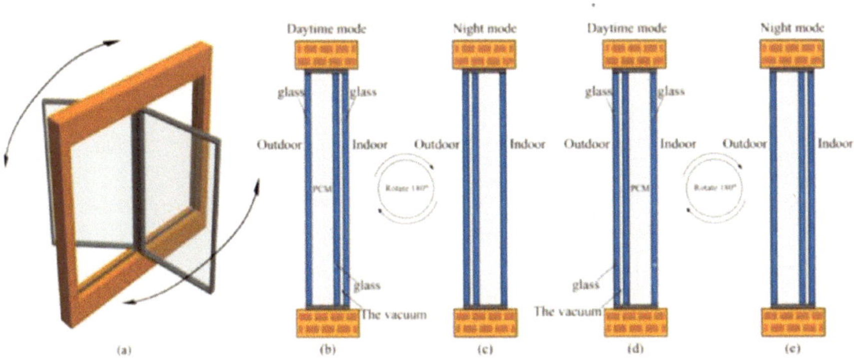

FIGURE 4.12 Rotating PCM window: (a) dynamic mode of winter condition, (b) night mode of winter condition, (c) daytime mode of summer condition, (d) night mode of summer condition, and (e) rendering of rotating PCM window [30]. (Permission from Elsevier.com.)

Their result indicates that the integration of 3 mm solid-solid PCMs with optimal properties in warm, mixed, and cold conditions can, respectively, save up to 17.2%, 14%, and 5.8% energy for the HVAC (heating, ventilation, and air conditioning) system, and 9.4%, 6.7%, and 3.2% energy for the whole building.

The results of the studies on the application of PCM in glazing windows reveal that up to 50% cooling load reduction can be achieved in comparison to the regular double-glazing windows. However, some challenges still exist such as failure in the phase transition due to insufficient heat transfer performance which leads to poor space lighting as well as absorption capacity.

An innovative double-glazing window incorporated with solid-solid PCM and silica aerogel is proposed by Ma et al. [32]. Parametric study has been performed using Energy Plus software with a focus on evaluating the implementation potential of using the window in a severe cold region of China. The numerical results of this study revealed that parameters such as transition temperature, latent heat, absorption coefficient, and refractive index are of remarkable relevance to the energy consumption and thermal performance of the buildings. When compared with 4 mm single-glazing window, the maximum energy saving of the building containing double glazing incorporated with PCM can be realized by 18.22% within the realistic range of PCM properties. The recommended thickness of silica aerogel in the optimum conditions is 10 mm.

4.3.5 PCM in Bricks

The application of PCM in gypsum and wallboards, concrete, tiles, and glass windows has been discussed in the previous sections. In this section, PCM application in bricks will be explained. Figure 4.13 illustrates the number of research

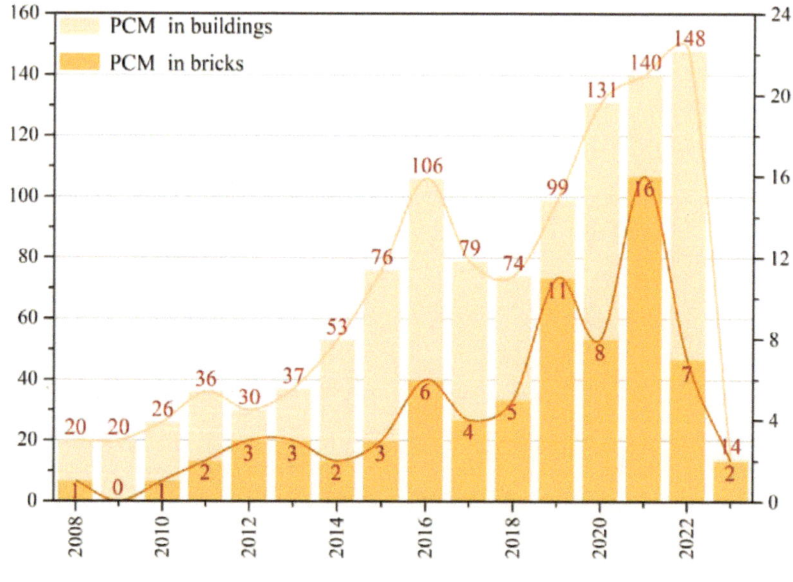

FIGURE 4.13 Literature frequency on the application of PCM in building materials after 2008 [33]. (Taylor & Francis publication.)

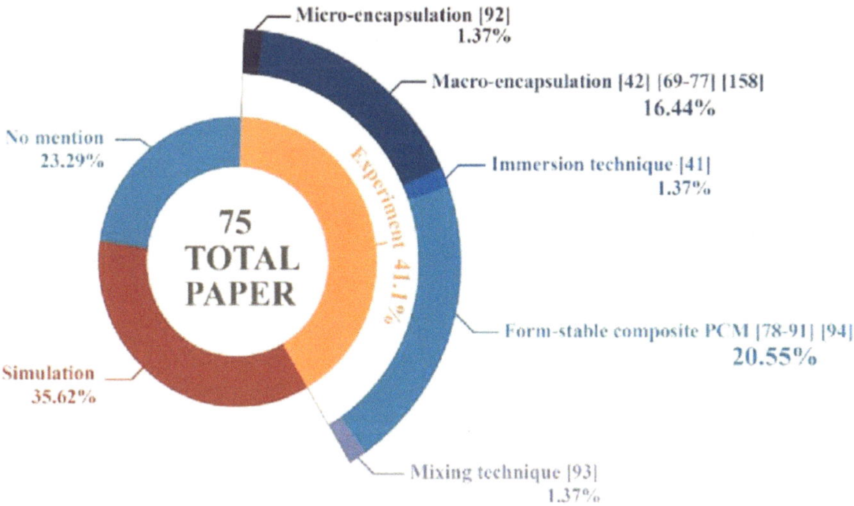

FIGURE 4.14 Different incorporation methods of PCM in bricks [33]. (Taylor & Francis publication.)

studies in the literature for applications of PCMs in building construction materials since 2008.

Due to the leakage and heat storage heat capacity loss for the solid-liquid PCM during phase transition which are incorporated in bricks PCM encapsulation is very important. The techniques include direct incorporation and indirect incorporation such as encapsulation and form-stable composite PCM. Figure 4.14 shows the percentage of PCM in bricks for different incorporation methods. Transparent acrylic tubes with a diameter of 45 mm and thermal conductivity of 0.19 W/m.K, filled with PCM and coated the tubes' surfaces with silicone, have been placed in the cylindrical holes of the bricks and studied for two days by Lai et al. [33].

The same experiment has been performed in six days by Saxena et al. [34] utilizing filled PCM in bricks with galvanized steel with dimensions of $15.5 \times 9 \times 1.5$ cm^3 and 0.4 mm thickness as illustrated in Figure 4.15.

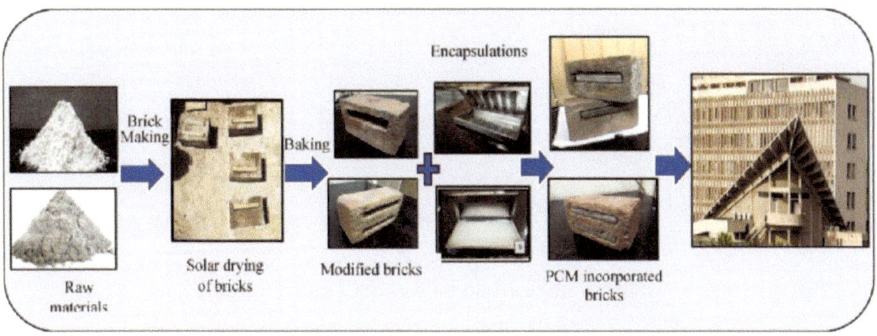

FIGURE 4.15 Schematic of PCM incorporation in bricks [34]. (Permission from Elsevier.com.)

The sides of the PCM bricks have been packed with thick layers of polystyrene to prevent leaking and heat leakage. However, polystyrene could not sustain the abrasiveness of the brick surface and leakage was the major issue during the experiments.

Due to the specific PCM properties, compatible materials can be incorporated into bricks to meet the needs of PCM thermal energy storage and performance enhancement of PCM in the bricks. Shape-stabilized PCM with mass percentage of 40% composed of paraffin, expanded graphite as thermal conductivity enhancement, and HDPE into cement mortar for the preparation of hybrid bricks of PCM has been investigated by Wang et al. [35]. They have faced a serious leakage in the thermal chamber during the experiments. It was concluded that the thermal, mechanical, and durability of PCM bricks have a direct impact on their application in building construction. Further research has been conducted by Wang et al. [36] for the optimization of mixing ratio of PCM in bricks using uniform and stable PCMs according to 70% of solidified mixed paraffin (30% of solid paraffin and 70% of liquid paraffin) and 30% of low-density polyethylene. The results reveal that the appropriate proportion of material could solve the problem of paraffin flow during the phase transition. Shape-stabilized PCMs have been incorporated into cement mortar for the preparation of PCMs bricks by Li et al. [37] as illustrated in Figure 4.16.

Shanghai in China has been selected for applying PCMs into building walls and thermal performance of the building has been studied during the summer. The results reveal an increase in the cooling load of 43.23% under continuous air conditioning, and an increase in heat transfer from the walls to the room by 15–25% under intermittent air conditioning in summer due to the liquid PCMs. Due to the narrow and fixed-phase transition temperatures, the thermal performance to PCMs brick wall was lower than of a common wall.

In general application of PCM-incorporated bricks in the building construction which has been widely proposed as a building passive cooling system, it proves to

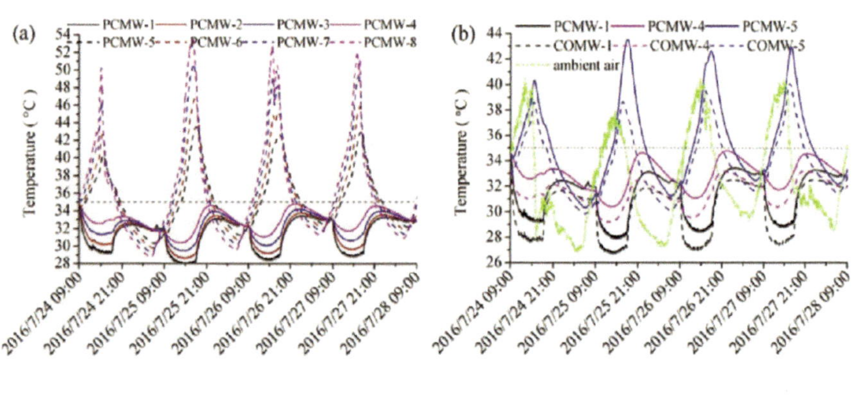

FIGURE 4.16 (a, b) Temperature distribution with AC system under summer day case [37]. (Permission from Elsevier.com.)

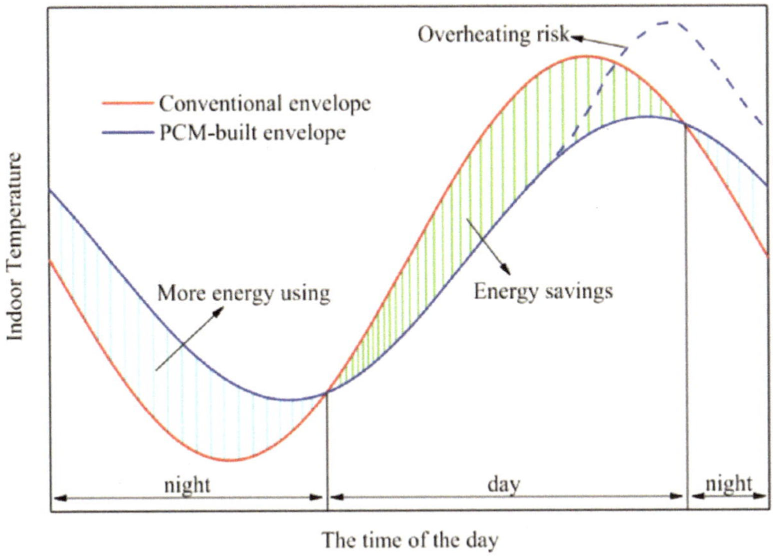

FIGURE 4.17 The comparison of temperature swings between a conventional and PCM-built envelope [38]. (Permission from Elsevier.com.)

be effective in the thermal performance of walls and indoor comfort due to the large amount of heat that can be stored and released during PCM phase transition. As illustrated in Figure 4.17, the heat stored in the PCM-built envelope is transferred to the room during night for better comfort temperature achievement.

In the study performed by Sama Ali Taj et al. [39] in Islamabad, Pakistan, with average summer temperature of 34–37°C, thermal performance of clay brick has been improved by incorporating a mixture of Lauric acid and Palmitic acid with melting temperature of 33–38°C as an organic eutectic PCM as a latent heat TES medium. Two experimental analyses have been conducted, one with PCM-incorporated bricks and the other with clay brick made and investigated under direct sunlight and in controlled conditions to analyse the temperature reduction and respective thermal load. The results indicate that a reduction of 4–5.5°C can be achieved in the inside temperature of the compartment with 32% reduction in thermal amplitude inside PCM compartment, with the overall reduction in heat flux of 25–30% inside PCM-incorporated bricks.

4.4 SUMMARY

Application of PCM in TES as a passive cooling media for building energy management has been widely used since 2008. A large amount of heat can be absorbed and released by PCMs during their phase transition periods to achieve the purpose of energy storage, indoor comfort temperature stability, and reduction of peak energy consumption in buildings. In this chapter, the incorporation of encapsulated PCMs in building materials has been discussed by different techniques such as form-stable (shape-stabilized) PCM, direct incorporation of PCM, and immersion methods. The incorporation of PCM in different construction materials like gypsum, concrete, tiles, glass windows, and bricks has also been presented. The studies prove that PCM-incorporated materials in the building construction can help for achieving comfort temperature in the building envelope and lead to energy consumption reduction and peak loads shifting in the residential and commercial buildings.

REFERENCES

1. UNEP, Global status report for buildings and construction: towards a zero-emission, efficiency (2021) Resilient Building Construction Sect., 1–105.
2. Christopher S., Vikram M.P., Bakli C., Thakur A.K., et al. (2023) Renewable energy potential towards attainment of net-zero energy buildings status - a critical review, J. Cleaner Prod., 405, p. 136942.
3. Hawes D.W. (1991) Latent heat storage in concrete, PhD Thesis, Concordia University, Montreal, Quebec, Canada.
4. Cabeza L.F., Castellon C., Nogues M., Medrano M., Leppers R., Zubillaga O. (2007) Use of microencapsulated PCM in concrete walls for energy savings, Energy Build., 39(2), pp. 113–119.
5. Feldman D., Banu D., Hawes D., Ghanbari E. (2022) Incorporation technology of bio-based PCMs for building envelope: a review, Energy Build., 260, p. 111620.
6. Feldman D., Banu D., Hawes D., Ghanbari E. (1991) Obtaining an energy storing building material by direct incorporation of an organic PCM in gypsum wallboard, Sol. Energy Mater., 22(2), pp. 231–242.
7. Ling T.C., Poon C.S. (2013) Use of PCMs for TES in concrete: an overview, Const. Build. Mater., 46, pp. 55–62.
8. Navarro L., Gracia D.E., Colclough S., Browne M., McCormack S.J., Griffiths P. et al. (2016) Thermal energy storage in building integrated thermal systems: a review, part 1, active storage systems, Renew. Energy, 88, pp. 526–547.
9. Flix Regin A., Solanki S.C., Saini J.S. (2009) An analysis of packed bed latent heat thermal energy storage system using PCM capsules: numerical investigation, Renew. Energy, 34(7), pp. 1765–1773.
10. Karaman S., Karaipekli A., Sari A., Bicer A. (2011) PEG/diatomite composite as a novel form stable PCM for thermal energy storage, Sol. Energy Mater. Sol. Cells, 95(7), pp. 1647–1653.
11. Sari A. (2014) Composites of PEG600 with gypsum and natural clay as new kinds of building PCMs for low temperature-thermal energy storage, Energy Build., 69, pp. 184–192.
12. Hasanabadi S., Sadrameli S.M., Soheili H., Moharami H., Heyhat M.M. (2019) A cost-effective form stable PCM composite with modified paraffin and expanded perlite for thermal energy storage in concrete, J. Therm. Anal. Calorimetry, 136(3), pp. 1201–1216.
13. Hasanabadi S., Sadrameli S.M., Sami S. (2021) Preparation, characterization, and thermal properties of surface-modified expanded perlite/paraffin as a form-stable phase change composite in concrete, J. Therm. Anal. Calorimetry, 144, pp. 61–69.
14. Sarcinella A., Aguiar J., Jesus C., Frigione M. (2023) Thermal properties of PEG-based form-stable PCMs incorporated in mortars for energy efficiency of buildings, J. Energy Storage, 67, p. 107545.

15. Zhang J., Ji Y., Zhang Y., Wu S., Wang D., Xu Z., Wei Q. (2024) Preparation and temperature regulation properties of TiO_2/CNTs-based form-stable PCMs, Colloids Surf. A, 685, p. 133278.
16. Voelker C., Kornadt O., Ostry M. (2008) Temperature reduction due to the application of PCMs, Energy Build., 40(5), pp. 937–944.
17. Hawes D.W., Feldman D., Banu D. (1993) Latent heat storage in building materials, Energy Build., 20(1), 77–86.
18. Shilei L., Neng Z., Guonui F. (2006) Impact of phase change wall room on indoor thermal environment in winter, Energy Build., 38(1), pp. 18–24.
19. Darkwa K., O'Callaghan P.W., Tetlow D. (2006) Phase change drywalls in passive solar building, Appl. Energy, 83(5), pp. 425–435.
20. Kuznik F., Vigone J., Roux J. (2008) Energetic efficiency of room wall containing PCM wallboard: a full-scale experimental investigation, Energy Build., 40(2), pp. 148–156.
21. Li C., Yu H., Song Y. (2019) Experimental investigation of thermal performance of microencapsulated PCM-contained wallboard by two measurement modes, Energy Build., 184, pp. 34–43.
22. Li X., Zhou Y., Zhang X., Zheng W., Chang C., Ren X., Zeng J., Hai C., Shen Y. (2018) Experimental investigation of thermal and mechanical properties of magnesium oxychloride cement with form stable PCM, Const. Build. Mater., 186, pp. 670–677.
23. Liu Y., Xie M., Gao X., Yang Y., Sang Y. (2018) Experimental exploration of incorporating form-stable hydrate salt PCMs into cement mortar for thermal energy storage, Appl. Therm. Eng., 140, pp. 112–119.
24. Gencel O., Hekimoglu G., Sari A., Sutcu M., Er Y., Ustaoglu A. (2021) A novel energy effective and carbon emission reducing mortars with bottom ash and PCM: physico-mechanical and thermal energy storage characteristics, J. Stor. Mater., 44, Part A, 103325.
25. Gencel O., Yaras A., Hekimoglu G., Ustaoglu A., Erdogmus E., Sutcu M., Sari A. (2022) Cement based thermal energy storage mortar including blast furnace slag/capric acid shape stabilized PCM: physical, mechanical, thermal properties and solar thermo-regulation performance, Energy Build., 258, p. 111849.
26. Alkan C., Alakara E., Aksoy S., Demir I. (2023) Cement mortar composites including 1-tetradecanol@PMMA Pickering emulsion particles for thermal energy management of buildings, Chem. Eng. J., 467, pp. 146843.
27. Ceron I., Neila J., Khayet M. (2011) Experimental tile with phase change materials (PCM) for building use, Energy Build., 43, pp. 1869–1874.
28. Li R., Zhou Y., Duan X. (2019) A novel composite phase change material with paraffin wax in tailing porous ceramics, Appl. Therm. Eng., 151, pp. 115–123.
29. Li D., Yang R., Arici M., Wang B., Tuncbilek E., Wu Y., Liu C., Ma Z., Ma Y. (2022) Incorporating phase change materials into glazing unit for building applications: current progress and challenges, Appl. Therm. Eng., 210, p. 118374.
30. Zhang X., Liu Z., Wang P., Li B. (2022) Performance evaluation of a novel rotatable dynamic window integrated with PCM and a vacuum layer, Energy Convers. Manag., 272, p. 116333.
31. Gao Y., Meng X. (2023) A comprehensive review of integrating phase change materials in building bricks: methods, performance and applications, J. Energy Storage, 62, p. 106913.
32. Ma Y., Li D., Yang R., Zhang S., Arici M., Liu C. (2022) Energy and daylighting performance of a building containing an innovative glazing window with solid-solid PCM and silica aerogel integration, Energy Convers. Manag., 271, p. 116341.
33. Lai C.M., Chiang C.M. (2006) How phase change materials affect thermal performance: hollow bricks, Build. Res. Inf., 34(2), pp. 118–130.

34. Saxena R., Rakshit D., Kaushik S.C. (2020) Experimental assessment of PCM embedded bricks for passive conditioning in buildings, Renew. Energy, 149, pp. 587–599.
35. Wang X., Yu H., Li L., Zhao M. (2016) Experimental assessment on a kind of composite wall incorporated with shape stabilized PCMs, Energy Build., 128, pp. 567–574.
36. Wang Q.H., Wang R.H., Wang C.P. (2014) Research on the insulation properties of composite phase change paraffin hollow block walls, New Build. Mater., 41(5), pp. 56–59.
37. Li L., Yu H., Liu R. (2017) Research on composite PCMs bricks in the west of room scale cubicle: mid-season and summer day cases, Build. Environ., 123, pp. 494–503.
38. Wang P., Liu Z., Zhang X., Hu M., Zhang L. (2023) Adaptive dynamic building envelope integrated with PCM to enhance the heat storage and release efficiency: a state-of-the-art review, Energy Build., 286, p. 112928.
39. Ali Taj S., Khalid W., Nazir H., Khan A., Sajid M., Waqas A., Hussain A., Ali M., Zaki S. (2024) Experimental investigation of eutectic PCM incorporated clay brick for thermal management of building envelope, J. Energy Storage, 84, p. 110838.

5 Phase Change Materials in Electrical Appliances

5.1 INTRODUCTION

Monthly high-energy consumption bills are a primary concern for residential homeowners which results in higher country's energy demand and increase in the greenhouse emissions. Therefore, cutting down the amount of energy consumption that is used for household electrical appliances while simultaneously satisfying the social and economic requirements of people is the primary goal of energy policies in many countries. Sustainable energy strategies can be planned through three major technological alternatives such as energy saving and management, efficiency enhancement, and substitution of fossil fuels to renewable clean energy sources. Waste to energy techniques has also been used as a solution to improve the efficiency in energy use and production in recent years [1–3]. This chapter discussed the application of PCM in the most common electrical appliances used in the residential and commercial buildings such as refrigerators, freezers, open-display refrigerators, and hot water heaters for energy efficiency enhancement and reduction of electrical energy consumptions. Application of PCM in heating, cooling, and ventilation (HVAC) will be discussed in a separate chapter.

5.2 HOUSEHOLD REFRIGERATORS

There is a very high-saving potential of the entire global number of household refrigerators due to the non-stop world demand, almost complete market penetration and typical continuous operation, although the power consumption of a single unit is low. If we consider in total, 1.5 billion household refrigeration systems are in use worldwide, this accounts for approximately 4% of global electricity consumption [4] and annual production of 480 million tons of CO_2 emission to the environment [5]. Therefore, refrigerators' efficiency enhancement towards electricity consumption reduction constitutes an important worldwide matter and extensive studies have focused on the performance improvement the systems to find new and optimum techniques to reduce energy consumption of these appliances [6].

One of the main and effective techniques of preserving food are refrigerators that contribute significantly to household energy consumption and therefore it is critical to enhance their energy efficiency. A conventional vapour compression in refrigeration cycle consists of a compressor, a condenser, and an evaporator which are illustrated in Figure 5.1.

The specific electrical energy consumption of an ordinary 165 L household refrigerator has been calculated to vary between 3.23 and 4.78 kWh/year/L [8]. There are several ideas for the efficiency improvement of refrigerators such as heat loss

DOI: 10.1201/9781003518280-5

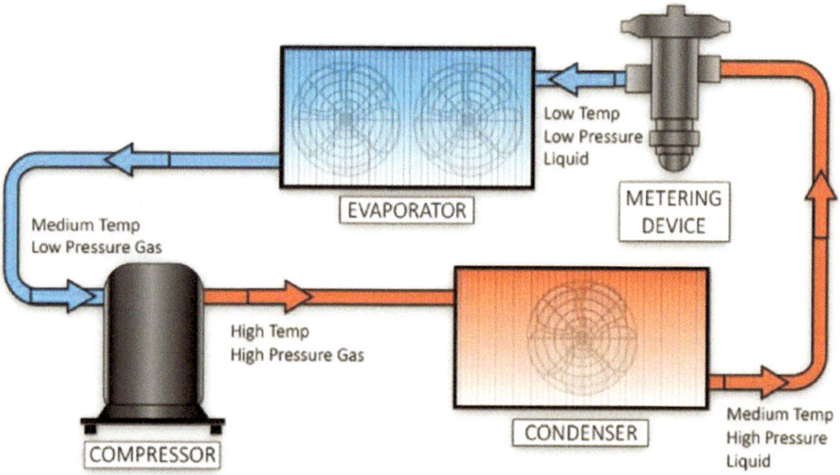

FIGURE 5.1 Conventional vapour compression refrigerator cycle [7]. (Permission from Violagroup.)

reduction by insulation improvement, using efficient compressors, and heat transfer enhancement in condenser and evaporator. The last technique is the most promising option for refrigeration efficiency enhancement by increasing heat transfer from the heat exchangers (condenser and evaporator) utilizing PCMs while minimizing environmental effects. Phase change materials can be used in the refrigerator compartment section, and in the heat exchangers such as condenser and evaporator as an energy storage system to absorb extra heat from the system and release it to the environment for efficiency enhancement.

5.2.1 Application of PCMs in Condenser

Performance improvement of vapour compression refrigeration system utilizing PCM and thermoelectric generator has been studied by Nandanwar et al. [9] using a single-door 190 L household refrigerator. A PCM container with dimensions of $26 \times 18 \times 7$ cm^3 has been placed just before expansion valve and at the condenser outflow in which the refrigerant helical tubes are installed and the heat stored in the PCM was supplied to the thermoelectric hot side while the cold side was served by an aluminium vessel carrying the condensed water yield coming from the evaporator coils as shown in Figure 5.2. Hexahydrate calcium chloride ($CaCl_2.6H_2O$) with melting point of 24°C, density of 1,470 kg/m^3, latent heat of fusion of 140 J/g, and solid and liquid thermal conductivities of 1.09 W/mK and 0.54 W/mK has been used as a PCM. They concluded that a hybrid system of PCM and thermoelectric in vapour compression system can increase the cooling effect and coefficient of performance (COP) by 7% and at the same time reduce the net amount of work required by 4%.

Eutectic mixtures of polyethylene glycol-1000 (PEG1000) and polyethylene glycol-600 (PEG600) with variable weight percentage as non-toxic, non-corrosive,

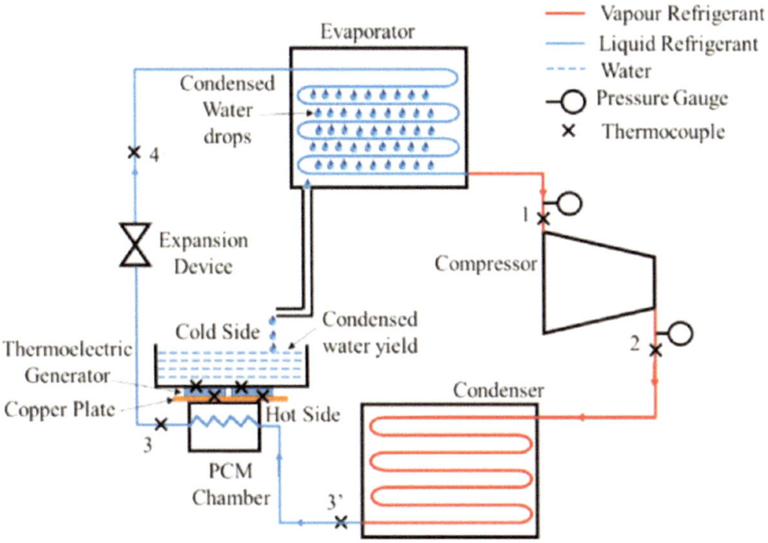

FIGURE 5.2 Experimental setup used by Nandanwar et al. [9]. (Permission from Taylor & Francis.)

and bio-compatibility PCMs have been used by Pirvaram et al. [6] for the thermal management of a household refrigerator model TDF320 Philver made by an Iranian company in a cascade condition. The properties of PCMs are listed in Table 5.1.

The refrigerator is a double door with capacity of 285 L with dimensions of $160 \times 62.5 \times 63$ cm^3 consists of two compartments for fresh food and frozen food having volume of 220 L and 65 L, respectively. For the simulation of frozen-food measurement packages (M-packs) containing 232 g of oxyethylmethylcellulose, 725 g of water, 43 g of sodium chloride, and 0.6 g of polyethyleneterphthalate with a freezing point of –5°C and dimensions of $50 \times 100 \times 100$ mm^3 have been used in the top compartment of the refrigerator. Four different cases have been included in their study, namely without measurement packs (10 kg M-packs)/without PCM, with M-packs/without PCM, with M-packs/single PCM on a condenser, and with M-packs/cascaded PCMs on a condenser and the results of condenser and interior compartments temperatures, compressor performance, and energy consumption of the refrigerator have been reported.

TABLE 5.1
PCMs Physical Characteristics [6]

Characteristics	PEG-1000	PEG-600
Phase change temperature (°C)	34–40	15–25
Latent heat of fusion (Cal g^{-1})	38	35
Density (g ml^{-1} at 20°C)	1.09	1.12
Thermal conductivity (W m^{-1} °C^{-1})	0.2–0.3	0.2–0.3
Liquid specific heat (Cal g^{-1} °C^{-1})	0.51	0.51

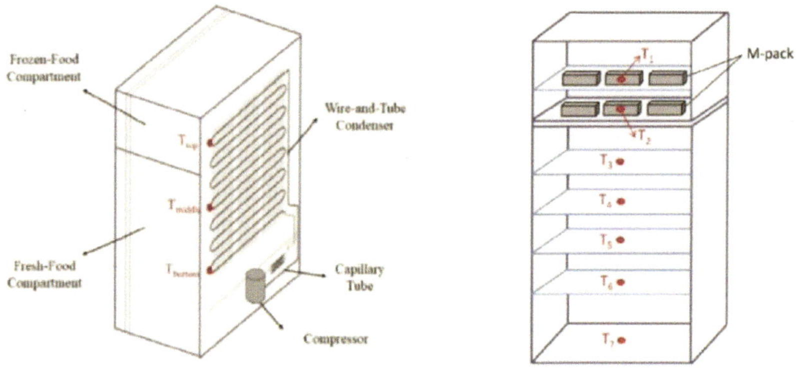

(a) (b)

FIGURE 5.3 (a) Schematic diagram of the refrigerator; (b) locations M-packs and thermocouples [6]. (Permission from Elsevier.com.)

All experimental runs have been performed inside the hot room in a Philver fridge and freezer manufacturing company in Tehran, Iran under standard conditions of 16–43°C temperature and relative humidity of 45–75% according to ISO 8187. Temperature measurements have been performed using T-type thermocouples at different locations of the refrigerator such as compressor, condenser, and internal cabinets as illustrated in Figure 5.3.

The higher melting point PCM has been located on the top half of the condenser while the lower melting point PCM pack has been fixed on the lower half of the surface based on the temperature distribution on the condenser surface during the operation as depicted in Figure 5.4.

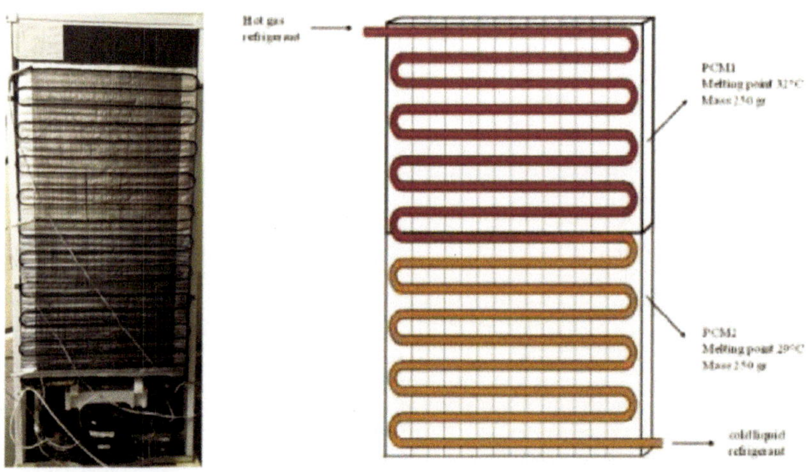

(a) (b)

FIGURE 5.4 (a) Wire-and-tube condenser with two PCMs; (b) placement of two eutectic PCMs [6]. (Permission from Elsevier.com.)

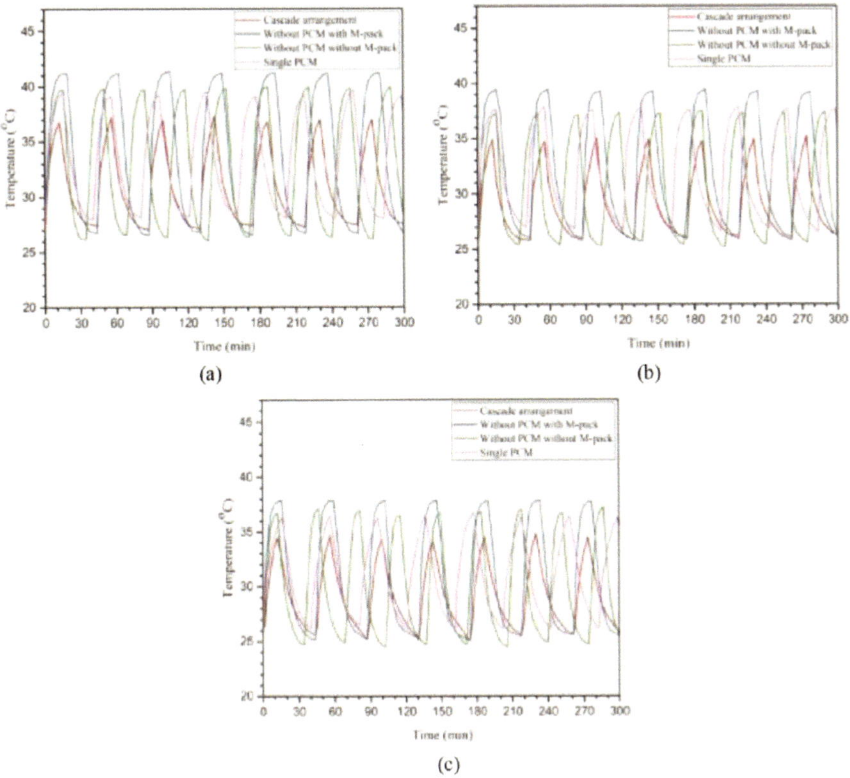

FIGURE 5.5 Surface condenser temperature distributions at (a) Top, (b) Middle, and (c) Bottom sections [6]. (Permission from Elsevier.com.)

The comparison between the condenser temperature distributions at four different cases is shown in Figure 5.5. Substantial reduction on the temperature distribution at all locations for the cascaded arrangement is very clear from the figures. In the case of no PCM with M-packs, (a) the condenser top temperature reached to the maximum of 41°C, while with single and cascaded PCMs, this reduces to 39°C and 37°C, respectively, which proves the heat transfer enhancement on the top of the condenser. The same trend is observed on the middle and bottom temperature distributions.

Figure 5.6 illustrates the impact of single and cascaded PCMs integrated into the backside of the condenser on the temperature distribution of the fresh-food compartment in the refrigerator when the thermostat settings were regulated to achieve the desired average fresh-food compartment temperature of below 5°C. The results indicated that for the refrigerator without PCM and with M-packs, the average temperature of the fresh-food compartment changes from 3.9°C to 4.8°C, while for the cases of single and cascaded PCMs, these variations are from 4.0°C to 4.7°C and from 4.1 to 4.6°C, respectively. Comparison of the results in the fresh-food part of the refrigerator reveals that temperature

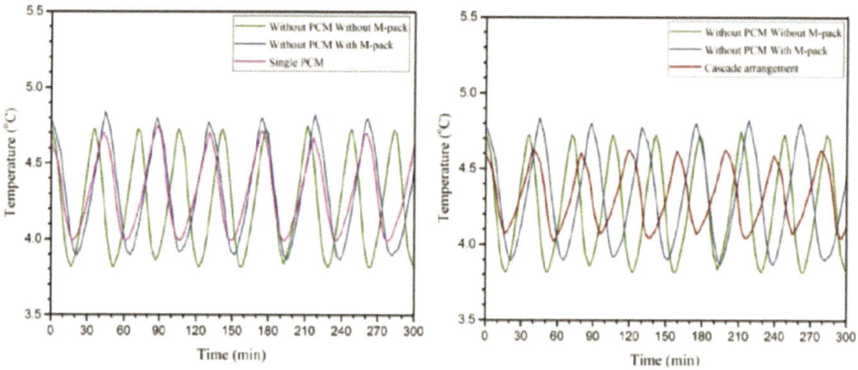

(a) (b)

FIGURE 5.6 Temperature distributions in fresh-food compartment (a) single PCM; (b) cascaded PCM [6]. (Permission from Elsevier.com.)

fluctuations decreased for the case of single PCM compared to no PCM and even more reduction is achieved with cascaded PCM in comparison to single PCM. The reduction in temperature fluctuations in the compartment preserves food quality.

Average temperature fluctuations in the food compartment for four different cases are illustrated in Figure 5.7 which proves that the temperature amplitude in the

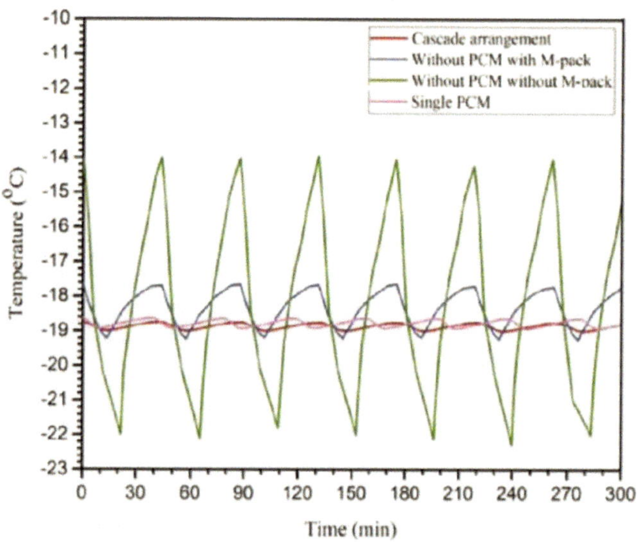

FIGURE 5.7 Temperature fluctuations in the frozen-food compartment [6]. (Permission from Elsevier.com.)

frozen-food section in the case of single PCM decreased from 1.4°C to 0.25°C, while for the cascade PCMs reached to 0.15°C.

In general, according to the results obtained during this study using two different PCMs on the condenser surface in a cascade arrangement, the compressor work time percentage has been reduced from 32.7 to 27.6 and energy consumption of the system with cascade PCM arrangement has been decreased by 13%.

A numerical calculation based on a physical model on the application of multiple PCMs in a household refrigerator has been performed by Ben Taher et al. [10] to study the energy efficiency of the system and their numerical results have been validated with experimental data from the literature. The investigated parameters were PCM thickness of 3–7 mm, internal load of 0–20 W, and external temperature of 15–30°C and concluded that PCM thickness has a significant effect on the system performance. The compressor downtime increases from 344 min for 3 mm water to 720 min using 7 mm. Their simulation results also indicated that the refrigerator air temperature is between 4°C and 6°C when utilizing multiple PCMs (water and eutectic solution).

Water-cooled system on the condenser of a 190 L household refrigeration system consists of hermetically sealed reciprocating compressor, air-cooled condenser, clinched-type evaporator, capillary tube, and capillary suction heat exchanger with technical specifications listed in Table 5.2, in the tropical region, located at the

TABLE 5.2
Technical Specification of the Refrigerator [11]

Component		Specification
Compressor power (W)	:	100 (hermetically-sealed reciprocating type)
Rated cooling capacity (W)	:	85
Displacement (cm^3)	:	4.49
Lubricant	:	Polyol ester oil (POE)
Speed (rpm)	:	1500
Air-cooled condenser	:	Mild Steel, Diameter – 4.7 mm, Length – 13 m (serpentine type, natural convection)
Water-cooled condenser	:	BPHE,10 plates of 0.3 mm thickness, a total heat transfer area – 0.048 m^2
Power of circulation pump (W)	:	15
Evaporator	:	Diameter – 6.35 mm, Length – 8 m, Aluminium pipes press fitted over plates (clinching type)
Capillary length (m)	:	CDR – 2.4 and MDR – 1.8
Capillary bore (mm)	:	0.7874
Length of CSHX (m)	:	0.9
Storage water tank capacity (*l*)	:	750, 1000, and 2000

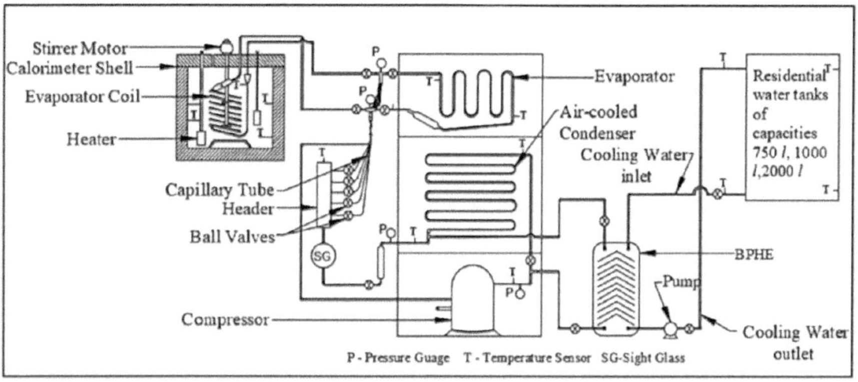

FIGURE 5.8 Experimental setup of water-cooled refrigerator system [11]. (Permission from Elsevier.com.)

latitude and longitude of 8.199° N and 77.384° E, respectively, and has been proposed by Raveendran et al. [11] as shown in Figure 5.8.

They concluded that the modified refrigerator with liquid-cooled condenser can decrease daily electricity consumption from 21% to 27% and compressor pull-down time by 45–70%, respectively. The heat released from the condenser can increase the water tank to a maximum of 5°C which does not require any heat recovery system.

Thermal performance of a household single-door and plate-type evaporator refrigerator with 165 L capacity in which paraffin wax with a melting point range of 57–60°C, as a PCM and enhanced paraffin with multi-walled carbon nanotube (MWCNT) as PCM/nanoparticles with properties listed in Table 5.3, has been integrated with a condenser has been studied by Manoj Kumar et al. [12] to predict the possible saving potential.

TABLE 5.3
Specification of Paraffin and MWCNT [12]

Paraffin Properties	**Values**
Melting point	57–60°C
Latent heat	140–145 kJ/kg
Thermal conductivity (solid 30°C)	0.245 W/mK
Thermal conductivity (solid 30°C)	0.161 W/mK
Specification of MWCNT	*Values*
Outside tube diameter	30–50 nm
Inside tube diameter	5–10 nm
Length	10–20 nm
Specific surface area	60 m^2/g
True density	2.1 g/cm^3

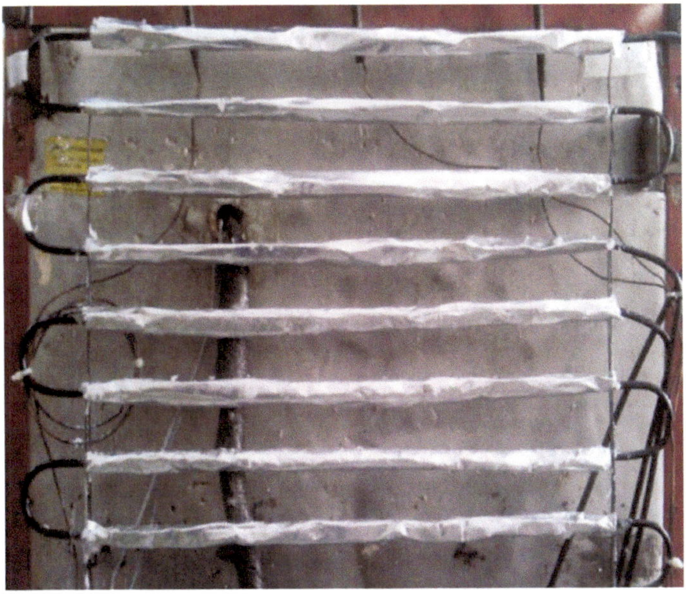

FIGURE 5.9 Photo view of PCM on a condenser [12]. (Permission from EBSCO.)

Aluminium foil tape has been used to fasten the TES along the condenser tubes as illustrated in Figure 5.9. Their results reveal that the integration of PCM and PCM/nanoparticle as TES on the condenser side decrease the ratio on cycle time to the total cycle time of the fridge by 8.26% and 17.95%, respectively, which lead to the saving potential of 13.06% with PCM and 18% with PCM/MWCNT-based TES systems.

Also, by utilizing PCM-based TES on the condenser side the condensation temperature has been reduced by 3°C and 2.1°C during on- and off-cycle periods, respectively. In the case of PCM/NWCNT, the condensation temperatures have been decreased by 4°C and 2.1°C during on- and off-cycle compared to the refrigerator with PCM only.

5.2.2 Application of PCMs in Evaporator

Performance improvement of a 190 L household refrigerator with R134a using water as a phase change material imbedded in the evaporator has been studied by Kappen et al. [13]. They concluded that using PCM in the refrigerator evaporator not only increases compressor cut-off time but also reduces the electricity consumption of the system with more flexibility with very good impact on the food preservation.

A dual-evaporator-based household refrigerator with PCM incorporated in the evaporator has been investigated by Subramaniam et al. for the COP enhancement by introducing a new subcooling routine [14]. Performance of a single-door household refrigerator in which water and eutectic PCM solutions with a melting point of –3°C have been in direct contact on the back side of the evaporator as illustrated

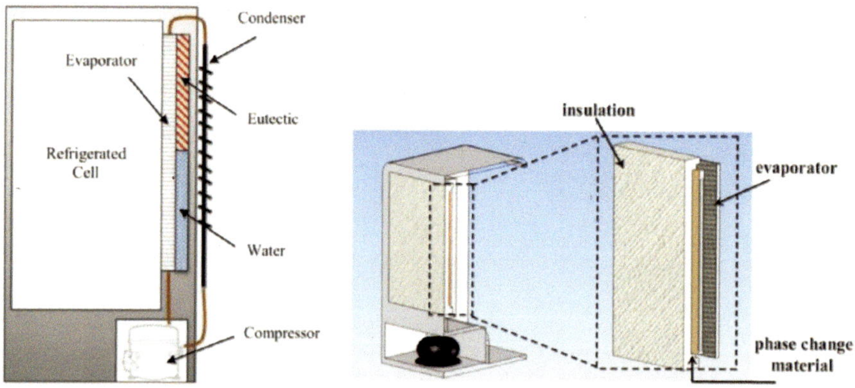

FIGURE 5.10 Refrigerator equipped with PCM [15]. (Permission from Elsevier.com.)

in Figure 5.10 has been studied by Azzouz et al. [15] to focus into how the operating conditions of the system affect its thermal performance of PCM. Influence of PCM amount on the cabinet temperature and compressor number of ON/OFF is shown in Figure 5.11 which proves that the percentage of compressor running time which is defined as a ratio of on-time to total cycle time is significantly lower than the case without PCM and less fluctuations in the refrigerated cell that allows 5–9 h of continuous operation without electrical supply.

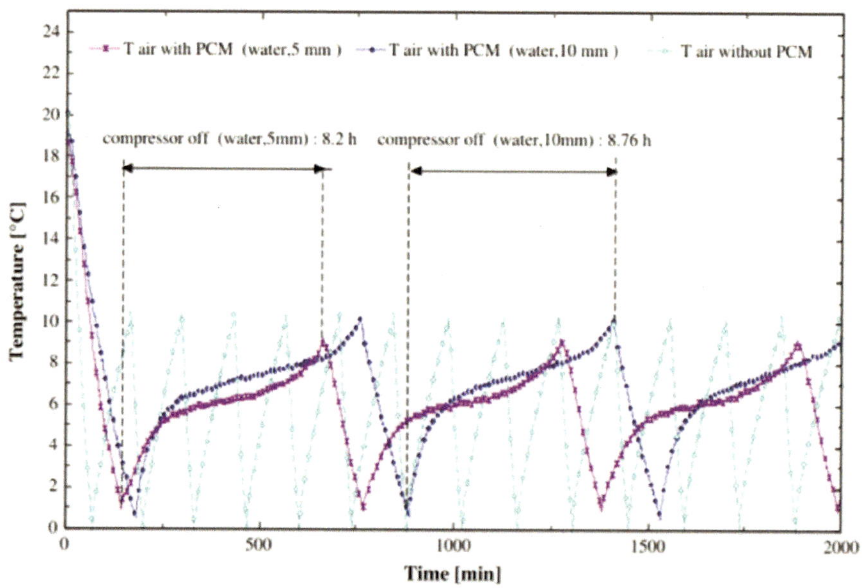

FIGURE 5.11 Time evolution of the average air temperature in the fridge [15]. (Permission from Elsevier.com.)

A recent review on the application of PCM for the performance improvement of refrigeration system consisted of different approaches for the incorporation of PCMs into the system has been presented by Nguyen et al. [16]. Effects of PCM geometry, heat transfer, and flow on the refrigeration system performance as well as prospective of PCMs in terms of enhancing energy saving have been discussed in this review paper which proves that PCMs can be a promising for electrical energy reduction contributing to operating costs reduction and decreasing CO_2 emissions.

5.2.3 Open-Display Supermarket Refrigerators

One of the common and widely used electrical systems in the groceries and supermarkets is open-display refrigerators for storing the goods at good quality over time. Performance enhancement of such systems can be achieved by using phase change materials. A dynamic model for a double cabinet open-display refrigerator incorporated with PCM has been developed and solved by Javeri et al. [17]. Parametric study has been performed on the model for the effect of PCM type and amount (1–8 kg), and location on the performance of the system. Their simulation results show the reduction of temperature fluctuations by 15.8%, number of compressor on/off by up to 5%, and energy consumption by 1.1%, respectively. If the evaporators are equipped with PCM with mass of 1–8 kg, the temperature fluctuations have been reduced by 6.6%, a decrease of 10% in compressor on/off times, and 1.5% reduction in energy consumption. The modelling results reveal that PCM can be efficiently used in open-display refrigerators for energy consumption purposes.

Electricity cost saving has been achieved for a cabinet refrigerator incorporating 15.6 kg of PCM in which the cycle operation has been optimized by Maiorino et al. [18] using a mathematical model. The model was based on the experimental data measured from a single-door commercial cabinet refrigerator with capacity of 520 L divided in six racks with PCM working with different hysteresis. Low-density polyethylene has been used as container for PCM and 2.4 kg of PCM covered each rack of the refrigerator excluding the upper rack where only 1.2 kg of PCM was placed due to the space limitation. The comparison results for on/off time and power consumption of refrigerator with and without PCM are listed in Table 5.4 in which τ^{ON}, τ^{OFF}, and P are the compressor on/off time and consumed power absorbed by the compressor, respectively.

TABLE 5.4
Measurement Data for Refrigerator with and without PCM [18]

	Δ_{exp} [K]	τ^{ON} [s]	τ^{OFF} [s]	P [W]
With PCM	1	1263	9112	215.2
	2	4322	32,337	218.0
	3	6040	48,420	221.9
Without PCM	1	233	1653	212.2
	2	638	4923	210.6
	3	1027	8325	210.5

The results of Table 5.4 clearly indicate that the incorporation of PCM in the system leads to increase the on/off time of the system especially for the higher hysteresis values; however, the power absorbed by the compressor increases as the hysteresis value increases which led to the increase in daily energy consumption with PCM up to 6.5%. The experimental results also indicate that the energy consumption of the refrigerator increases with PCM at higher hysteresis.

5.3 HOUSEHOLD FREEZERS

One of the common household appliances that has a significant impact in keeping frozen food at a safe and stable temperature below −18°C for a long time is a freezer. Due to the colour and texture change of meats in the freezers caused by temperature fluctuations, many researchers focused their studies to reduce temperature fluctuations in the freezers which are produced by door openings, heat generated during defrosting, and power failure which can significantly change the freezer temperature. Application of PCMs in freezers for improving compressor efficiency and optimizing heat transfer from condenser and evaporator has been a topic for the research in last decades.

Efficiency enhancement of household freezers using environmental friendly and eutectic PCM of PEG200 and PEG300 with the composition listed in Table 5.5 has been studied by Abdolmaleki et al. under standard conditions of ISIRI 13700 [19]. The parameters under study were power supply, electrical energy consumption, temperature fluctuations in the freezer, and the compressor operating cycles.

Experimental tests have been carried out on two 262 L vertical freezers capacity model TUP 310 N Philver with dimensions 63 × 60 × 156 cm^3 and power of 402 kWh, one equipped with PCMs which have been stored in aluminium packs located in every tray and one without PCM as shown in Figure 5.10. Measurement packages (M-packs) consisted of 232 g of oxyethylmethylcellulose, 725 g of water, 43 g of sodium chloride, and 0.6 g of polyethyleneterphthalate with dimensions of 5 × 10 × 10 cm^3 and freezing point of −5°C and total weight of 60 kg have been located on the trays to simulate frozen foods in the freezer as shown in Figure 5.12. Both freezers have been tested under the same standard conditions according to ISIRI 13700.

Experimental tests have been performed inside a test room with temperature range of 16–43°C and humidity of 45–75% to simulate a hot kitchen environment. The comparison of power consumption is depicted in Figure 5.13. As can be observed

TABLE 5.5
Composition of PCM Used in the Freezer [19]

PCM	Melting Point (°C)	PEG200 (wt.%)	PEG300 (wt.%)
A	−10	4	96
B	−15	20	80
C	−20	30	70

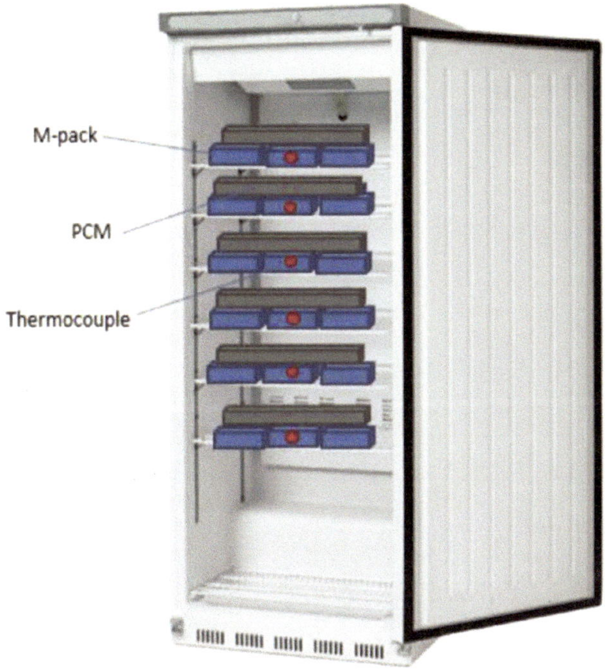

FIGURE 5.12 Schematic of the freezer [19]. (Permission from Elsevier.com.)

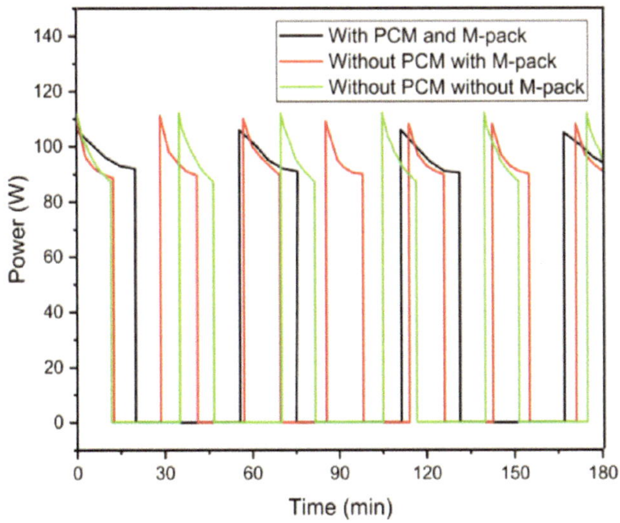

FIGURE 5.13 Power consumptions of the freezers [19]. (Permission from Elsevier.com.)

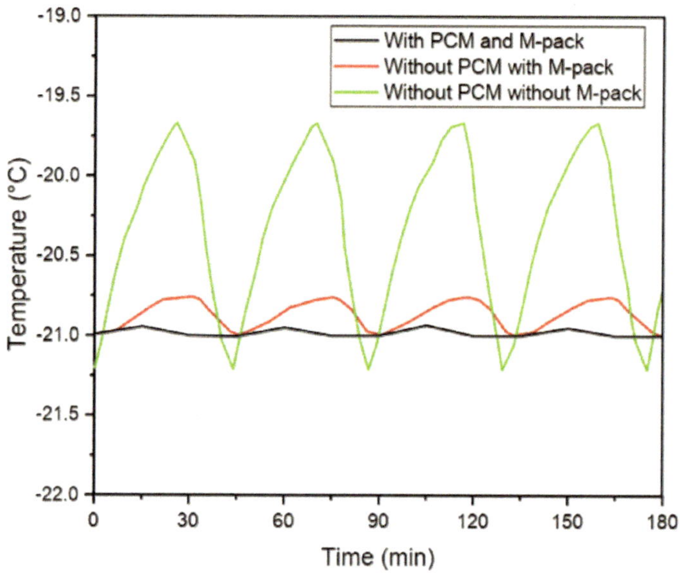

FIGURE 5.14 Temperature fluctuations of the freezers [19]. (Permission from Elsevier.com.)

from the figure, the power consumptions of the freezers after 24 h, with PCM and M-pack, without PCM and with M-pack, and without PCM and M-pack were 0.89 kWh, 0.93 kWh, and 0.94 kWh, respectively, which indicate that freezer with PCM could consume less electrical energy due to the compressor longer off-time.

As can be seen from the temperature fluctuations of three cases in Figure 5.14, there is a dramatic decrease in the temperature fluctuations in the presence of PCM in the freezer.

They finally concluded that the maximum energy reduction occurs with PCM melting point and amount of −20°C and 1.5 kg, respectively.

Further investigations on the application of eutectic PCMs in cascade arrangement for household freezers have been performed by Pirvaram et al. [20] in which optimization of energy consumption and temperature fluctuations have been studied. Experimental runs have been conducted on the same system reported in [19] under two conditions (loaded with and without M-packs) equipped with two eutectic PCMs arranged in a cascade-like configuration to find the effect of thermal load on the system performance as illustrated in Figure 5.15.

A comparison of the energy consumptions during 10-h experiments for two freezers, one with cascaded PCM and one without PCM, is shown in Figure 5.16. Energy consumption of the system without PCM and M-packs has been measured to be 0.9401 kWh, whereas for the freezer with M-packs and no PCM, this is reduced to 0.9382 kWh. It is clearly obtained from the energy consumption results that since storing PCMs in the freezer increases the compressor OFF-time, a large amount of energy is saved. Experimental runs with using 9 kg of single PCM with a melting temperature of −20°C leads to reduction of energy consumption by 8.37%, while when using two PCMs in a cascade condition, the energy consumption further

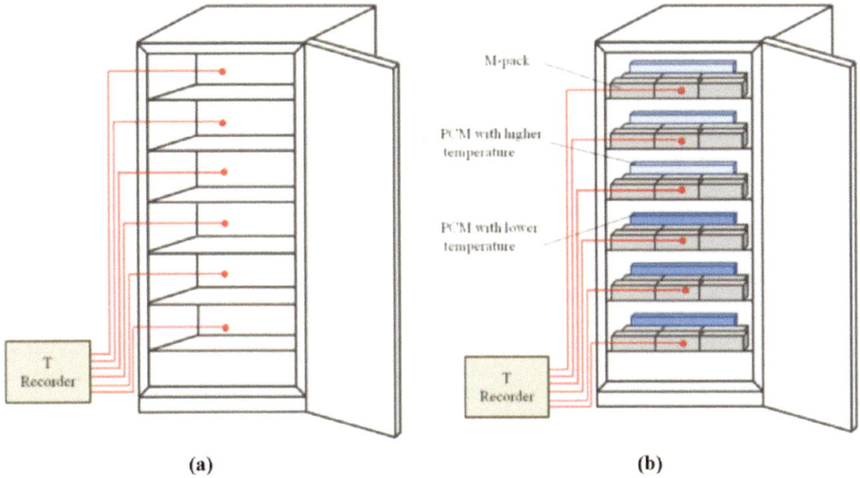

FIGURE 5.15 Schematic diagrams of household freezers: (a) thermocouple locations; (b) cascade arrangement of two PCMs [20]. (Permission from John Wiley.com.)

reduces to 13.42% which proves the significant role of cascade PCMs in energy saving of the freezers.

An extensive review on the thermal management and performance enhancement of household refrigerators and freezers using PCM consists of effect of PCM's configuration, geometry, orientation, and location on the performance under different thermal loads, including PCM, ambient temperature, evaporation temperature, door openings, defrosting, and power failure has been published by Omara et al. [21].

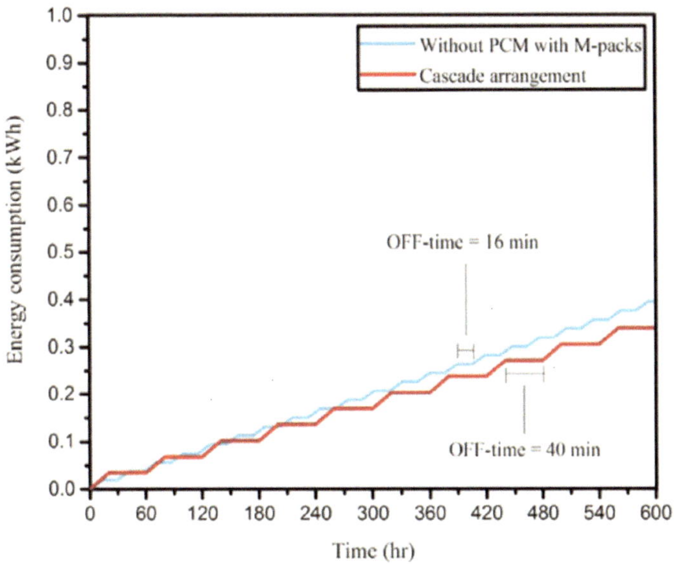

FIGURE 5.16 Comparison of the energy consumption [20]. (Permission from John Wiley.com.)

5.4 HOT WATER HEATERS

A large part of the total annual energy consumption is due the energy consumption of domestic hot water heaters which accounts for 21% of the total energy consumption in Danish buildings [22]. Existing buildings in the Netherlands consume more than 23% of the gas usage that will be increased to 50% in the new buildings [22]. The figure is higher for South Africa, where the water heating process consumes up to 40% of the total energy [23]. Water heating can be produced using different techniques such as classical gas or electrical heaters, solar heaters, heat pump water heaters, geothermal water heating, etc. Several parameters affect the energy consumption of water heaters which includes geographic location, number of occupants, occupants' behaviour, and season of the year. Application of PCM as a short-term energy storage system by phase transition in domestic water heater systems has been very effective in energy consumption reduction. For domestic hot water heaters, PCMs with phase transition temperatures between 40°C and 80°C such as paraffins, fatty acids, salt hydrates, and alcohols are commonly used.

Application of a homogenous mixture of 25% animal fat and 75% paraffin wax as lower-cost PCMs in domestic hot water heater to enhance energy efficiency has been studied by Djeffal et al. [24]. The melting points and heating values have been measured by a DSC analysis (Figure 5.17) to be 35.58°C and 180 kJ/kg for animal fat and 62.58°C and 210 J/g for paraffin wax, respectively.

Water production at a setpoint temperature range of 50–75°C and the energy consumption of the water heater is shown in Table 5.6 in which the energy cost was assumed at 10.54 DZD/kWh (1 US$= 136.5 DZD) as per Algerian utility costs.

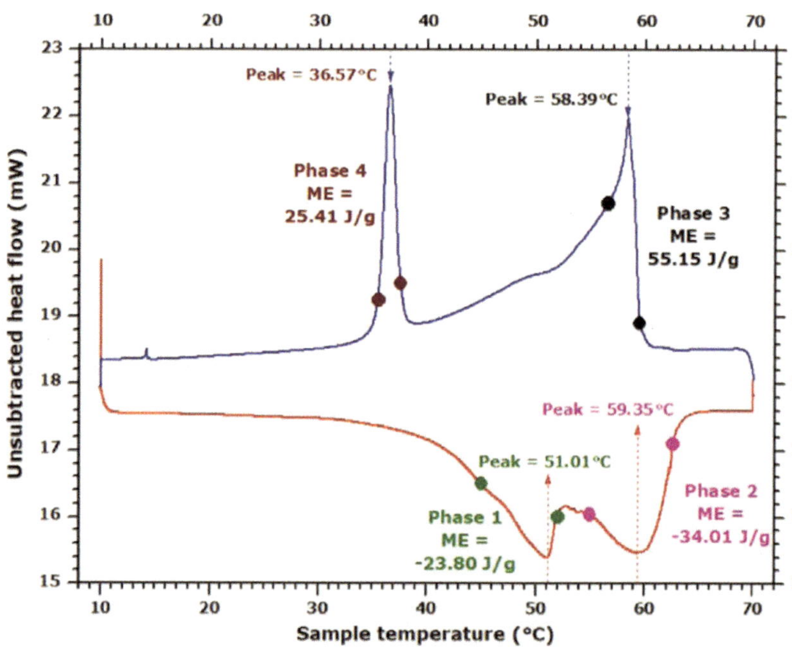

FIGURE 5.17 DSC curves of PCM during phase transition process [24]. (Permission from Elsevier.com.)

TABLE 5.6
Heater Energy Consumption at Different Temperatures [24]

Thermostat Temperature (°C)	Without PCM		With PCM	
	Production (Litre)	Power Consumption (kWh)	Production (Liter)	Power Consumption (kWh)
50	75	2.8	75	2.7
55	83	3.35	112	3.25
60	98	4.1	156	3.95
65	117	4.45	198	4.4
70	131	5.25	259	5.25
75	143	6.25	317	6

As can be seen from the table hotter water is produced from PCM-embedded tank than the conventional type when the thermostat setpoint temperature is increased above 50°C (above PCM melting temperature).

In addition, the PCM-embedded tank utilizes slightly less energy than the conventional tank. Their experimental results also indicated that the PCM-embedded water tank produces one third to two times more hot water than the conventional tank based on the thermostat setting and consumes less energy by 3.94% and 56.88% at 50°C and 75°C, respectively. A composite PCM consists of sodium acetate and 10% graphite as a thermal conductivity enhancer encapsulated in 100 cylindrical aluminium tubes of 16 mm in diameter and 48.33 cm in height, utilized in a 270 L domestic hot water heater by Nkwetta et al. [25], as depicted in Figure 5.18, to study the influence of PCM amounts on the energy efficiency and system performance. Based on the PCM density of 1,360 kg/m^3, the total weight of PCM has been calculated to be 13.2 kg. Properties of the hot water tank are listed in Table 5.7.

TABLE 5.7
Properties of Hot Water Tank [25]

Parameters	Values
Tank capacity	270 l
Height	1.45 m
Number of heater	2
Height of low heater	0.20 m
Height of thermostat controlling low heater	0.25 m
Dead band	5°C
Power supplied	4.2 kW
Height of top heater	1.02 m
Height of thermostat controlling top heater	1.07 m
Dead band	10°C
Power supplied	4.2 kW
Tank lost coefficient	1.05 W/m^2 K

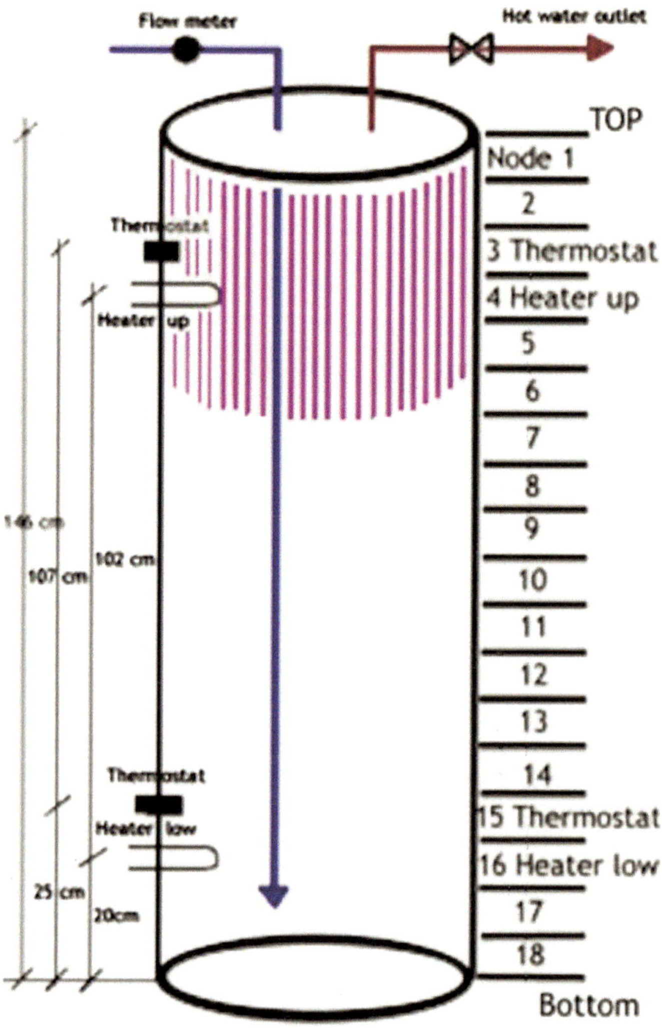

FIGURE 5.18 PCM used on a water heater [25]. (Permission from Elsevier.com.)

The temperature distribution of hot water discharge from the tank with a constant draw-off of 15 L/h utilizing different amounts of PCMs in the tank is illustrated in Figure 5.19. It is observed form the temperature profiles shown in the figure that application of PCM in the tank can increase the total hot water discharging time when the water is used at 45°C which is due to the additional amounts of energy stored. Also, by increasing the PCM amounts on the system the energy stored and the discharging time is increased and at the same time can help to shift the peak power demand in the range of 3, 6, and 8 h corresponding to 13 kg, 26 kg, and 52 kg of PCM, respectively.

An experimental analysis of a full-scale PCM thermal energy storage system has been designed for a domestic hot water system by Pakalka et al. [26] in which a fin-and-tube-type heat exchanger and a polypropylene storage tank filled with

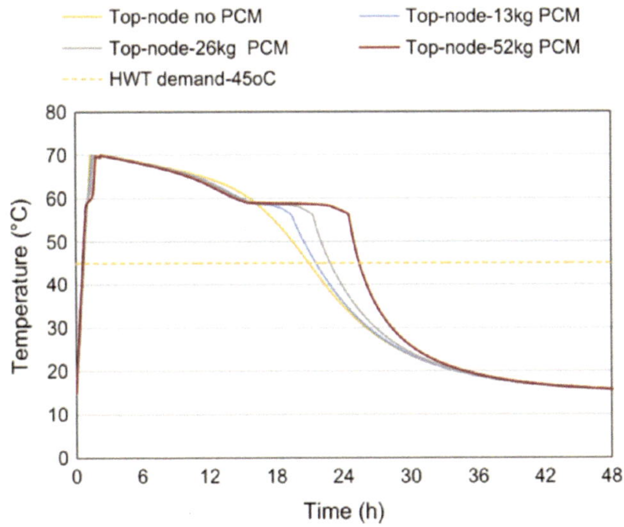

FIGURE 5.19 Temperature distribution of hot water at different PCM weight [25]. (Permission from Elsevier.com.)

commercial organic PCM RT58 (Rubitherm) have been used. The experimental setup is depicted in Figure 5.20.

The heat exchanger was made of copper tubes with aluminium fins that had two separate circuits for heat transfer fluid as shown in Figure 5.21.

The results obtained from the prototype unit show that the hot water system can deliver 99.1 and 156.2 L of hot water at 40°C and 37°C at constant mass flow rate of 0.1 kg/s for up to 15 min and 39 s, respectively.

FIGURE 5.20 Schematic of experimental setup [26]. (Permission from Elsevier.com.)

(a) (b)

FIGURE 5.21 General view of the finned tube heat exchangers: (a) heat exchanger; (b) heat exchanger integrated into storage tank [26]. (Permission from Elsevier.com.)

A domestic hot water system incorporated with a new lower-cost PCM consisted of animal fat and parafin wax with melting point of 35.58°C and 62.58°C and heating values of 180 and 210 J/g has been developed by Djeffal et al. [27] for the efficiency enhancement of the system. The system has been experimentally tested under different conditions of hot water volume, energy consumption, and associated energy cost. The regression model for the conventional water tank system and the PCM-embedded water tank at different thermostat setpoints is illustrated in Figure 5.22. As shown in

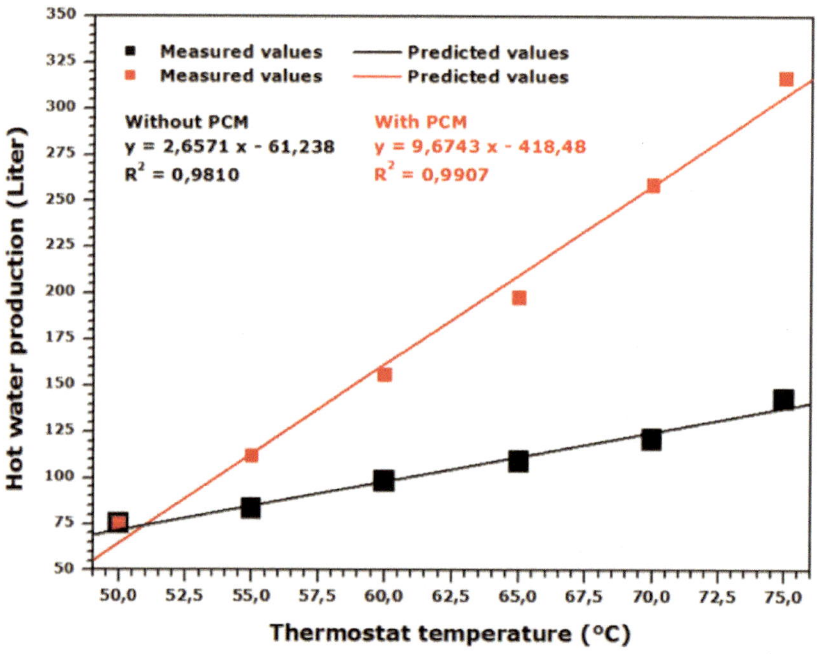

FIGURE 5.22 Regression models for the hot water production system [27]. (Permission from Elsevier.com.)

the figure, the experimental data for the conventional and PCM-embedded systems are best fitted with linear regression model with an R^2 more than 98% and 99%, respectively. It is also clear from the figure that hot water system with PCM produce one third to two times more hot water than the conventional type depending on the thermostat setpoint. The energy cost has also been reduced by 3.94% and 56.88% at 50°C and 75°C, respectively, which proves that the higher temperature setpoint are more economical.

An extensive review on the application of PCMs in solar domestic hot water systems presented by Douvi et al. [28] in which many challenges such as different techniques related to the solar energy storage and utilizing the latent heat content of PCMs to produce hot water have been discussed.

5.5 SUMMARY

Application of PCMs in household appliances for effective handling and control of the thermal demands under different climate conditions has been discussed. The results of the experimental data obtained for the household refrigerators and freezers demonstrated that the PCM's behaviour in relation to design variables like system configuration, PCM thickness, shape, orientation, and location are strongly impacted the selection of PCM and thermal load. Water and eutectic solutions of PCMs were mostly used in the compartment and evaporator parts of refrigerators while organic PCMs were mainly considered for the condenser part. The results reveal that PCMs could be promising to energy savings, contributing to lowering the operating and maintenance costs and as a result decreasing CO_2 emission. However, the PCM characteristics such as cycling stability, corrosion impact, and the choice of PCM container material are among the primary parameters limiting the utilization of PCM in various household appliances. More research in the future is required such as nucleating agent development for non-organic materials and thermal conductivity enhancement for the application of PCMs in electrical appliances.

REFERENCES

1. Bahadorian A., Sadrameli S.M., Pahlavanzadeh H., Nabi Ilani Kashkouli M. (2023) Effects of particle size in linseed biodiesel production via in-situ transesterification and slow pyrolysis of obtained linseed waste, Renew. Energy, 203, pp. 10–19.
2. Mohseni Roodbari S., Seyed Mousavi S.A.H., Sadrameli S.M., Pahlavanzadeh H. (2022) Thermal pyrolysis of linseed waste to produce a renewable biofuel using response surface methodology in a fixed bed reactor, Appl. Anal. Pyrol., 168, 105701, pp. 1–11.
3. Seyed Mousavi S.A.H., Sadrameli S.M., Saeedi A.H. (2022) Catalytic pyrolysis of municipal plastic waste over nano MIL-53 (Cu) derived @Y zeolite for gasoline, jet fuel and diesel range fuel production, Process Saf. Environ. Protect., 164, pp. 449–467.
4. Coulomb D., Dupon J.L., Pichard A. (2015) The role of refrigeration in the global economy. http://www.iifiir.org
5. Barthel C., Gotz T. (2012) The overall worldwide saving potential from domestic refrigerators and freezers, Wuppertal (Germany). http://www.gigee.net

6. Pirvaram A., Sadrameli S.M., Abdolmaleki L. (2019) Energy management of a household refrigerator using eutectic environmental friendly PCMs in cascade condition energy, Energy, 181, pp. 321–330.
7. https://master-bilt.com/general-information/news/refrigeration-u-the-basic-refrigeration-cycle
8. Kumar V., Shrivastava R., Nandan G. (2016) Energy saving using PCM in refrigerating, 3rd Int. Conf. on Manufacturing Excellence, MANFEX 2016, Amity University, Uttar Pradesh, India.
9. Nandanwar Y.N., Walke P.V., Kalbande V.P., Mohan M. (2023) Performance improvement of vapor compression system using PCM and thermoelectric generator, Int. J. Thermofluids, 18, p. 100352.
10. Ben Taher M.A., Ahachad M., Mahdaoui M., Zeraouli Y., Kousksou T. (2022) Thermal performance of domestic refrigerator with multiple PCM: numerical study, J. Energy Storage, 55, p. 105673.
11. Raveendran P.S., Sekhar S.J. (2020) Experimental studies on the performance improvement of household refrigerator connected to domestic water system with a water-cooled condenser in tropical regions, Appl. Therm. Eng., 179, p. 115684.
12. Manoj Kumar P., Elakkiyadasan R., Sathishkumar N., Ashwin Probhu G., Balasubramanian T. (2020) Performance enhancement of a domestic refrigerator with nanoparticle enhanced PCM over the condenser side, FME Trans., 48, pp. 620–627.
13. Kappen A., George T.J., Vinay V.N. (2016) Performance improvement of a household refrigerator by use of a phase change material, Int. J. Scient. Eng. Res., 7(3), pp. 58–68.
14. Subramaniam P.R., Tulapurkar C., Thagamani G., Thiyagarajan R. (2010) Phase change materials for domestic refrigerators to improve food quality and prolong compressor off time, International refrigeration and air conditioning conference. USA: School of Mechanical Engineering, Purdue University.
15. Azzouz K., Keducq D., Gobin D. (2009) Enhancing the performance of household refrigerators with latent heat storage: an experimental investigation, Int. J. Refrig., 32, pp. 1634–1644.
16. Nguyen V.N., Le T.L., Duong X.Q., Le V.V., Nguyen D.T., Nguyen P.Q.P., Rajamohan S., Vo A.V., Le H.S. (2023) Application of PCMs in improving the performance of refrigeration systems, Sustain. Energy Technol. Assess., 56, p. 103097.
17. Javeri Shahreza I., Fakhroleslam M., Sadrameli S.M. (2023) Application of PCMs for performance enhancement of open-display supermarket refrigerators: numerical simulation and parametric study, J. Energy Storage, 66, p. 107506.
18. Maiorino A., Duca M., Mota-Babiloni A., Aprea C. (2020) Achieving a running cost saving with a cabinet refrigerator incorporating a PCM by the scheduling optimization of its cycle operations, Int. J. Refrig., 117, pp. 237–246.
19. Abdolmaleki L., Sadrameli S.M., Pirvaram A. (2020) Application of environmentally friendly and eutectic phase change materials for the efficiency enhancement of household freezers, Renew. Energy, 145, pp. 233–241.
20. Pirvaram A., Sadrameli S.M., Abdolmaleki L. (2021) Optimization of energy consumption and temperature fluctuations for a household freezer using non-toxic and non-flammable eutectic phase change materials with a cascade arrangement, Int. J. Energy Res., 45(2), pp. 1775–1788.
21. Omara A.A.M., Mohammedali A.A.M. (2020) Thermal management and performance enhancement of domestic refrigerators and freezers via PCMs: a review, Innov. Food Sci. Emerg. Technol., 66, p. 105522.
22. Bertrand A., Mastrucci A., Schuler N., Aggoune R., Marechal F. (2017) Characterization of domestic hot water end-uses for integrated urban thermal energy assessment and optimization, Appl. Energy, 186, pp. 152–166.

23. Hohne P.A., Kusakana K., Numbi B.P. (2019) A review of water heating technology: an application to the South African context, Energy Rep., 5, pp. 1–19.
24. Djeffar R., Cherier M.K., Bekkouche S.M., Younsi Z., Hamdani M., Al-Saadi S. (2022) Concept development and experimentation of a PCM enhanced domestic hot water, J. Energy Storage, 51, p. 104400.
25. Nkwetta D.N., Vouillamoz P.E., Haghighat F., Mankibi M.E., Desai K. (2014) PCMs in hot water tank for shifting peak power demand, Sol. Energy, 107, pp. 628–635.
26. Pakalka S., Doneliene J., Rudzikas M., Valancius K. (2024) Development and experimental investigation of full-scale PCM thermal energy storage prototype for domestic hot water applications, J. Energy Storage, 80, p. 110283.
27. Djeffal R., Cherier M., Bekkouche S., Younsi Z., Hamdani M., Al-Saadi S. (2022) Concept development and experimentation of PCM enhanced domestic hot water, J. Energy Storage, 51, p. 104400.
28. Douvi E., Pagkalos C., Dogkas G., Koukou M.K., Stathopoulos V.N., Caouris Y., Vrachopoulos M. (2021) PCMs in solar domestic hot water systems: a review, Int. J. Thermofluids, 10, p. 100075.

6 Phase Change Materials and Lithium-ion Batteries

6.1 INTRODUCTION

One of the best alternatives energy sources for fossil fuels in the world is Lithium-ion batteries (LIBs) which have usage in a variety of applications. They have been utilized in small electronics like mobile, camera, laptops and in transport technologies such as scooters, battery electric vehicles, plug-in hybrid electric cars, and hybrid electric vehicles as a key component for fossil fuel savings potential. However, they are facing some challenges like limited world lithium sources, life span ability, thermal degradation which leads to battery capacity reduction and overheating which causes fire in critical situation. Different cooling techniques have been offered in the active and passive modes of operation utilizing air, water, refrigeration and PCMs to solve overheating issue in the batteries from which PCMs has more advantages to other cooling method. This chapter discusses recent developments on the application of PCM as a thermal energy storage (TES) on thermal management systems (TMSs) in LIBs, including thermal efficiency of different PCMs, techniques, numerical and experimental methods for the simulation of the system and the challenges that are faced when using PCMs in such systems.

6.2 LIB STRUCTURES

Lithium metal has some outstanding features, such as high thermal conductivity, light weight, and high electrochemical equivalence, that makes it a very attractive metal for the anode electrode in manufacture of LIBs. They have been used in some military applications since the 1970s, but their applications were limited due to the suitable cell structure, formulations, and safety considerations [1]. LIB cells are manufactured from of two electrodes with porous structure in the cathode which is made of different materials based on the manufacturers and anode which is mainly graphite with silicon and titanium-based materials, with one porous structure separator made of polymeric membrane in the electrolyte solution which is an organic solvent with dissolved lithium salt. Figure 6.1 shows different structures of the LIB cells.

Charged lithium ions are migrated back and forth between positive and negative electrodes, passing through the separator and electrolyte during the charge and discharge as shown in Figure 6.2 [2]. Based on the materials of the cathode and anode, the involved electrochemistry is different. For a LIB cell with positive electrode of $LiFeSO_4$ and negative electrode made from graphite, the reactions at the cathode and anode are shown in equations (6.1) and (6.2) [2].

$$LiFePO_4 \rightleftharpoons xLi^+ + xe^- + Li_{(1-x)}FePO_4 \tag{6.1}$$

$$6C + xLi^+ + xe^- \rightleftharpoons Li_xC6 \tag{6.2}$$

DOI: 10.1201/9781003518280-6

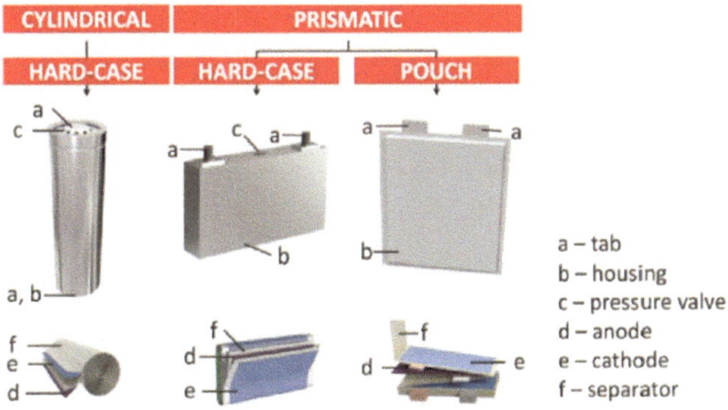

FIGURE 6.1 Different structures of LIB: (a) cylindrical, (b) prismatic, and (c) pouch [1]. (Permission from Elsevier.com.)

In general, LIBs can be classified to two main categories: lithium iron phosphate (*LFP*, $LiFePO_4$) and metal oxides (NCM, NCA, Cobalt, Manganese). Table 6.1 shows the characteristics of two different chemistry types of batteries.

Despite the outstanding features of LIBs in comparison to *NiCd* and *NIMH* cells, there are some challenges for the application of LIBs such as instability due to the decomposition of electrodes at high temperature, overheating, which is caused by heat dissipation during change and discharge processes, and other safety issues.

Among above drawbacks, high surface temperatures during charging and discharging processes are very critical affecting the safety, performance, and life span deterioration of the cells. In large-scale battery packs in which many cells are in series or parallel, the amount of heat dissipation is much higher and the temperature difference between the cells may cause performance failure or in the critical situation may lead to an explosion or fire in the system as illustrated in Figure 6.3 [4].

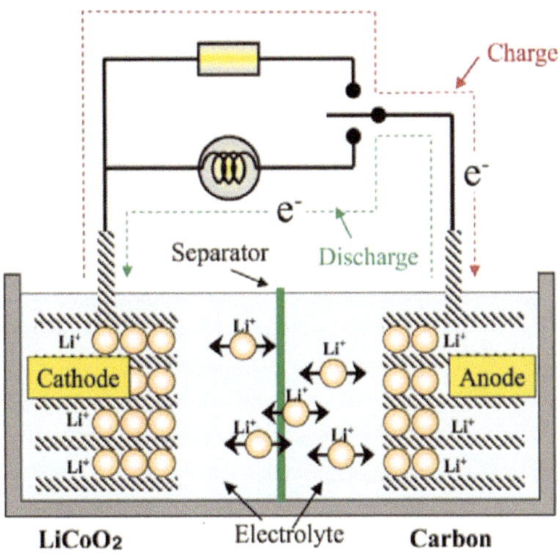

FIGURE 6.2 LIB working mechanisms [2]. (Permission from Elsevier.com.)

TABLE 6.1
Lithium-ion Subcategory Comparison [3]

	LFP	LiNCM
Voltage	3.3 V nominal (2-3.6 V/cell)	3.7 V nominal (2.7-4.2 V/cell)
Energy density	300 Wh/L	735 Wh/L
Specific energy	128 Wh/kg	256 Wh/kg
Power	1000 W/kg	512 W/kg
Cycle life	2,000 @ 100% DoD 3,000 @ 80% DoD	750 @ 100% DoD 1,900 @ 80% DoD
Calendar life	6 years	8 years
Max recommended temperature	40°C	55°C
Safety	High	Moderate
Commercial suppliers	A123, Valence, BAK, BYD, K2, Lishen, and many Chinese vendors	Sanyo, Panasonic, Samsung, DowKokam, Sony, LG Chem, and Moli

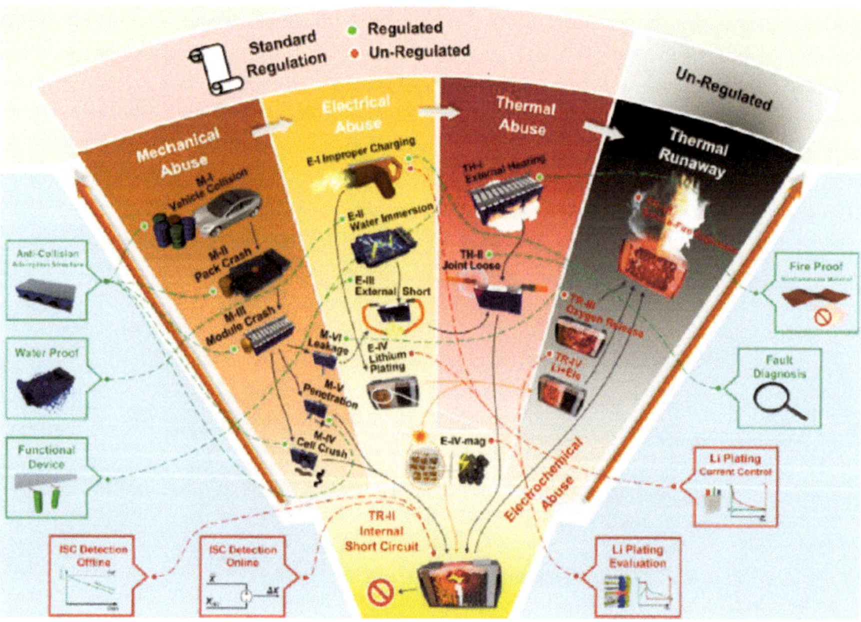

FIGURE 6.3 Thermal runaway conditions in LIBs [4]. (Permission from Elsevier.com.)

6.3 THERMAL MANAGEMENT OF LIBs

It has been recognized over the years that the optimum operating temperature for LIBs is between 25°C and 40°C with temperature difference within cell and among cells of not greater than 5°C [5]. Thermal management of LIBs is crucially important since inappropriate environmental conditions can deteriorate systems performance and in critical situations cause thermal runaway that sometimes leads to explosion. TMS consists of heating/cooling control systems which can not only keep the battery within allowable optimal temperature range but also provide temperature uniformity between cells in a pack.

6.3.1 Heating Systems

In an electrical vehicle, the low performance of the battery due to the slower electrochemical reactions at low temperature can affect other components of the vehicle. Therefore, a proper heating system consists of anticipated heating time, required power for heating and initial cost, considering weight, size, and space limits, is necessary to maintain the battery at optimum conditions [6]. The heating techniques can be classified in to three types: internal, external, and hybrid. A summary of the alternative heating systems for electrical vehicle battery packs is shown in Figure 6.4. In internal heating methods, the battery is heated with the heat generated locally inside the battery shell and therefore there is no need for an external agency to heat the cells, while in external heating except for the application of PCMs, an extra power supply is required for the heating systems.

Application of PCMs for heating is a passive approach for maintaining the battery pack temperature at optimum temperature in a cold environment by transferring PCM's heat of fusion to the battery [7]. PCM-based TMS is the least affected by environmental conditions and temperature differences among

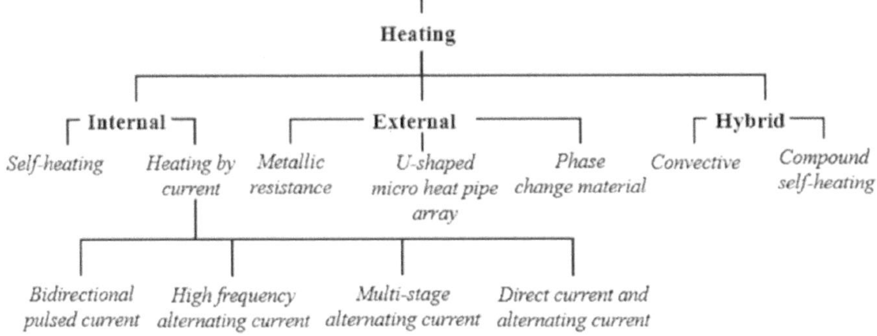

FIGURE 6.4 Heating techniques for battery thermal management systems [4]. (Permission from Elsevier.com.)

TABLE 6.2
Comparison of Alternative Heating Systems [4]

Heating Method		Advantages	Disadvantages
Internal	Self-heating	Simple concept	Needs to be implemented prior to the battery shell's fabrication
	Heating by current	Wide range of heating rates	Needs a complex control circuit
External	Metallic resistance	Good pack thermo-consistency	Slow heating
	Heat pipes	Good heating efficiency	Complex structural design
	PCM	Consumes no heating power	Relatively low heating efficiency
Hybrid	Convective	Short heating time	Needs extra actuators
	Compound self-heating	Good electro-thermal conversion	Heating efficiency is greatly influenced by the design of the heaters

other heating alternatives. Table 6.2 summarizes the advantages and disadvantages of heating systems for battery packs.

6.3.2 Cooling Systems

Heat is generated in the batteries during charging and discharging due to the transformation of chemical into electrical energy and vice versa, and consists of heat of reaction, Joule heating, and polarization heating. In hot climate regions, extra heat from the environment is added to the generated heat and a suitable cooling system is required to keep the battery pack within optimum temperature range. The cooling techniques for batteries can be categorized into internal, external, and hybrid as for the heating systems. In the internal methods, the generated heat by the reactions has been minimized during the charging and discharging processes while in an external cooling system either active using air or water or passive, including PCMs, an external agency is involved. A detailed configuration of cooling techniques is shown in Figure 6.5. A conventional active cooling system using air is illustrated in Figure 6.6.

Application of PCMs as a cooling method for batteries has been studied by many researchers during the last two decades. The role of PCM is to absorb the dissipated heat form the electrochemical reactions during charging and discharging process. Therefore, melting point, latent heat, and density of PCMs are among

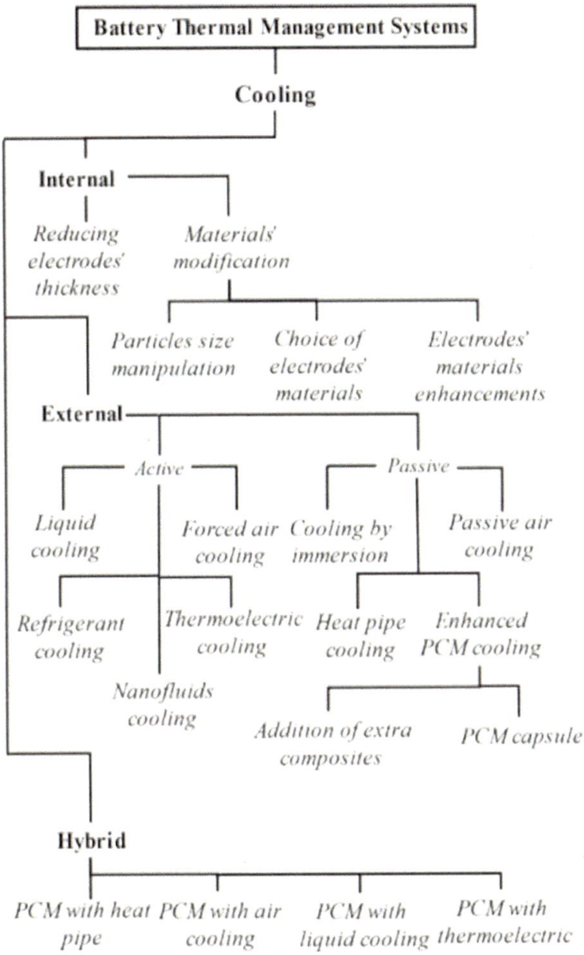

FIGURE 6.5 Cooling techniques in batteries [4]. (Permission from Elsevier.com.)

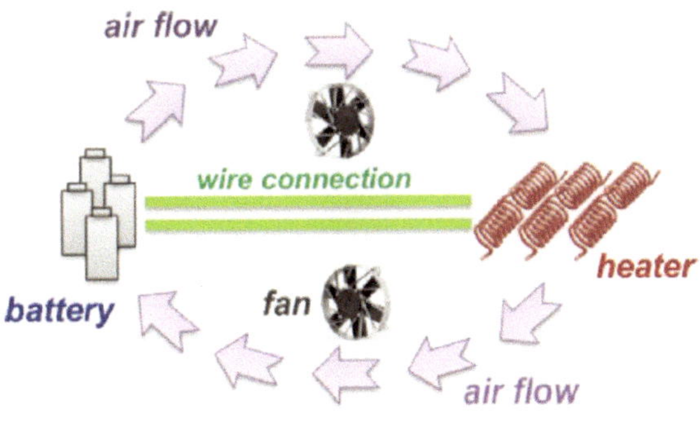

FIGURE 6.6 Active cooling by air convection [7]. (Permission from Elsevier.com.)

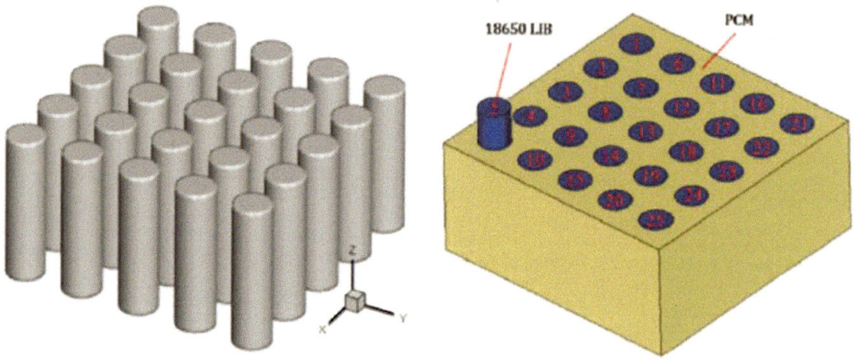

(a) (b)

FIGURE 6.7 A PCM cooling system for LIB: (a) before packing and (b) after packing [8]. (Permission from Elsevier.com.)

the significant parameters for the selection of PCMs. A conventional cooling system using PCM is shown in Figure 6.7 in which the cells are mounted in direct contact with PCM.

Despite the advantages of utilizing PCM as TMS for LIBs, there are some challenges in practical applications of PCMs such as heat leakage, poor thermal conductivity of PCM, and low cycle performance. PCM low thermal conductivity can be cured by adding nanoscale and metallic particles, porous materials, and other carbon-based materials into PCMs.

Application of aluminium sheets as thermal conductivity enhancer of PCM has been used by Azizi et al. [9] for the thermal management of LIBs in moderate, hot, and cold climate regions. PEG1000 with physical properties shown in Table 6.3 has been selected as a PCM.

A complete battery pack consist of six cylindrical shape LIBs of type $LiFePO_4$ model 38120 with 38 mm in diameter and 122 mm in length has been fitted inside a plexiglass container with dimensions of $17 \times 12 \times 18$ cm^3 filled with PCM and

TABLE 6.3
Physical Properties of PCM [9]

Average M.W.	950–1050
Viscosity at 100°C, cSt	17.2
Solubility in water at 20°C, by wt.%	80
ρ (kg/m^3) @ 60°C	1092.7
k (W/m K)	0.23
ΔH_m (J/kg)	158,992
ΔT_m (°C)	35–40
C_p (J/g °C)	2.142

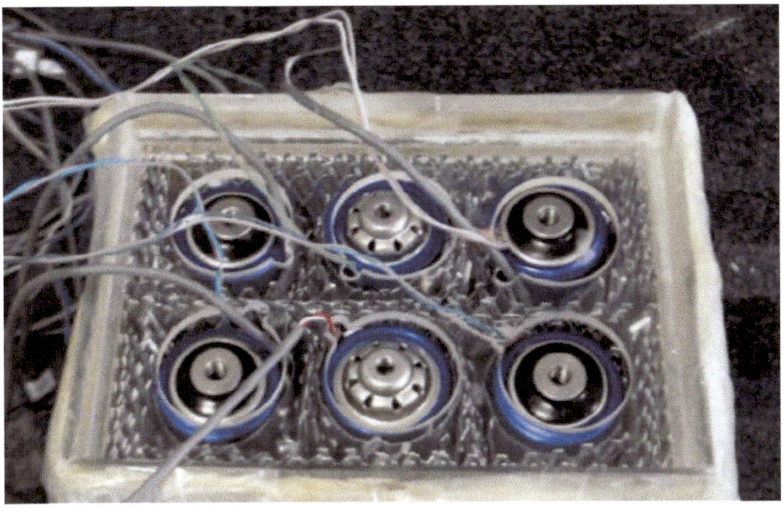

FIGURE 6.8 Configuration of TMS for LIBs [9]. (Permission from Elsevier.com.)

aluminium sheets are placed between the cells inside PCM as illustrated in Figure 6.8. An oven model UFE400 and a freezer model 20SMT have been used for the hot and cold temperature environments, respectively.

The charge of the batteries has been done using Kimia-Stat instrument made by an Iranian company for 5–6 hours and has been continued until the voltage difference of the cells has been reached to 3.7 V. The batteries have been discharged at the rates of 1°C to 4°C using a GB cell model AL2505010K with 1–50 V, 1–250 A, and continuous power of 10 kW as depicted in Figure 6.9.

FIGURE 6.9 GB cell for charge and discharge of LIBs [9]. (Permission from Elsevier.com.)

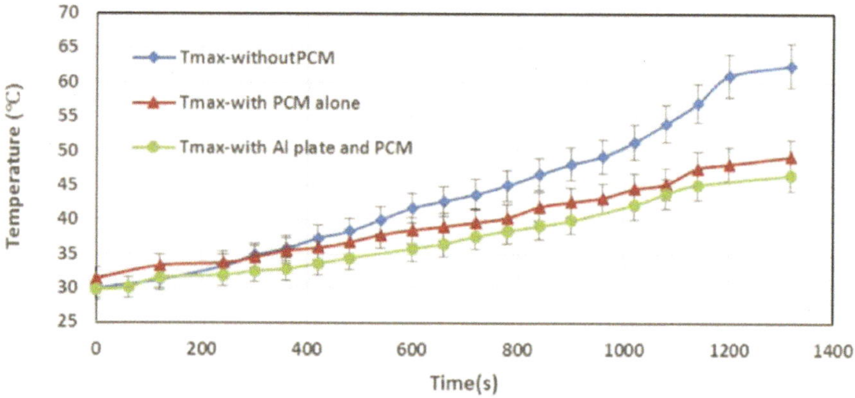

FIGURE 6.10 Temperature distribution of LIBs at rate 3C [9]. (Permission from Elsevier.com.)

Temperature distribution of cells at rate 3C (60A) in an ambient temperature for three cases of without PCM, with PCM, and with PCM/aluminium sheets is shown in Figure 6.10 in which the maximum temperature of the cells without PCM has reached to 62.2°C. Using PCM and PCM/AS can reduce the maximum temperatures to 49.2°C and 46.5°C, respectively, which is in the safe mode of cells operation.

Maximum allowable temperature difference for the cells at rate 2C, which indicates the temperature uniformity during the operation, is illustrated in Figure 6.11 which shows that the maximum temperature differences reach to 2.6°C, 4°C, and 6.7°C for systems with TMS, PCM only, and without TMS, respectively. This proves that their TMS system can provide better uniformity and higher performance for the LIB cells.

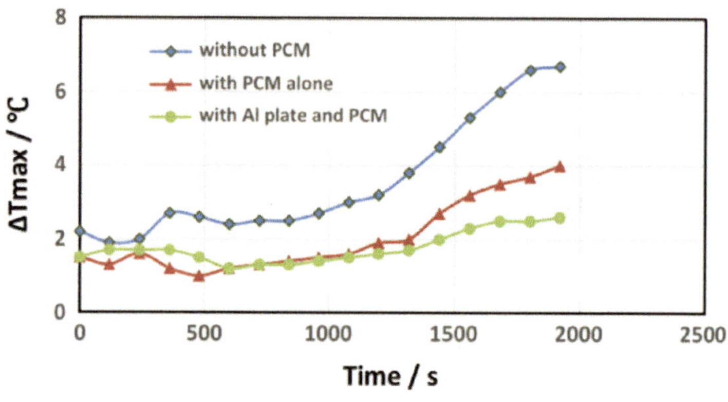

FIGURE 6.11 Maximum temperature difference at discharge rate of 2C [9]. (Permission from Elsevier.com.)

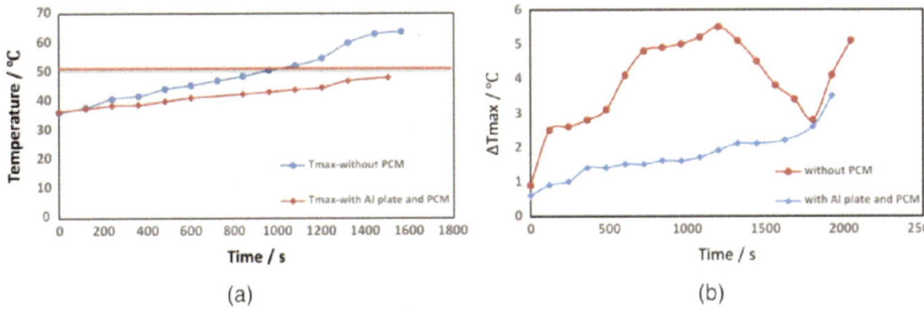

FIGURE 6.12 Cell temperature distributions (a) and maximum temperature differences at hot temperature (b) [9]. (Permission from Elsevier.com.)

Temperature profiles at rate 2.5C and maximum temperature differences at rate 2C for the cell at hot temperature of 50°C are shown in Figure 6.12. The maximum temperature for this case has reached to 63.8°C which is very critical in the operation of such batteries and has been decreased to 48.2°C when operates under TMS which represents that the system is in much more safer and efficient conditions. The maximum temperature differences have also reduced from 5.2°C to 3.5°C for the systems without and with TMS, respectively.

An extended work of Azizi et al. [9] for the cold climate region conditions of 0 to –20°C has been presented by Sadrameli et al. [10] in which the same configuration of TMS has been utilized for cold temperature conditions using a freezer. Temperature distributions at rate 1°C and maximum temperature differences of LIBs at cold temperature of –20°C are illustrated in Figure 6.13. As seen in Figure 6.13.a, the temperature distributions of the systems without and with TMS are very similar which is due the environment condition in which PCM is still in the solid form and no heat is absorbed by the PCM at low temperature. The maximum allowable temperature differences have been reduced considerably from 5.8°C to below 4°C when using TMS.

The impact of critical PCM thermal properties such as latent heat (100–250 J/g), thermal conductivity (0.2–23 W/mK), melting temperature, and PCM thickness on the LIB pack thermal performance during discharge rate of 25 A (3.2C) using detailed CFD analysis has been developed by Napa et al. [11]. Each battery pack consists of several individual cylindrical cells with 2.6 Ah, 3.7 V and size of 18 mm

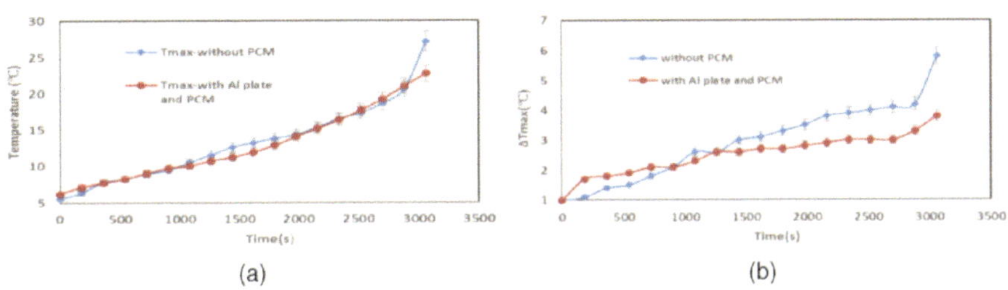

FIGURE 6.13 Cell temperature distributions (a) and maximum temperature differences at cold temperature (b) [10]. (Permission from Elsevier.com.)

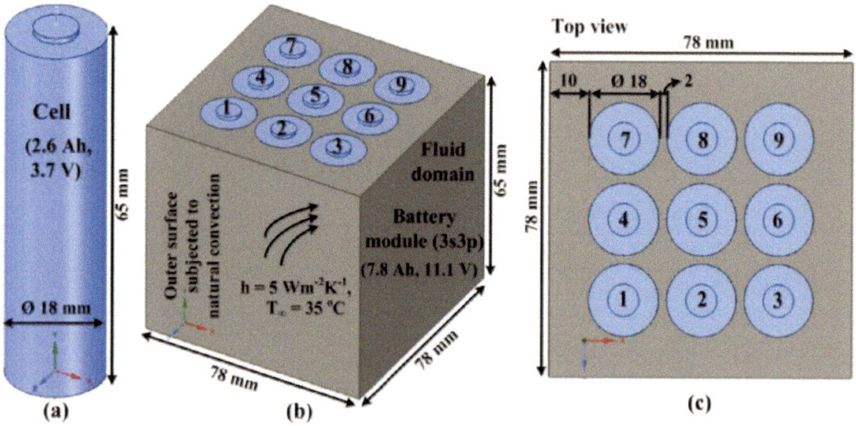

FIGURE 6.14 Schematic of (a) single battery cell, (b) battery pack, and (c) top view of battery pack [11]. (Permission from Elsevier.com.)

in diameter and 65 mm in length (Figure 6.14a) which are connected in series and parallel (three in series and three in parallel) arrangement to get the pack voltage of 11.1 V and 7.8 Ah as illustrated in Figures 6.14b. The PCM thickness between the cells is kept at 2 mm and the gap between the cell and wall is 10 mm with an ambient temperature of 35°C as illustrated in Figure 6.14c.

The results reveal that PCM with higher latent heat of greater than 175 J/g, with thermal conductivity larger higher than 3 W/mK and thickness of more than 3 mm with a melting point range of 41–44°C has the highest heat dissipation for the high discharge rates by reducing the maximum temperature to 44.4°C and temperature difference of 2.4°C for cells of 25 A and ambient temperature of 36°C.

A comprehensive review on the recent applications of PCMs on TMS or LIBs, including thermal performance of different PCMs, and numerical and experimental models has been presented by Nasajpour-Esfahani et al. [12].

6.4 MODELLING AND SIMULATION

Performance improvement of LIB passive cooling process using copper fins and paraffin wax with melting temperature of 40°C and latent heat of 195 J/g, as a PCM has been presented by Alghassab [13]. Numerical simulation of the system has been performed using ANSYS FLUENT's solidification and melting solver in an axisymmetric configuration. The results reveal that in comparison with conventional air cooling which produce the maximum battery temperature around 55°C after one hour, PCM passive cooling maintained lower temperature during the melting process if PCM is solid. Using 13 mm PCM thickness on the sides and 10 mm on top the maximum temperatures were up to 15°C lower than air cooling for the first 2 hours. Further improvement can be achieved by using copper fins with thermal conductivity of 385 W/mK that produce 8°C lower peak temperatures than PCM without fins over two hours.

A simplified two-dimensional mathematical model in cylindrical geometry for one of the battery cells, including PCM jacket has been developed by Azizi et al. [9] using COMSOL 5.1 to obtain surface temperature distribution of the cell.

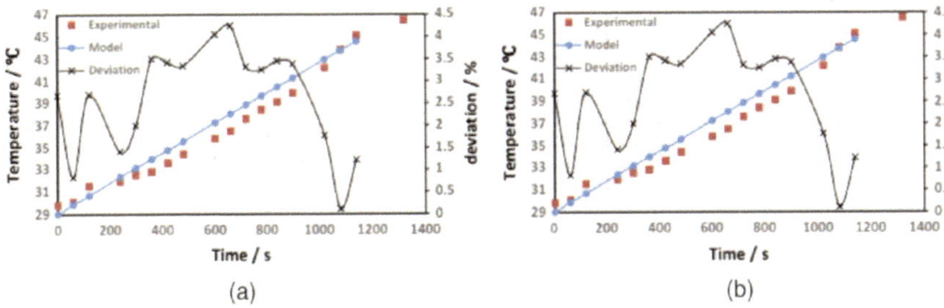

FIGURE 6.15 Comparison of results for the cell with PCM (a) and with PCM/aluminium sheet (b) [9]. (Permission from Elsevier.com.)

The generated heat during discharge process in 20 minutes in the cell has been considered as a heat source in one solid cylinder in W/m^3 in which the size of the solid cylinder and its PCM jacket are the same as real battery cell. Three cases have been examined for the cell without PCM, cell with PCM, and cell with PCM and aluminium wire mesh plate. The comparison of temperature distributions of the cell at discharge rate of 3C, with PCM only and with PCM/aluminium wire mesh plate are illustrated in Figures 6.15.a and 6.15.b, respectively. As can been seen from the figures, there is a good agreement between the model and experimental results.

Three-dimensional representation of the temperature distribution after 900 min is depicted in Figure 6.16.

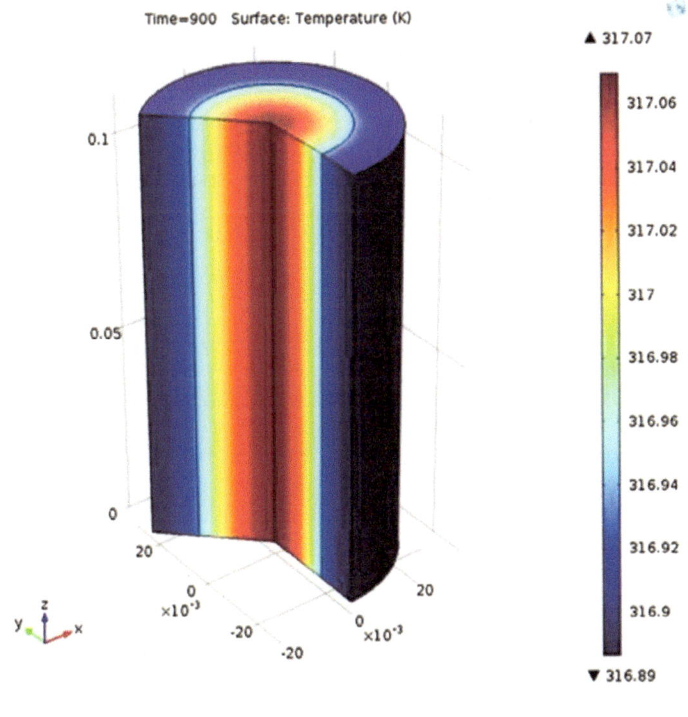

FIGURE 6.16 3-D Temperature distribution of the cell [9]. (Permission from Elsevier.com.)

FIGURE 6.17 Schematic diagram of TMS [14]. (Permission from Elsevier.com.)

A numerical study on a thermal management of a battery pack using a hybrid system of encapsulated PCM with thickness of 0–3 mm and convective air cooling of inlet velocity (U_{in} = 0.05 to 0.2 m/s) has been presented by Singh et al. [14] in which commercial SONY 18650 type LIB with nominal capacity of 3 Ah, nominal voltage of 3.7 V, and cell diameter of 18 mm were taken in two arrangements of inline and staggered as depicted in Figure 6.17.

N-eicosane with properties listed in Table 6.4 has been used as PCM. A two-dimensional, incompressible, and laminar flow of air passing through cylindrical LIBs cells covered with encapsulated PCM has been considered for the modelling purposes as shown in Figure 6.14.

The electrochemical-thermal system has been simulated by a coupled model using COMSOL Multiphysics v. 5.3. The air velocity magnitude contours inside the battery pack, including eight cells and different PCM thickness in two inline and staggered arrangements at velocities of 0.05 and 0.2 m/s is shown in Figure 6.18.a. It is obvious from the results that for both velocities the maximum velocity exits for 3 mm PCM thickness which is due to the mass conservation which sharply increases the velocity up to 10-12 times of the inlet air velocity in constricted annular regions and the maximum velocity is higher for inline arrangement in comparison to the

TABLE 6.4
Properties of n-eicosane [14]

PCM Property	Solid Phase	Liquid Phase
T_m (K)	309.55 ± 1	
C_p (J/kg·K)	1926	2400
ρ (kg/m^3)	839.5	
μ (kg/m·s)	–	3.88×10^{-3}
k (W/m·K)	0.423	0.146
L_f (J/kg)	248×10^3	–

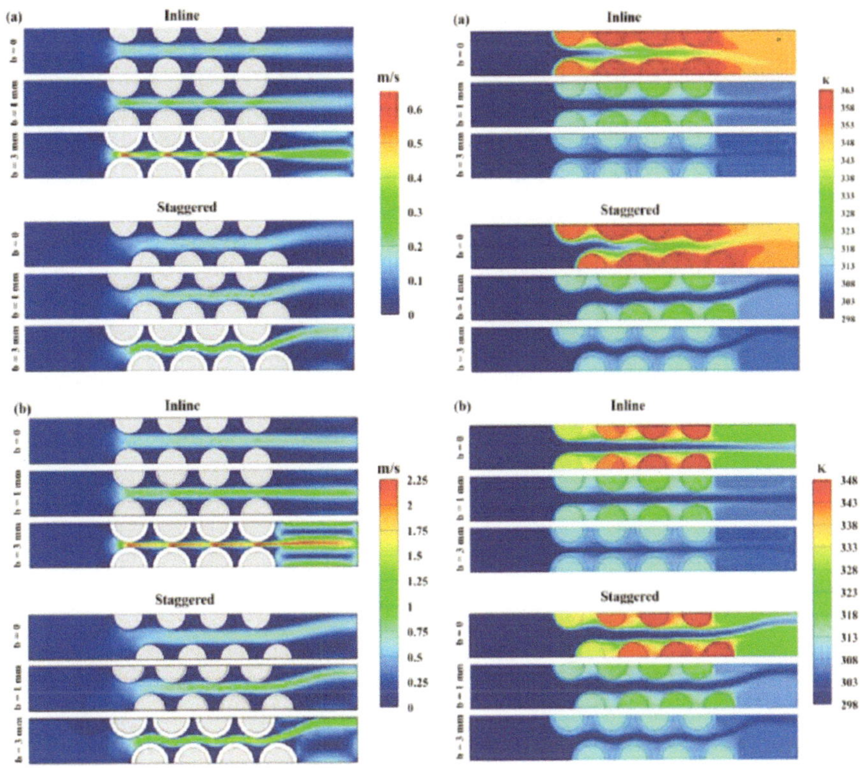

(I) (II)

FIGURE 6.18 Velocity (I) and temperature (II) contours for N = 8, (a) U_{in} = 0.05 m/s; (b) 0.2 m/s [14]. (Permission from Elsevier.com.)

staggered cell configuration. Temperature distributions with similar conditions are illustrated in Figure 6.18.b. It can be seen from the temperature results that the maximum cooling effect can be observed when air is in contact with the first cell which the minimum happened for the last cell which is caused by relatively lower temperature gradient at the exit.

A new cooling system of battery pack, including five prismatic Li-ion battery each of size 14 × 100 × 180 mm^3 sandwiched between cold plates of phase change materials has been studied experimentally and numerically by Mousavi et al. [15]. N-eicosane with thermal properties listed in Table 6.5 has been used as a PCM.

Numerical optimization for application paraffin wax as PCM and expanded graphite as a thermal conductivity enhancer for PCM in TMS for LIB has been done by Wang et al. [16] to investigate the effect of composite PCM with changing melting temperature on the thermal performance of a battery pack. They included a one-dimensional electrochemical reaction and a 3-D thermal module in the model

TABLE 6.5
Thermophysical Properties of n-eicosane Used as PCM [15]

	Solid PCM (at 298.15 K)	Liquid PCM (at 323.15 K)	Water
Density, ρ(kg/m^3)	910	769	997.5
Latent heat of fusion, h_{sl} (kJ/kg)	248	–	–
Melting point, T_{melt} (K)	309.55	–	–
Specific heat capacity, c_p (J/(kg.K))	1926	2400	4179
Thermal conductivity, k (W/(m.K))	0.423	0.146	0.613
Thermal expansion coefficient, β (1/K)	–	8.161×10^{-4}	–

with the mesh grids shown in Figure 6.19. The electrochemical model developed for the diffusion process of Li+ for variety of phases and the reaction process of the electrode material particle surface as illustrated in Figure 6.20.

Maximum surface cell temperature T_{max} distributions and maximum temperature difference of the cell, ΔT_{max} vs. time at rate 3C, which has a greater fire risk for different convective heat transfer coefficient, h in the range of 5–40 W/m^{2}K are shown in Figure 6.21. They concluded that increasing h has opposite effects on the maximum cell surface temperature and the maximum temperature differences as increasing h causes an increase in ΔT_{max} while decreasing the maximum surface temperature which is due to the faster heat dissipation at higher convection.

Capacity of PCMs with different thermal-physical properties for the thermal management of LIBs using the CFD model has been examined by Wang et al. [17]. Effects

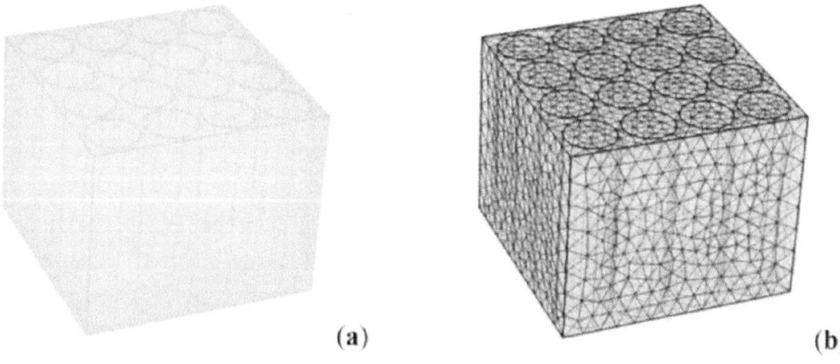

FIGURE 6.19 Thermal model (a) and mesh grids (b) [16]. (Permission from Elsevier.com.)

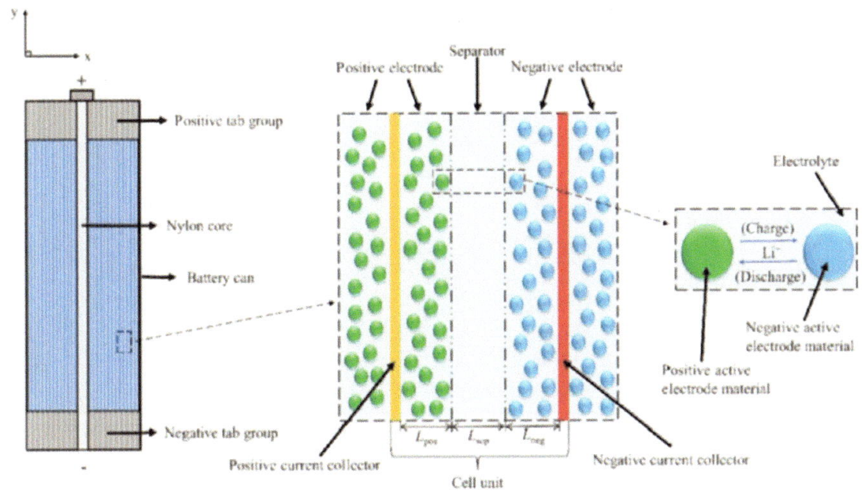

FIGURE 6.20 Battery axial structure, 2-D model, and diffusion process [16]. (Permission from Elsevier.com.)

of system parameters such as shell material, heat transfer coefficient, PCM fill volume, and shape of the battery pack on the thermal performance of the system have been studied. The results show that at an ambient temperature range of 20–30°C, PCM of RT-35 have the best thermal performance and control ability for the proposed system, while the PCM of RT-50 has the better performance at an ambient temperature of 40°C. The maximum temperature of the battery and the temperature difference decrease by increasing the heat transfer coefficient and shell material thermal conductivity. This is due to the better convective and conductive heat transfer from the battery surface. As far as the geometry of the batteries is concerned, the rectangular

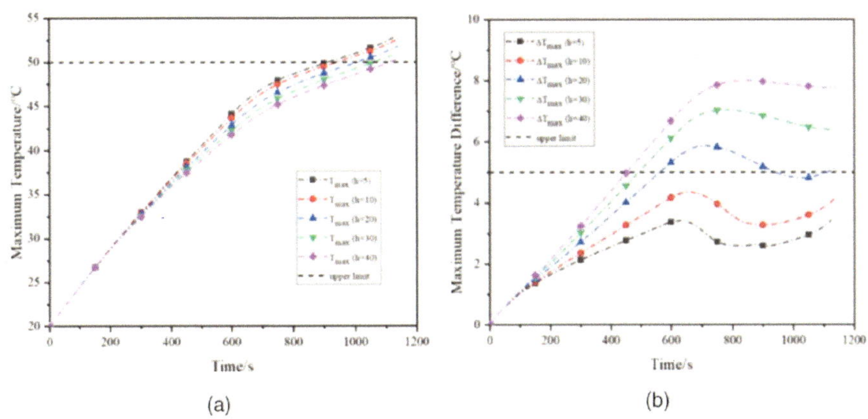

FIGURE 6.21 T_{max} (a) and ΔT_{max} (b) distributions vs. time [16]. (Permission from Elsevier.com.)

LIB with PCM has a lowest maximum temperature and the smallest temperature difference in comparison with the circular and square shapes.

The above modelling and simulation studies prove the benefits of simulation techniques for the design, parametric study, and optimization of the LIB pack with phase change materials.

6.5 SUMMARY

One of the promising solutions for energy storage in different applications such as mobile electronic devices, electric vehicles, scooters, and building facilities are LIBs. However, they are facing many challenges and one of most important challenges is their surface temperature increase during charging and discharging processes. Therefore, a suitable TMS is required to keep the batteries at safe and optimum environments.

Phase change materials with great characteristics such as low viscosity, variable melting points, high latent heat of fusion for absorbing more dissipated heat from the cells, and good wettability on the surface for better heat transfer have been applied for the thermal management of LIBs. This chapter discusses the structure of LIBs, and the main challenges involve during charging and discharging processes of the systems. Recent developments on the application of phase change materials and thermal management technologies for both high- and low-temperature environments have been discussed and compared in terms of complexity, performance, and types and sizes of batteries to which they applied. Different TMSs for the optimum operation of the cells, including heating and cooling techniques in active and passive modes have been explained. Finally, mathematical modelling and simulation of the system have been discussed using COMSOL Multiphysics software for the optimization of temperature distribution on the cell surfaces utilizing PCMs and some of the results have been presented. The main goal of the chapter was to provide an extensive overview the PCMs part in the thermal management of LIBs.

REFERENCES

1. Schroder R., Aydemir M., Seliger G. (2017) Comparatively assessing different shapes of lithium-ion battery cells, Proc. Manuf., 8, pp. 104–111.
2. Wang X., Wu P. (2019) One step photo-mediated grafting of poly(methyl methacrylate) on to fluorinated carbon nanotube for the enhanced thermal conductive property of polymer composites, Chem. Eng. J., 269, pp. 272–279.
3. Albright G., Al Hallaj S. (2012) A comparison of lead acid to lithium ion in stationary storage applications, All Cell Technologies LLC.
4. Feng X., He D., Ouyang M. (2020) Mitigating thermal runaway of LIBs, Joule, Cell Press, 4(4), pp. 743–770.
5. Pesaran A.A. (2002) Battery thermal models for hybrid vehicle simulations, J. Power Sources, 110, pp. 377–382.
6. Wang Y., Zhang X., Chen Z. (2022) Low temperature preheating techniques for LIBs, recent advances and future challenges, Appl. Energy, 313, p. 118832.
7. Verma A., Rakshit D. (2022) Performance analysis of PCM-fin combination for heat abatement of LIB pack in electric vehicles at high ambient temperature, Therm. Sci. Eng. Progr., 32, p. 101314.

8. Ji Y., Wang C.Y. (2013) Heating strategies for LIBs operated from subzero temperatures, Electrochim. Acta, 107, pp. 664–674.
9. Azizi Y., Sadrameli S.M. (2016) Thermal management of a LiFePO4 battery pack at high temperature environment using a composite of PCMs and aluminium wire mesh plates, Energy Convers. Manag., 128, pp. 294–302.
10. Sadrameli S.M., Azizi Y. (2023) Application of PCMs for thermal management of $LiFePO_4$ battery pack in a cold temperature environment, JOM.
11. Napa N., Agrawal M., Tamma B. (2024) PCM properties identification for the design of efficient thermal management system for cylindrical lithium-ion battery module, J. Energy Storage, 99, p. 113241.
12. Nasajpour-Esfahani N., Garmestani H., Rozati M.R., Fadhil Smaisim G. (2023) The role of PCMs in LIBs: a brief review on current materials, thermal management systems, numerical methods, and experimental models, J. Energy Storage, 63, p. 107061.
13. Alghassab M.A. (2024) Investigating the performance improvement of lithium-ion battery cooling process using copper fins and PCMs, Case Stud. Therm. Eng., 59, p. 104473.
14. Singh L.K., Kumar R., Gupta A.K., Sharma A.K., Panchal S. (2023) Computational study on hybrid air-PCM cooling inside lithium-ion battery packs with varying number of cells, J. Energy Storage, 67, p. 107649.
15. Mousavi S., Siavashi M., Zadehkabir A. (2021) A new design for hybrid cooling of LIB pack utilizing PCM and mini channel cold plates, Appl. Therm. Eng., 197, p. 117398.
16. Wang, S., Zhang D., Li C., Wang J., Zhang J., Cheng Y., Mei W., Cheng S., Qin P., Duan Q., Sun J., Wang Q. (2023) Numerical optimization for PCM based LIB thermal management system, Appl. Therm. Eng., 222, p. 119839.
17. Wang H., Guo Y., Ren Y., Yeboah S., Wang J., Long F., Zhang Zh., Jiang R. (2024) Investigation of the thermal management potential of PCM for lithium-ion battery, Appl. Therm. Eng., 236, p. 121590.

7 Application of Phase Change Materials in Textiles Thermal Management

7.1 INTRODUCTION

Materials that sense and react to environmental conditions or stimuli, like mechanical, thermal, electrical, chemical, and magnetic effects, are called smart materials and can be divided into three categories based on their reaction to the environment, namely passive smart materials that only sense the external conditions, active smart materials that sense and react to the conditions, and very smart materials that can sense, react, and adapt themselves [1]. Silk thread, including a shape memory, was the first textile material that was named as a smart textile. Waterproof and breathable textiles have been developed by attaching membranes on textile substrates and fabrics. Smart fabrics that change, radiate, or erase colour depending on the environmental conditions such as light, heat, electricity, pressure, liquid, or electron beam are called chromic textile. This chapter reviews the suitable types of PCM which are utilized in textile applications, their incorporation methods, and parameters affected on the thermal management of textiles and fibres using PCMs.

Many researchers studied the application of PCMs in clothes, textile, and accessories for maintaining comfort in extreme hot and cold environments. PCM is added to the filaments of textile to absorb excessive heat from environment and prevent from reaching the human body in hot weather, while in the cold environment incorporated PCM in the textile increases the thermal resistance between the body and external ambient and prevents heat flow to exit to the external ambient.

Incorporation of PCMs into textile structures was first applied in the early 1980s at the US National Aeronautics and Space Administration (NASA) as a research study to improve the thermal performance of space suits to provide the astronauts with comfort temperature at extreme temperature conditions in the space. Although it has not been practically utilized in the space suits, but the technology has been further extended into terrestrial applications with thermoregulation ability which means keeping the body temperature at comfort level in diverse environments.

Body systems performance is less when the body temperature is beyond the thermo-neutral zone which is 37 ± 1°C and can even stop working when pushed to extreme conditions. This can be solved by using smart textiles containing PCMs which can act as a transient thermal barrier, which controls the heat flux. Melting

DOI: 10.1201/9781003518280-7

point of PCM is one of the most significant parameters that can affect the textile thermoregulation. There are other important criteria for the selection of PCM for smart textile such as PCM type, heat capacity, maximum allowable loading, fabric structure, and garment structure that must be considered with designing PCM-based smart textiles for thermoregulation.

7.2 PCM IN TEXTILES

The balanced status between total heat gain and release in the microclimate around wearer's body skin using textiles is called thermal comfort. Comfort or discomfort is also defined as the mental satisfaction or dissatisfaction, respectively, with the thermal environment. Thermal dissatisfaction may be caused by hot and cold discomfort of the body or local discomfort due to the undesired cooling/heating which is also known as Predicted Mean Vote (PMV) and Predicted Percentage of Dissatisfied (PDD) indices. Textile materials in contact with the skin can also be a reason for thermal comfort/discomfort of the body and is a complex phenomenon that depends on the integration of thermal (hot and cold), wetness and tactile (contact). The level of perceived sensation can be affected by whether conditions such as humidity, temperature, and wind, levels of physical activity, and the physical and physiological status of individuals and the textile properties.

All mechanisms of heat transfer such as conduction through the textile thickness, convection, and radiation from body to textile, and from textile to ambient and vice versa, are involved in thermoregulation of textiles as illustrated in Figure 7.1.

Controlling the microclimate between the clothing and skin, which significantly affects thermal comfort through temperature and humidity is one of the most important mechanisms for the personal thermal management. The main heat loss from the body in typical indoor environments occurs through mid-infrared (MIR) radiation. The heat loss can be reduced by wearing a low thermal conductivity material in colder environments, while increasing the thermal conductivity

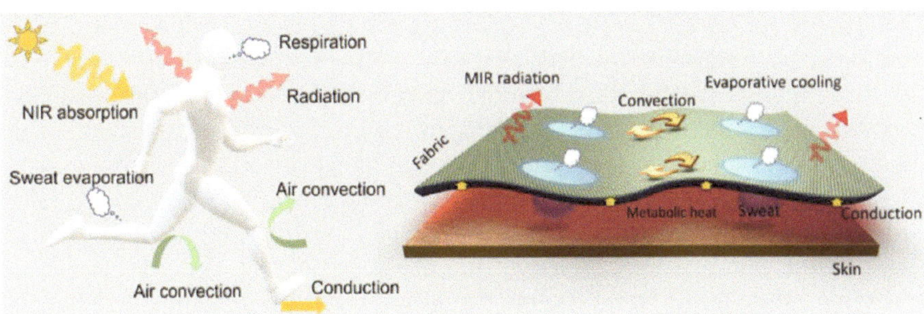

FIGURE 7.1 Heat exchange mechanisms from a naked human body and skin covered by a fabric [2]. (Permission from Elsevier.com.)

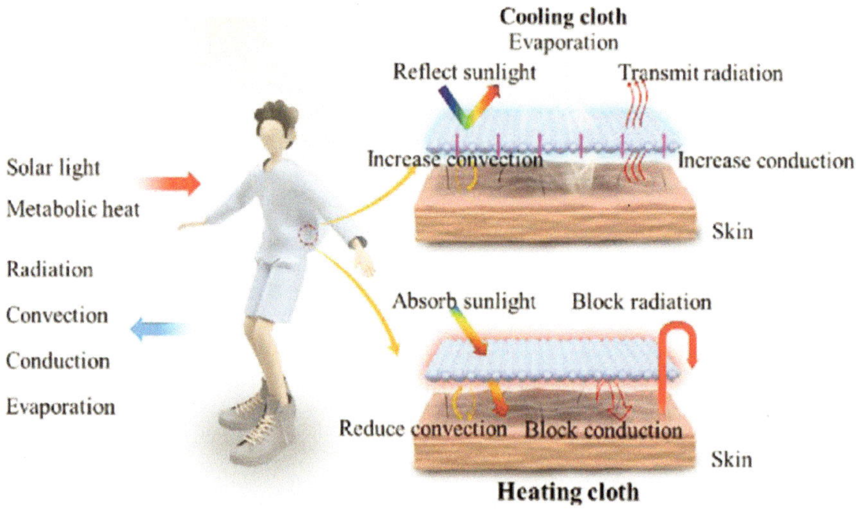

FIGURE 7.2 Schematic of heat input and output in a human body in outdoor environment [3]. (Permission from Elsevier.com.)

of the textiles for cooling can increase the heat transfer from the body to the environment as illustrated in Figure 7.2.

Thermal comfort is the most important factor especially in a hot climate for human being since its poor management can lead to heat stress, health problems, poor performance, and even at critical conditions death [4]. Thermoregulating features to textiles and clothing can be added with different mechanisms such as PCM integration in coatings and fibre-spinning dopes, evaporative cooling, reflective nanoparticles, conductive ingredients into coating solutions, and spinning and attachment of portable heaters and fans to clothing [2]. Thermal management textiles can be classified based on their preparation techniques and the type of used substrate materials as illustrated in Figure 7.3.

7.2.1 Cooling Methods in Textiles

Traditional and innovative methods of textile cooling have been studied to address the challenges posed by warm external environment. Active cooling methods, including air and liquid cooling, are among the traditional strategies to enhance heat dissipation, while thermal management through radiative and conductive means is classified as advanced cooling technologies (Figure 7.3).

In the air-cooling process, devices such as electrical fans are used to enhance sweat evaporation and air convection in a dual layer of textiles for efficient heat extraction, while in the liquid cooling textiles, water and glycol are utilized to circulate through tubes to facilitate heat convection. Liquid cooling is especially used for high-temperature environments where effective heat dissipation is low. One of the challenges in active cooling is excessive skin cooling which makes the textile

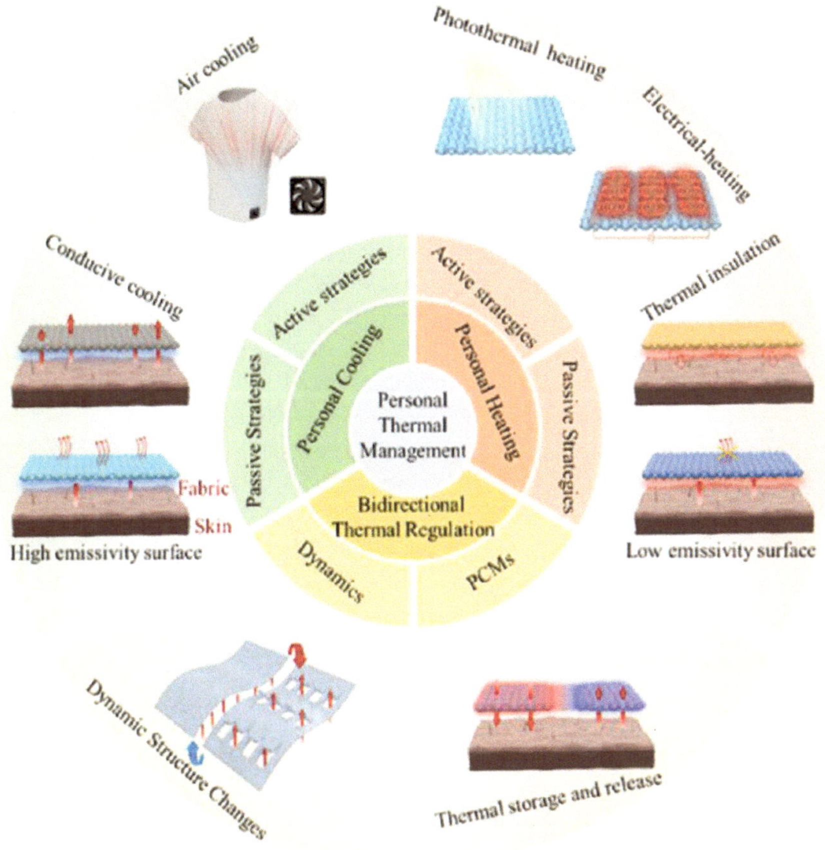

FIGURE 7.3 Summary of active/passive cooling/heating strategies for thermoregulating textiles [3].

disconformable. Therefore, passive cooling which focuses on thermal radiation and conduction management as an advance technique can balance cooling effects with more user comfort.

7.2.2 Heating Methods in Textiles

Active heating techniques can be achieved by electrical or chemical heating principles, while in the passive methods, heating is performed with thermal radiative/conductive management as depicted in Figure 7.3. Thermal comfort in cold environment can be obtained by electrical heating elements inside textiles using the Joule-heating effect which converts electrical energy into thermal energy.

In the chemical method, heat is generated within the textile by chemical reaction. This method is not extensively used as electrical heating techniques due to the excessive temperature increase beyond the comfortable skin temperature and leakage of the chemicals during the heating process.

Textile passive heating methods can be divided into lowering MIR emissivity and thermal conductivity through managing thermal radiation and thermal conduction in the textiles.

There are some challenges on the application of PCMs in textiles such as low thermal conductivity, PCM supercooling or subcooling, and PCM leakage. Low thermal conductivity can be treated by adding a variety of thermally conductive materials like metals and metal oxide nanoparticles, nanowires, carbon fibres, carbon nanotubes, graphene, and graphene oxide to PCM. Effects of supercooling can be reduced by adding nucleating agents to the PCM to maintain consistent temperature during transition phases. Leakage of PCM materials during melting is another limitation that can be prevented by microencapsulation of PCMs which is produced by depositing a thin layer of polymer coating onto small solid PCM particles or liquid drops as illustrated in Figure 7.4. One of the most important issues in encapsulation process is obtaining high enthalpy as possible for the microcapsule, whereas the microcapsules should not coalesce during thermal cycles of phase changes. This can be achieved by using a higher core to shell ratio to improve the thermoregulating performance of PCM on textiles substrate. The common diameter for the microcapsules is in the range of 40 μm [5].

Microencapsulation techniques have been classified into three categories, namely physical methods, chemical methods, and physicochemical methods as reviewed by Jamekhorshid et al. [6].

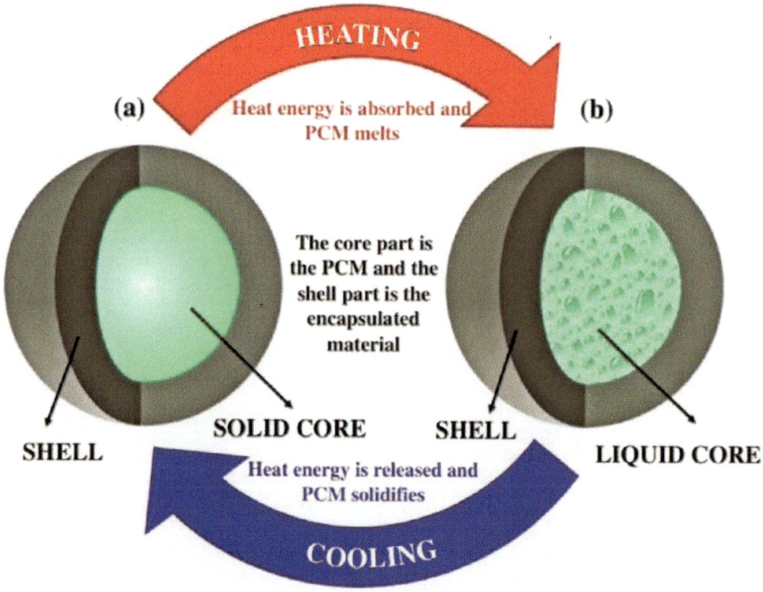

FIGURE 7.4 Working principles of encapsulated PCMs: (a) heat absorption; (b) heat release [4]. (Permission from Elsevier.com.)

Microcapsules made of biopolymers instead of synthetic polymers are very attractive for textile and fibre applications due to their lower production costs, greater accessibility, and biodegradability which lead to sustainable textiles.

A short review on the application of biopolymers such as chitosan, gelatine, alginate, and cellulose in microencapsulation of sustainable textile has been presented recently by Peng et al. [7]. Paraffins, salt hydrates, and fatty acids with melting temperatures in the range of 18–38°C are the most suitable PCMs in textile applications. Paraffin PCMS can be used purely with purity of 95% or as eutectic to maintain the specific transition temperature for specific applications. They are noncorrosive, non-toxic, and non-hygroscopic with stable thermal behaviour under permanent application and reasonable cost. Salt hydrates are alloys of inorganic salts and water with high latent heat of fusion and non-combustible which are suitable for fire-resistance textiles and fibres. However, subcooling behaviour, which results in the lack in reversible melting and freezing, is one of the drawbacks of salt hydrated that makes them unsuitable for permanent use in textiles. This property can be treated by adding stabilizing agents to the PCM before encapsulation process. Some of the common PCMs are summarized in Table 7.1.

7.2.2.1 PCM Integration Methods into Textiles

One of the main challenges in the application of latent heat storage materials in textile is the useful life of the PCM's container system and the number of cycles it can withstand before degradation of its functional properties. Therefore, cycle tests of the selected PCMs are required before incorporation into the textiles and fibres. There are different ways of applying PCMs into textiles such as mixing/blending into the fibre-spinning solution/melt, incorporation of PCM into hollow fibre cores, and adding the PCM to the polymer chain during synthesis. An alternative technique especially in research studies is electrospinning that provides a textile with thermoregulation. In this method, PCM such as paraffin wax is encapsulated into fibres to enhance the thermal management capabilities of the textile through efficient encapsulation of PCMs for better thermal buffering. Adding PCMs to the fabrics,

TABLE 7.1
Some of the PCMs Which Are Utilized in Textiles [7]

Phase Change Material	Melting Temperature, °C	Latent Heat Storage, J/g
Paraffins:		
Hexadecane	18.5	237
Heptadecane	22.5	213
Octadecane	28.2	244
Nonadecane	32.1	222
Eicosane	36.1	247
Salt hydrates		
Calcium chloride hexahydrate	29.4	170
Sodium hydrogen phosphate dodecahydrate	36.0	280

by impregnation during finishing, embedded in a coating compound or mixed into a foam to be applied to the fabrics by lamination. The coating materials of the encapsulated PCM must be compatible to textile materials before applying PCM to a textile substrate to prevent dissolution during liquid phase. Another method used to create core-shell fibrous structures integrating PCMs into various polymer shells is coaxial electrospinning.

7.2.2.1.1 Mixing/Blending Methods

Synthetic filaments are produced via melt and solution-spinning processes. In melt spinning process the polymer granules or chips are first melted in an extruder, squirted through the nozzle into the quenching chamber and finally spin finish is applied and the filament bundles are wound on tube rolls. In solution-spinning technique, the polymer is dissolved in a solvent, and the solution is forwarded to the spinning process while the solvent is separated from the polymer by heating (dry spinning) or by coagulation in another solvent compatible fluid (wet spinning). PCM with high thermal stability can be added to melt spinning, while in solution spinning, PCMs must have higher chemical stability [8].

7.2.2.1.2 Coating Method

Incorporation of PCM into textile structure by coating method includes wetted microcapsules containing PCMs dispersed throughout the polymer binder such as polyurethane or acrylic mixture, a surfactant, dispersant, antifoaming agent, and thickener. The coating composition is made of wetted microsphere PCMs in a dispersion of water solution containing surfactant, dispersant, antifoaming agent, and polymer binder and then applied to the textile substrate by suitable coating methods such as knife-over-roll, knife-over-air, pad-dry-cure, gravure, dip coating, and transfer coating [9]. The substrate materials can be classified into the following categories [3]:

- Polymeric Materials such as polyvinyl alcohol, polydimethylsiloxane and cellulose can be utilized for enhanced thermal management properties.
- Aerogels of silica, Kevlar, cellulose, and polyimide can be used for their exceptional insulating properties for lightweight passive cooling textiles.
- Phase change materials can also be used to store and release thermal energy during the process of melting and freezing in textile thermoregulation.
- Nanomaterial such as carbon nanotubes, graphene, and boron nitride nanosheets can also be added into textiles for thermal conductivity enhancement.

Super hydrophobic silica aerogel nanoparticle with percentage of 2–8% incorporated in 65/35 wool-Aramid blended fabric of around 230 g/m^2 by coating method using acrylic binder for fire fighter's protective clothing has been synthetized and thermo-physiological comfort was analysed by Shaid et al. [10]. The results show that 2%, 4%, and 8% coating of aerogel nanoparticle can reduce the fibre permeability by up to 45.46%, 61.76%, and 91.36%, respectively. The results also prove that moisture absorption rate in normal uncoated state was 332% and 0% at the top and bottom of the surface, respectively, while after coating this has been increased to 380% and

4.15% at the top and bottom surfaces, respectively. The conclusion suggests that 8% coating fibres can be used as an alternative covering for fire fighter's clothing.

Thermoregulating textiles using microcapsules of polystyrene containing paraffin wax by coating techniques has been synthesized by Sanchez et al. [11] using suspension polymerization. Their results indicated that the coating fabric with 35 wt.% added microcapsules had the energy storage density of 7.6 J/g and a difference of 8.8°C for 6 s observed with high durability and adequate stability after washing, rub fastness, and ironing treatments when compared with a coated one without microcapsules.

7.2.2.1.3 Lamination Process

PCM can also be incorporated into the thin polymer film of textiles and fibres and applied to the inner side to the fabric system by lamination to improve the thermo-physiological wear comfort of non-woven protective garments [12]. Strength of films adhesion to the textile substrate which may lost due to the mechanical and/or extreme environmental condition affect the lamination performance. PCM laminated textile can be utilized in chemical protective suits and can improve the thermo-physiological wearing comfort of other protective garments such as surgical gowns, uniforms or garments worn in clean rooms [12].

A composite aerogel fibre with highly porous structure and good mechanical strength with thermal conductivity of 5.34 W/mK and heating density of 125.8 J/g, consisted of polyimide/boron nitride (PI-BN) has been prepared by freeze-spinning technique by Zhang et al. [13] in which the pores of the textile have been filled with polyethylene glycol (PEG) by vacuum impregnation. PI-BN fibres have been knitted and immersed in the molten PEG at 80°C under vacuum and kept overnight for absorption of PEG in the textile structure as illustrated in Figure 7.5.

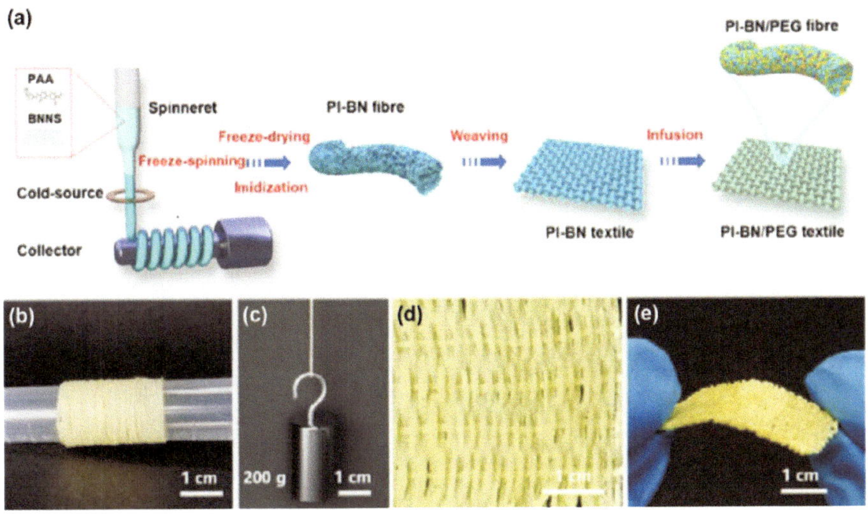

FIGURE 7.5 (a) Textile preparation, (b) aerogel fibre, (c) 200 g strength, (d, e) optical images of textile [13]. (Permission from Elsevier.com.)

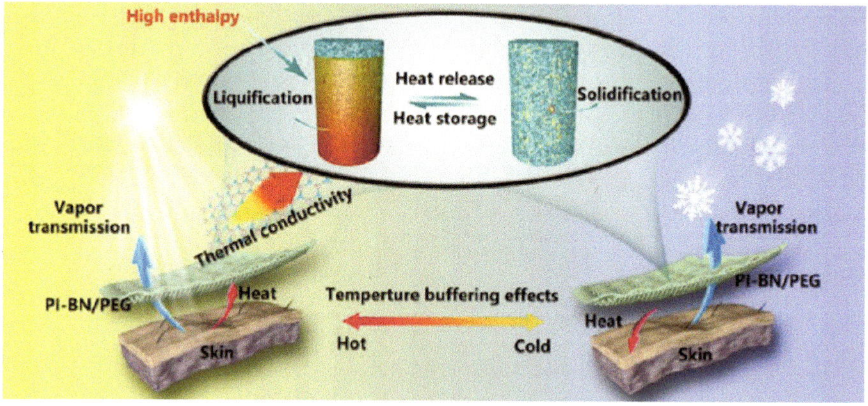

FIGURE 7.6 Thermoregulating performance of PI-BN/PEG textile [13]. (Permission from Elsevier.com.)

PI-BN/PEG textile has been coated by polydimethylsiloxane (PDMS) for waterproofness of the textile by immersing in 60 mL of anhydrous ethanol containing 2 g of PDMZ precursor with curing agent (0.2 g) and maintaining at 60°C for 1 h.

In a hot environment, PEG in the composite can be melted by storing heat energy during the melting process, while in a very cold condition, the PCM would be solidified in the textile to release heat and prevent significant temperature decrease as shown in Figure 7.6.

They concluded that the prepared PI-BN/PEG textile has a rapid thermal response to the environmental conditions, exhibit excellent cycling stability after 100 cycles and has a great potential for the next generation of smart textiles.

Performance of a cooling vest containing Glauber's salt ($Na_2SO_4.10H_2O$) as a PCM with melting temperature of 28°C, and latent heat of 251 J/g, on human physiological parameters, including heart rate, systolic and diastolic blood pressure, thermal sensation, and sweating rates of 18 volunteers in hot and humid conditions of Persian Gulf region with temperature of 40°C and humidity of 45% has been studied by Yousefi et al. [4]. The cooling vest was made of polyester fabric containing 16 PhaseCore^{TM} cooling elements six in the front and ten in the back with dimension of each $0.8 \times 6.6 \times 11.4$ cm^3 and weight of 100 g as illustrated in Figure 7.7.

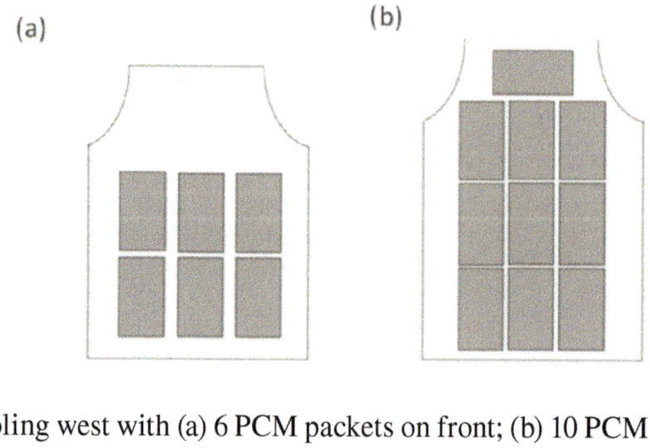

FIGURE 7.7 Cooling west with (a) 6 PCM packets on front; (b) 10 PCM packets on the back [4]. (Permission from Elsevier.com.)

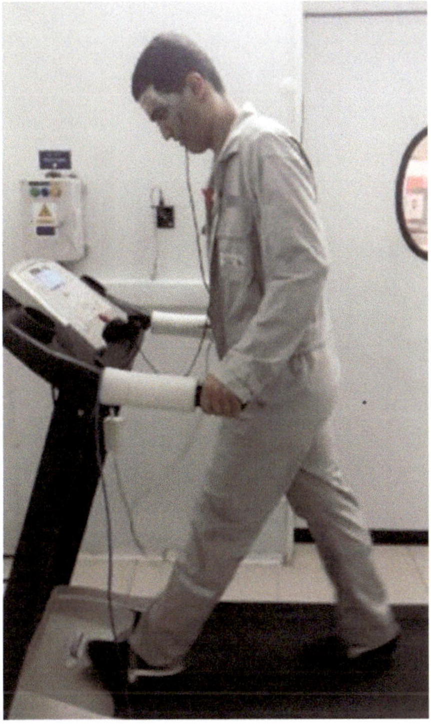

FIGURE 7.8 One volunteer on treadmill [4]. (Permission from Elsevier.com.)

The vest has been kept at 20°C for solidification before and after the test. The clothes of the participants were a T-shirt, cooling vest jacket and a boiler suit over the cooling jacket. The participants were 18 non-athlete healthy male students with average height, weight and age of 178.17 ± 4.75 cm, 68.62 ± 12.58 kg, and 21 ± 1 years, respectively, with no drinking alcohol and no smoking. Each volunteer was ready to walk for 30 min at a speed of 5 km/h in the climatic chamber on a treadmill and heart rate was measured at 1 min intervals, as shown in Figure 7.8. Modelling and numerical analysis have been performed to simulate the effects of the heat generated by human's body on the cooling vest using Ansys Fluent software version 19.0.

Individual sweating activities with and without vest are illustrated in Figure 7.9, which shows that wearing cooling vest has a significant effect for increasing sweat production which is since heavy weight and high thickness of the vest prevent air circulation and sweat evaporation.

Their results prove that the vest did not affect the mean decrease in heart rate and blood pressure over time; however, it affects people thermal sensation.

Four types of paraffins with melting points of 25°C, 30°C, 35°C, and 42°C have been selected and coated on the fabrics with intelligent bidirectional thermal regulation by Su et al. [14] for absorbing excessive thermal energy before the skin burn,

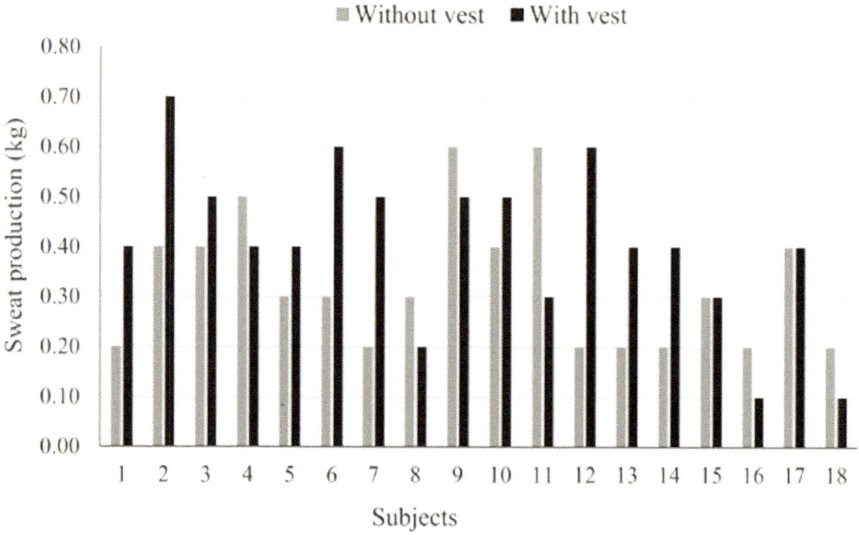

FIGURE 7.9 Perspiration rate with and without cooling vest [4]. (Permission from Elsevier.com.)

thus preventing the skin burn. PCM coated fabric was prepared by well-pressing of substrate into the fabric. The coating slurry was prepared using microcapsule suspension, thickener, adhesive, and water as illustrated in Figure 7.10.

PCM and skin temperature distributions over time are illustrated in Figure 7.11 which indicates that skin temperature for the bare fabric is still far more than the one coated with PCM although the skin temperatures for all PCM layers are the same which shows that change of melt temperature does not have any impact on skin temperature. The PCM layer with melt temperature of 35 and 42°C has a higher PCM temperature which is due to the higher thermal energy stored in the PCM.

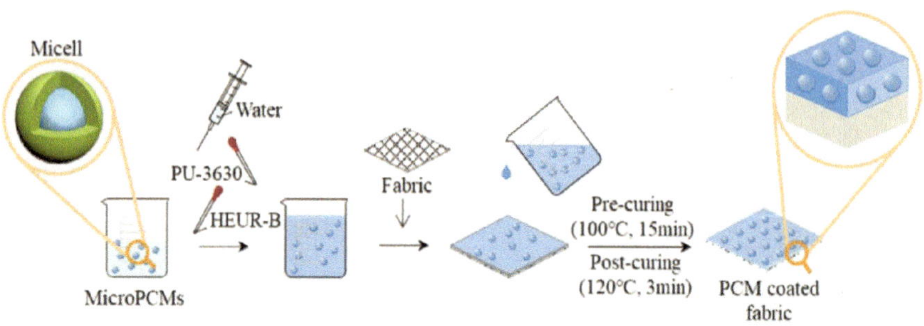

FIGURE 7.10 PCM coated fabric preparation steps [14]. (Permission from Elsevier.com.)

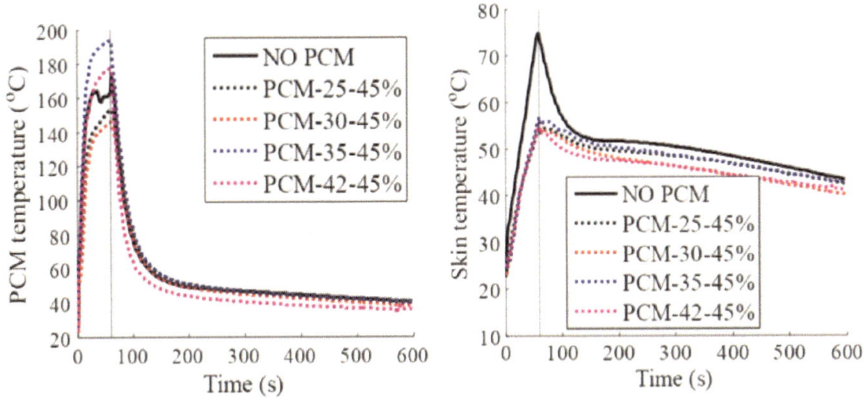

FIGURE 7.11 PCM and skin temperature distributions vs. time [14]. (Permission from Elsevier.com.)

One of the main challenges for healthcare workers during the COVID-19 period was wearing of Personal Protective Equipment (PPE) for a long time which was essential to protect them from the disease and with traditional cooling methods the does not meet the human body comfort. Therefore, incorporation of PCM in cooling clothing (PCM-CC) for comfort requirement of healthcare workers performing nucleic acid sample collection outdoors has been proposed by Hu et al. [15]. Each cooling vest was made of polyester containing individual pockets filled with 72 PCM packets filled with n-octadecane with melting temperature of 25°C and total weight of 1.48 kg as depicted in Figure 7.12.

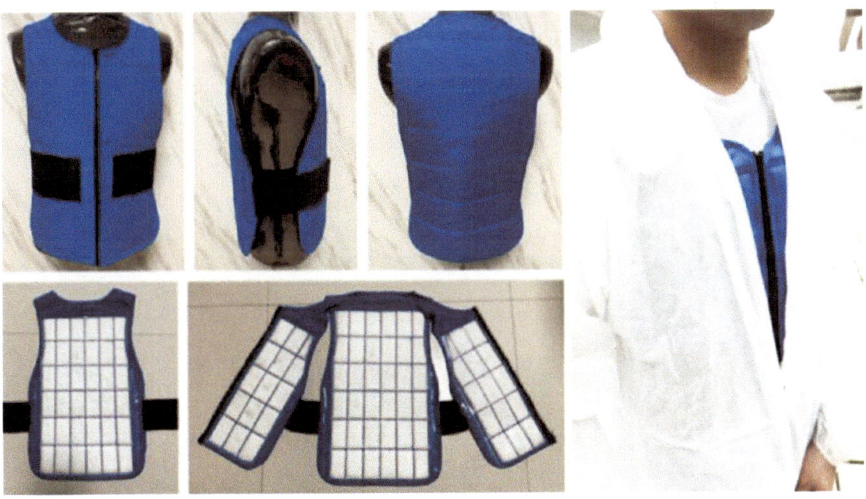

FIGURE 7.12 Cooling vest for healthcare workers [15]. (Permission from Elsevier.com.)

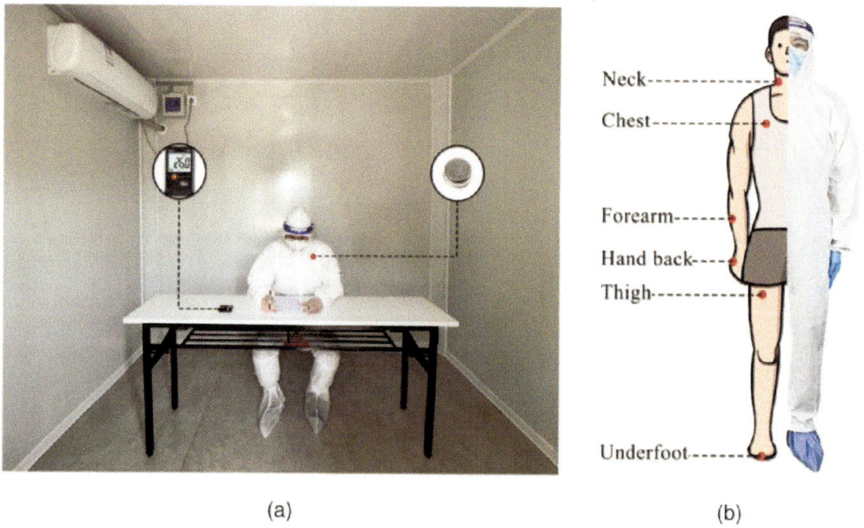

FIGURE 7.13 Experimental test room (a); measurement points (b) [15]. (Permission from Elsevier.com.)

A real experimental room with dimensions of 3.0 × 3.0 × 2.8 m^3 was constructed using steel frame structure for walls and roof, including 2.5 mm steel layer, 90 mm rock wool insulation, and 2.5 mm steel layer as shown in Figure 7.13.a. Temperature measurements have been performed using skin thermometers on six body positions, including the chest, forearm, neck, back of the hand, thigh, and underfoot (Figure 7.13.b).

The body discomfort sites distribution for with and without cooling vest are presented in Figure 7.14 at 26°C and 32°C environmental temperatures, which shows that the maximum discomfort sites are facial and head which were associated with direct human contact with the cloth and face masks. More than 75% of the participants without cooling vest at 26°C felt head and facial discomfort, while these

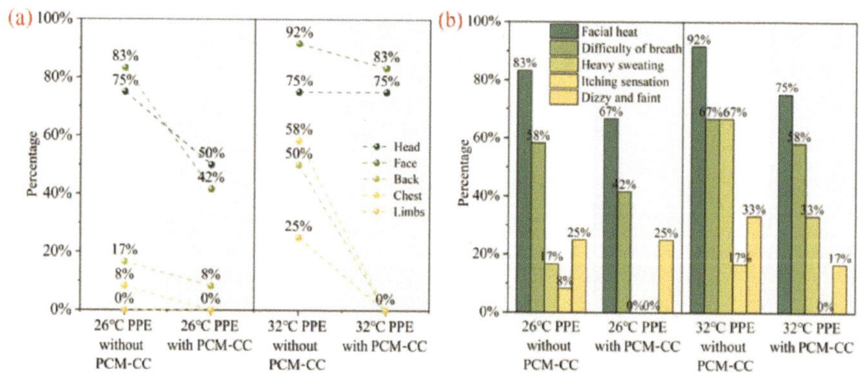

FIGURE 7.14 Body discomfort sites distribution (a) and participants discomfort symptoms (b) [15]. (Permission from Elsevier.com.)

figures for head and facial reduce to 25% and 41% for volunteers with cooling vests, respectively, although the effect on the back was not significant. When utilizing the system at 32°C, the discomfort percentage was reduced sharply by wearing cooling jackets and its values were lowered from 92% to 83%, from 58% to 0% and from 25% to 0% for face, chest, back and limbs, respectively. They concluded that using the cooling vest at 32°C working environment, improve the comfort conditions significantly on the back of the body, chest, and trunk with limited comfort improvement on the head and face.

Regulation of heat exchange between the human body and the surrounding environment can be achieved by personal management of textiles. An air-permeable phase change fluorinated polyurethane-silicon nitride membrane with hierarchically micro-/nano-structure consisting of polyurethane glycol core and polyurethane/ Si_3N_4 shell fabricated by Ma et al. [16] using coaxial electrostatic spinning. The fabricated textile had solar reflectance of 91%, infrared emissivity of 92%, thermal conductivity of 4.5 W/mK and latent heat of 43.3 J/g. When using the textile in hot and humid conditions with ambient temperature of 28.2°C and humidity of over 50% the regulated textile realizes a daytime temperature drop of 10.7°C for human skin.

7.3 SUMMARY

Human body thermal comfort which is an integral part of the overall human comfort in various environmental conditions and is especially important for medical textiles can be achieved by wearing suitable clothing. The body can be covered by textile to protect the body from extreme environmental conditions by maintaining optimum thermal state of the body created by the balance between the heat generation by the body and heat loss to the surrounding.

Better thermal and comfort control for a human body can be achieved through incorporation of PCMs as an energy storage medium in clothes and textiles. There are different techniques for adding PCMs into textiles such as coating, lamination, and mixing and blending methods. Encapsulated PCMs can also be kept between the layers of the clothes and garments as an energy storage medium. As an example, portable head cooling system using PCMs, if made easy to use, light weighted, ergonomic, and cost-effective can successfully lead to the brain temperature control.

The studies also indicated that temperature increase from 20 to 28°C can take 13 s in common silk, 20 s for Outlast silk and 37 s for treated fabric utilizing PCMs.

Generally, incorporated PCM clothes and textiles can reduce the body heat load, maintaining lower and more comfort temperature for the body in the hot climate regions and preventing heat loss from the body to the surroundings and maintaining a higher skin temperature during the cold situations.

REFERENCES

1. Tao X. (2001) Smart technology for textiles and clothing-introduction and overview. In: Tao I.X., editor. Smart fibers, fabrics and clothing. Woodhead Publishing, Sawston, Cambridge, U.K. pp. 1–6.
2. Pakdel E., Wang X. (2023) Thermoregulating textiles and fibrous materials for passive radiative cooling functionality, Mater. Des., 231, p. 112006.

3. Zhang Q., Cheng H., Zhang Sh., Li Y., Li Z., Ma J., Liu X. (2024) Advancement and challenges in thermoregulating textiles: smart clothing for enhanced personal thermal management, Chem. Eng. J., 488, p. 151040.
4. Yousefi S., Jamekhorshid A., Tahmasebi S., Sadrameli S.M. (2021) Experimental and numerical performance evaluation of a cooling vest subtending phase change material under extremely hot and humid environment, Therm. Sci. Eng. Progr., 26, p. 101103.
5. Kizildag N. (2023), Smart textiles with PCMs for thermoregulation. Multifunctional PCMs, Chapter 11. Elsevier Ltd. Kinga Pielichowska and Krzysztof Pielichowski, Amsterdam, Netherlands
6. Jamekhorshid A., Sadrameli S.M., Farid M. (2014) A review of microencapsulation methods of phase change materials (PCMs) as a thermal energy storage (TES) medium, Renew. Sustain. Energy Rev., 32(3), pp. 531–542.
7. Peng X., Umer M., Pervez A., Hasan K.M., Habib A., Islam S., Lin L., Xiong X., Naddeo V., Cai Y. (2023) Biopolymers-based microencapsulation technology for sustainable textiles development: a short review, Case Stud. Chem. Environ. Eng., 7, p. 100349.
8. Pause B. (2010) Phase change materials and their application in coating and laminates for textiles. In: Smart textile coatings and laminates. Woodhead Publishing Limited. Sawston, Cambridge, U.K.
9. Ho C.P., Fan J., Newton E., Au R. (2011) Improving thermal comfort in apparel. In: Song I.G., editor. Improving comfort in clothing. USA: Woodhead Publishing, pp. 165–181.
10. Sheid A., Furgusson M., Wang L. (2014) Thermophysiological comfort analysis of aerogel nanoparticle incorporated fabric for fire fighter's protective clothing, Chem. Mater. Eng., 2(2), pp. 37–43.
11. Sanchez P., Sanchez-Fernandez M.V., Romero A., Rodriguez J. (2010) Development of thermo-regulating textiles using paraffin wax microcapsules, Thermochim. Acta, 498, pp. 16–21.
12. Pause B. (2004) Nonwoven protective garments with thermo-regulating properties, Asian Textile J., 13(4), pp. 62–64.
13. Zhang Q., Xue T., Tian J., Yang Y., Fan W., Liu T. (2022) Polyimide/boron nitrite composite aerogel fiber-based-changeable textile for intelligent personal thermoregulation, Compos. Sci. Technol., 226, p. 109541.
14. Su Y., Zhu W., Tian M., Wang Y., Zhang X., Li J. (2020) Intelligent bidirectional thermal regulation of PCM incorporated in thermal protective clothing, Appl. Therm. Eng., 174, p. 115340.
15. Hu C., Wang Z., Bo R., Li C., Meng X. (2023) Effect of cooling clothing integrating with PCM on the thermal comfort of healthcare workers with personal prospective equipment during the COVID-19, Case Stud. Therm. Eng., 42, p. 102725.
16. Ma C., Gao Y., Cao Y., Yang Y., Wang W., Wang J. (2024) Hierarchically core-shell nanofiber textiles for personal cooling in hot and humid conditions, Nano Energy, 123, p. 109400.

8 Application of Phase Change Materials in Solar Cells and Solar Collectors

8.1 INTRODUCTION

Fossil fuels can be replaced by renewable energy sources such as solar and wind for world energy demand to reduce carbon emission and hence minimize greenhouse gas (GHG) emissions [1]. Solar photovoltaic (PV) power plant is one of the most common alternatives of energy sources in the last decades due to its lower maintenance, easy installation, and reasonable energy conversion into electricity. Another application of solar energy is solar collectors which are devices in which radiation energy from the sun is converted to heat energy and can be applied as solar water desalination or solar water heating. Direct conversion of solar radiation into electrical energy utilizing the benefits of radiative heat is performed using PV panels but their surface temperature affects the panel efficiency which is defined as a relationship between the supplied electrical energy and the incident solar irradiation [2]. Electrical efficiency of the solar panels decreases by 0.45% for a 1°C surface temperature increase [3]. Active and passive cooling techniques can be used to overcome this problem and increase the panel's electrical efficiency. One of the common passive methods of panel cooling is utilizing phase change materials (PCMs) with high latent heat on the panel as an energy storage (ES) system to absorb extra heat from the panel during the day and release them to the environment during the night. This chapter reviews the application of PCMs as a passive cooling technique for solar panels for electricity and heat generation and solar collectors. PCM selection criteria, PV cells performance indicator variables, and PCM heat transfer enhancement will also be discussed.

8.2 THERMAL ENERGY STORAGE (TES) SYSTEMS

Continuous utilization of solar energy can be attained through an optimized thermal energy storage (TES) system to be used at night and cloudy days to achieve energy peak regulation. TES systems can be classified into three categories which are sensible TES, latent TES, and chemical reaction TES. In sensible TES system which is one of most widespread forms of ES and utilization, rocks, water, clays, and diverse metallic materials can be utilized as a thermal storage medium, and energy is stored in the materials when it is available and released to the environment when it is required. In the latent TES system, the same phenomena are done on latent heat materials such as PCMs with a common form of liquid to solid and vice versa as an

DOI: 10.1201/9781003518280-8

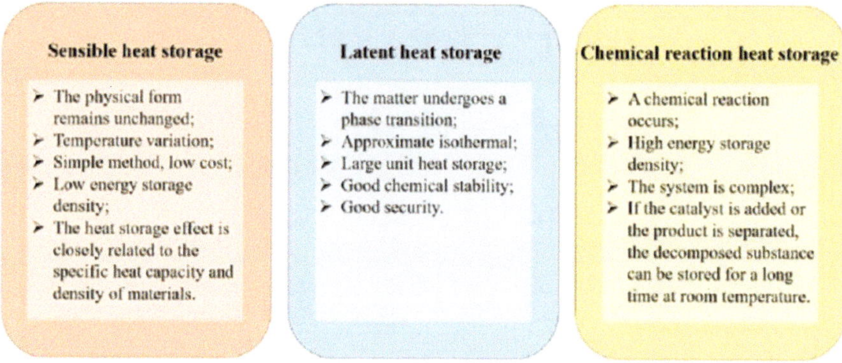

FIGURE 8.1 The characteristics of different TES methods [4]. (Permission from Elsevier.com.)

exchange energy medium. In the third type, i.e., thermochemical thermal storage, energy of forward and backward chemical reactions with heat energy as the catalyst can be stored and released in chemical bonds lattices [4]. The characteristics of the three TES systems are demonstrated in Figure 8.1. Details of all TES systems have been explained in Chapter 3.

8.3 SOLAR PANELS

Solar as a source of renewable energy has been utilized since ancient times. There are various application areas for solar energy which is a clean, reliable, and non-polluting source due to its availability, cost effectiveness, accessibility, capacity, and performance features. It can be utilized for domestic hot water production, vegetable and fruits drying, water desalination, air conditioning, heat and electricity production, irrigation, heat generation, and photochemical reactions [5].

The first application of solar energy to electricity through PV has been found by Alexander Henri Becquerel who discovered exposure of semiconductors to sunlight can generate electricity [6]. The first discovery of the PV cell in history was manufactured by Charles Fritts in 1883 with 1% efficiency. The first commercial production of solar panels with 2% efficiency and 14 mW power was built by Hoffman Electronics in 1955. Application of solar panels in the Explorer 6 increases their publicity to the adaptation of solar electricity [7]. This has been constantly developed with higher performance and it is predicted to be one of the major renewable energy sources in the future.

8.3.1 PV Panels

Conversion of solar radiation to electricity by PVs has some advantages such as no moving parts, no noise and gas emissions, long lifetime, and direct conversion process. The most important components of the PV panels are illustrated in Figure 8.2.

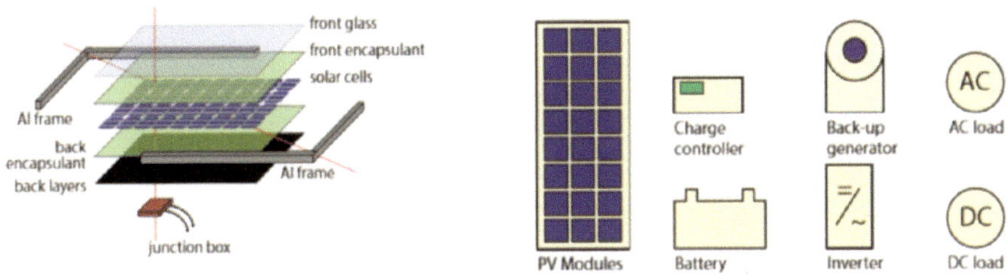

Components of a PV panel. Components of a PV system.

FIGURE 8.2 Components of a PV panel and PV system [8]. (Permission from Elsevier.com.)

Conversion of solar energy to electricity can be performed on the PV cell surface where the radiation energy is absorbed by the cells from sunlight and is converted to direct current (DC) as shown in Figure 8.3.

Monocrystalline silicon, polycrystalline silicon, thin film, flexible and transparent solar panels are the common types of PV cells [9]. Although the performance of monocrystalline silicon cells can reach up to 24% which is the highest efficiency among other solar panel types, they are also more expensive than other kinds. However, polycrystalline solar cells with efficiency of 15% and the cheapest cost-efficiency rate are the most used product in the market. Thin film solar panels with low efficiency of about 7%, and transparent solar panels are not widely used in the market application areas. Flexible solar cells with durable structure, more flexibility, and being strong have more usage areas in the market.

Due to the absorption and conversion of sunlight to electricity in PV modules, there are some parameters that affect the cells performance such as environmental factors, PV structure, installation, and operation and maintenance factors. Surface fouling because of rainfall, hurricane or powdering may cause efficiency loss due

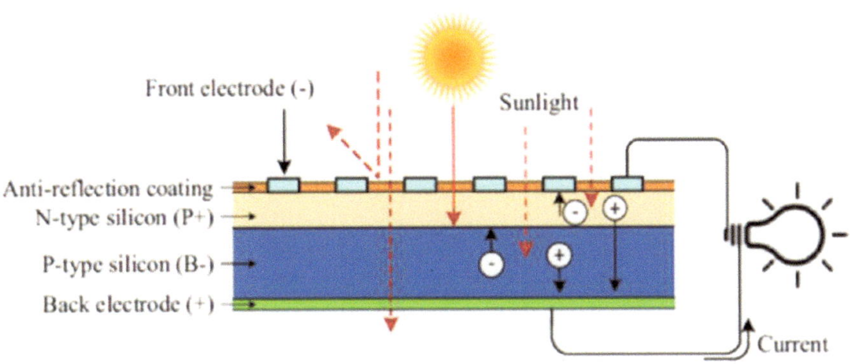

FIGURE 8.3 Conversion of solar energy to electricity by a PV cell [8]. (Permission from Elsevier.com.)

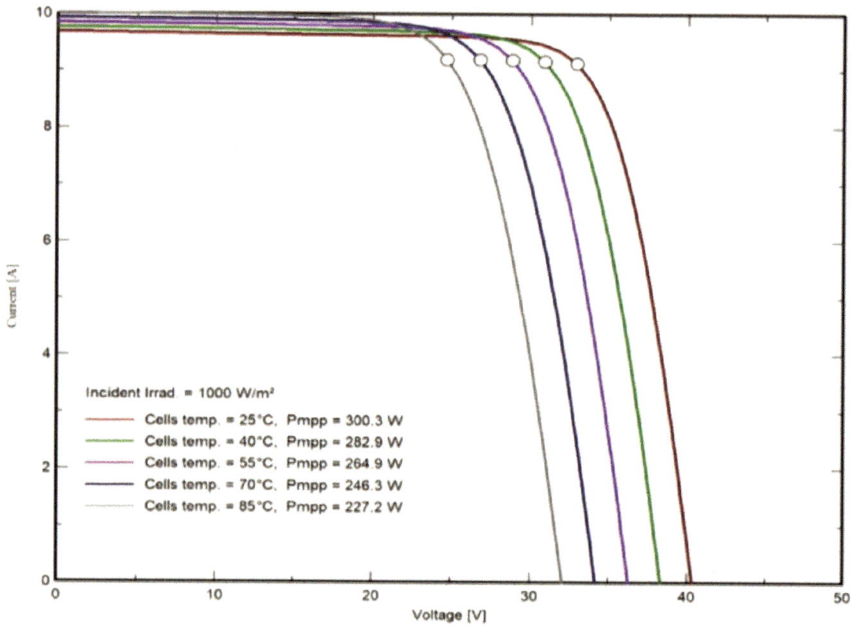

FIGURE 8.4 Deficiency of solar panels with temperature [10]. (Permission from Elsevier.com.)

to the reduction of sunlight absorption on the surface that decreases the electricity generation. Another important factor is the position of movable panels which gives the maximum performance when they are faced to the south for the southern hemisphere and vice versa for the northern hemisphere. Shadow over the panel caused by mountains, forests, trees, buildings, etc., can also decrease the cell efficiency. One of the drawbacks of the solar panels which is common for all types is the deficiency of the panel during overheating, especially in hot climate regions as demonstrated in Figure 8.4 and, therefore, different cooling techniques have been offered by the researchers for cell efficiency enhancement.

8.3.1.1 PV Cooling Techniques

There are two main types of cooling methods namely active and passive with some subclassifications as shown in Figure 8.5 [8]. An extensive review on the cooling techniques, including air-based, water-based, PCM-based, and hybrid cooling systems using PCM, nanomaterials, and nanofluids to decrease panel surface temperature is reported by Shoaib Sheik et al. [10]. In this chapter, only application of PCMs on the solar panel' will be reviewed as a passive method of cooling.

When PCMs are integrated into the back of a solar PV cell, extra heat from the panel surface will be absorbed by the cell to provide uniform cooling of the PV cell and utilizes the heat for domestic applications. Therefore, the efficiency of the panel will be improved by decreasing the surface temperature by incorporation of PCMs on the panel.

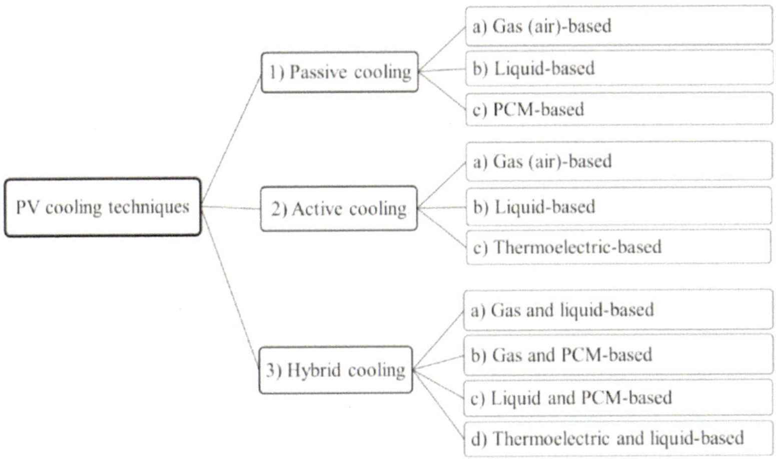

FIGURE 8.5 Classification of solar panel cooling techniques [8]. (Permission from Elsevier.com.)

Application of composite PCMs integrated with solar panels for four different modules of a reference panel, solar panel with paraffin jelly (PJ) and expanded graphite (EG), solar panel with PJ, expanded perlite (EP), and solar panel with PJ and expanded vermiculite (EV) with melting points and heating values demonstrated in Figure 8.6 has been studied by Govindasamy et al. [11].

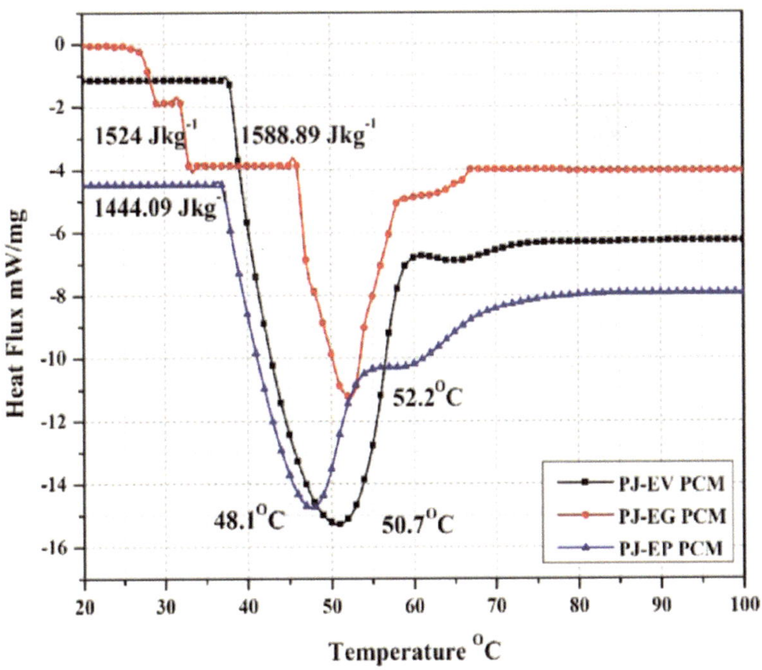

FIGURE 8.6 DSC results of composite PCMs [11]. (Permission from Elsevier.com.)

TABLE 8.1
Specification of Solar Panels [11]

Solar PV Panel Type	Polycrystalline
Maximum power (P_{max})	40 Wp
Maximum power voltage (V_{mp})	19.25 V
Maximum power current (I_{mp})	2.08 A
Short circuit current (I_{sc})	2.22 A
Open circuit voltage (V_{oc})	22.5 V
Number of cells	36
Weight	3 kg
Dimensions	430 × 665 × 35 mm

Their experimental setup consisted of four polycrystalline cells with specification listed in Table 8.1, tilted in 13.7°, three of them equipped with different composite PCMs and the other with no PCM as shown in Figure 8.7.

The maximum surface temperature of the reference panel with no PCM and PV panels with PJ-EG, PJ-EP, and PJ-Ex has been increased from 32.9°C to 47.2°C,

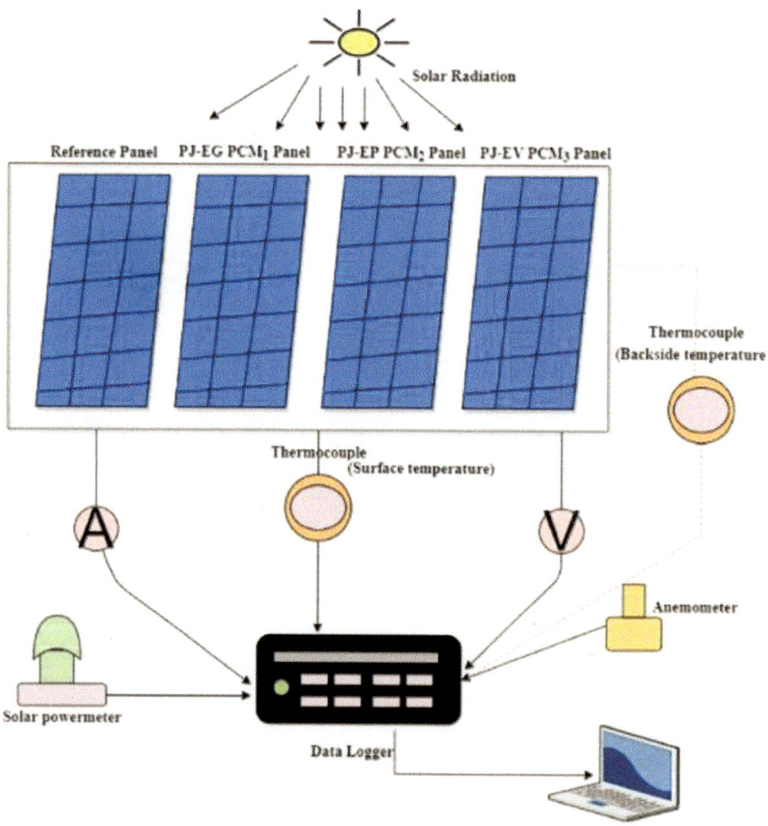

FIGURE 8.7 Schematic of experimental setup [11]. (Permission from Elsevier.com.)

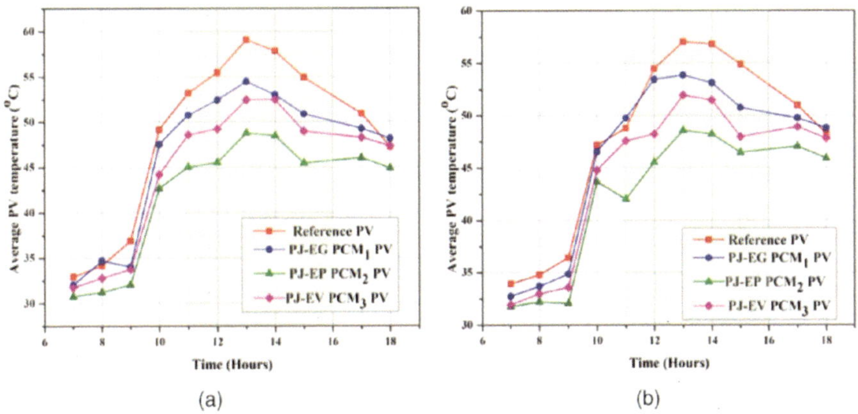

FIGURE 8.8 Average solar panel temperature during (a) June and (b) July [11]. (Permission from Elsevier.com.)

from 32°C to 48.1°C, from 30.7°C to 44.9°C and 31.7 to 47.3°C, respectively. The temperature difference between the reference panel and PV cell incorporated with composite PCMs was obtained to be 4°C, 4.59°C, 10.29°C, and 6.56°C, respectively, as illustrated in Figure 8.8.a. The same temperature levels have been achieved during the month of July (Figure 8.8.b).

The maximum power output for the reference and other panels with composite PCMs during the months of June and July are shown in Figure 8.9.a and Figure 8.9.b, respectively. As can be seen from the results, the lowest and highest power output values are 28.95 W and 36.99 W for the PV panels without PCM and the one with PJ-EP during the month of June (Figure 8.9.a) and 28.65 W and 33.51 W for the month of July (Figure 8.9.b), respectively.

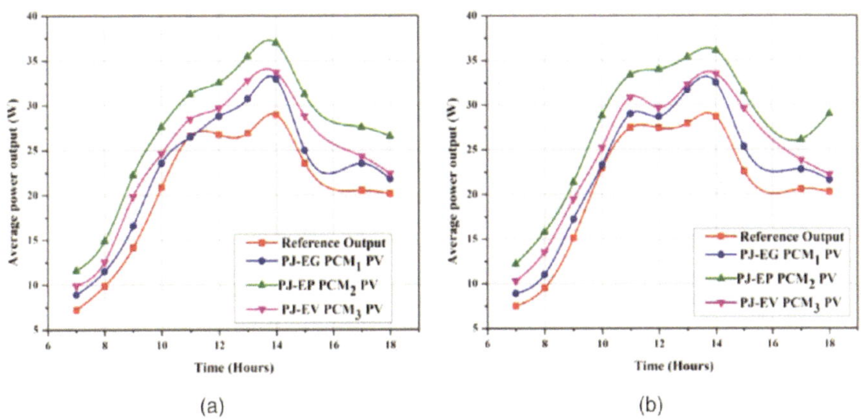

FIGURE 8.9 Average power output of the PV panels in (a) June and (b) July [11]. (Permission from Elsevier.com.)

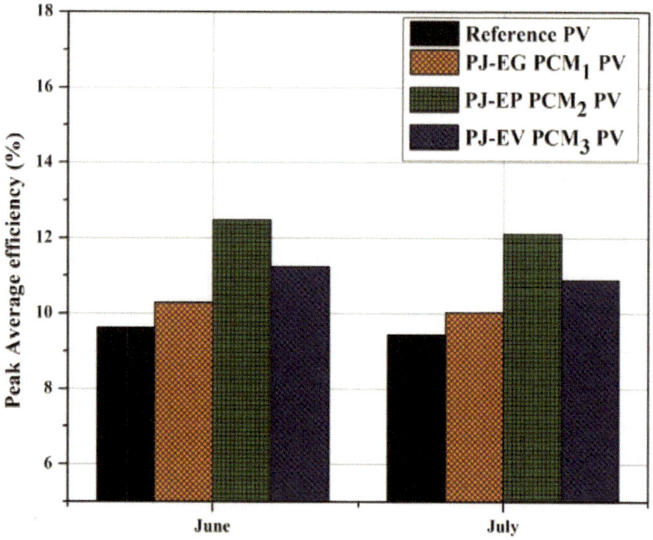

FIGURE 8.10 Overall average maximum efficiency of the panels [11]. (Permission from Elsevier.com.)

The efficiencies of the panels with and without composite PCMs during the months of June and July are illustrated in Figure 8.10 with the minimum efficiency of 5.2% and 12.74%, respectively.

They finally concluded that integration of composite PCMs to the panels not only increases their efficiencies significantly but can also decrease the surface panel temperature that causes lifetime increase of the panels.

Integration of polyethylene glycol 1000 (PEG1000) as a PCM incorporated into a PV solar panel of 40 W and zero and 15° inclination has been studied by Mousavi et al. [2]. The PCM embedded in one of the two selected panels with the same size and power outputs manufactured by Max Volta, an Iranian Company as shown in Figure 8.11.

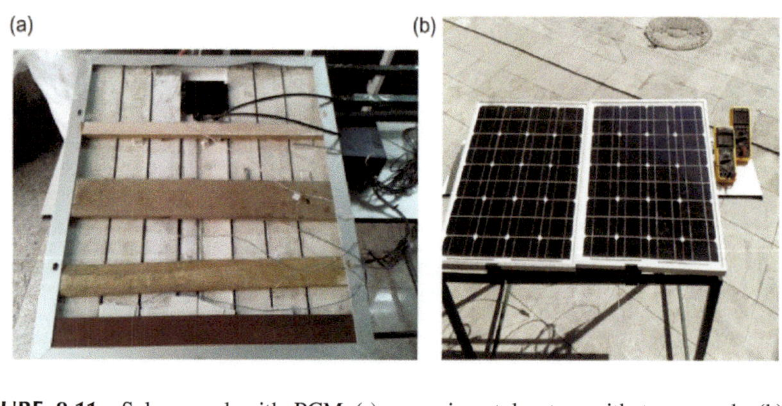

FIGURE 8.11 Solar panel with PCM (a); experimental setup with two panels (b) [2]. (Permission from Elsevier.com.)

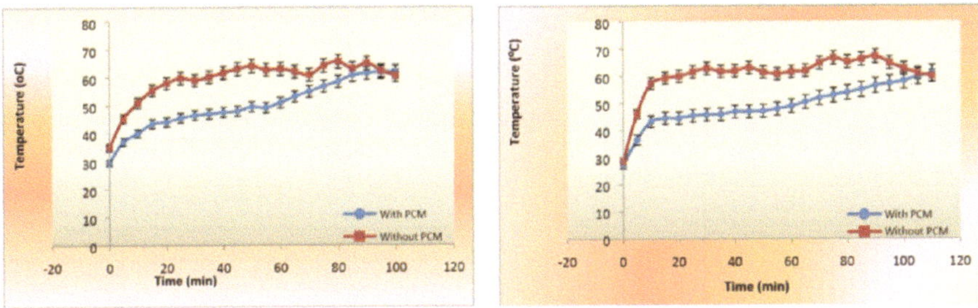

(a) (b)

FIGURE 8.12 Temperature profiles of PV panels with 0° (a) and 15° (b) inclination [2]. (Permission from Elsevier.com.)

Temperature profiles of two panels with and without PCM at zero and 15° inclination are illustrated in Figure 8.12a and 8.12.b, respectively. As observed from the figures the panel temperature with PCM increases sharply at the start of the experiment and reaches to the steady temperature of 60°C and then fluctuates caused by different convection coefficient of ambient air by wind. Temperature distribution in the panel with PCM also increases after the experiment with a lower slope in comparison to other panel without PCM and reaches to the constant PCM melting temperature (40°C) and continues to increase to match the profile in non PCM panel.

The maximum power outputs from the panels with and without PCM at zero and 15° inclinations are demonstrated in Figure 8.13.a and 8.13.b, respectively, which shows maximum power of 23.8 W, and 24.85 W can be achieved by the panels with PCM at zero and 15° inclination, respectively.

They finally concluded that the efficiency of the PV cell with PCM passive cooling can be increased to 8.5% and this can even be enhanced by water cooling inside the PCM. A comprehensive review on the utilization of composite PCM and nanoparticles, including influences of PCM thermo-physical properties such as melting temperature, heat of fusion, viscosity, and supercooling degree, has been presented by Maghrabie et al. [12].

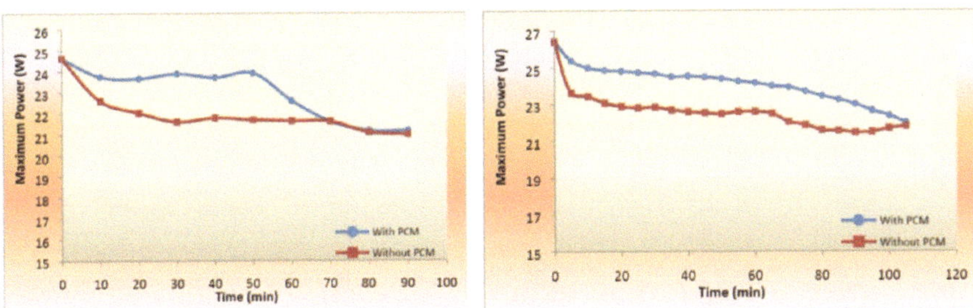

(a) (b)

FIGURE 8.13 Maximum power output for horizontal panel (a) and angled panel (b) [2]. (Permission from Elsevier.com.)

8.3.2 Thermal Panels

Performance of the solar water heater can be enhanced by incorporation of PCM into the system.

Experimental investigation of solar water heating system efficiency in a shell and tube TES utilizing paraffin wax as illustrated in Figure 8.14 has been presented by Mahfuz et al. [13].

According to the experimental data, energy efficiency of solar water heating system can be increased by 63.88% and 77.41% for the water flowrate of 33 and 167 gr/min, respectively. Incorporation of nanoparticles of CuO with concentration from 0.25% to 1.0% into paraffin wax and utilization of a composite PCM in a solar water heater during the night in an experimental setup shown in Figure 8.15 has been studied by Mandal et al. [14]. Maximum thermal efficiency of 179.2% has been achieved by utilizing composite PCM as an ES media.

An experimental analysis of a double-channel solar air heater, including two identical prototypes with and without PCM as TES, has been performed by Palacio et al. [15]. Their experimental setup was instrumented for hot air temperature and velocity, solar radiation, temperature probes in the glass cover, absorber plate, PCM storage system, and a weather station to record the environmental conditions as illustrated in Figure 8.16.

Performance enhancements of a PCM-based flat plate solar collector utilizing metal foams, nanoparticles, and wavy wall-Y-shaped surface have been investigated by Nematpour Keshteli et al. [16]. A Computational Fluid Dynamics (CFD) model has been developed to study different heat transfer performance solutions of PCM for

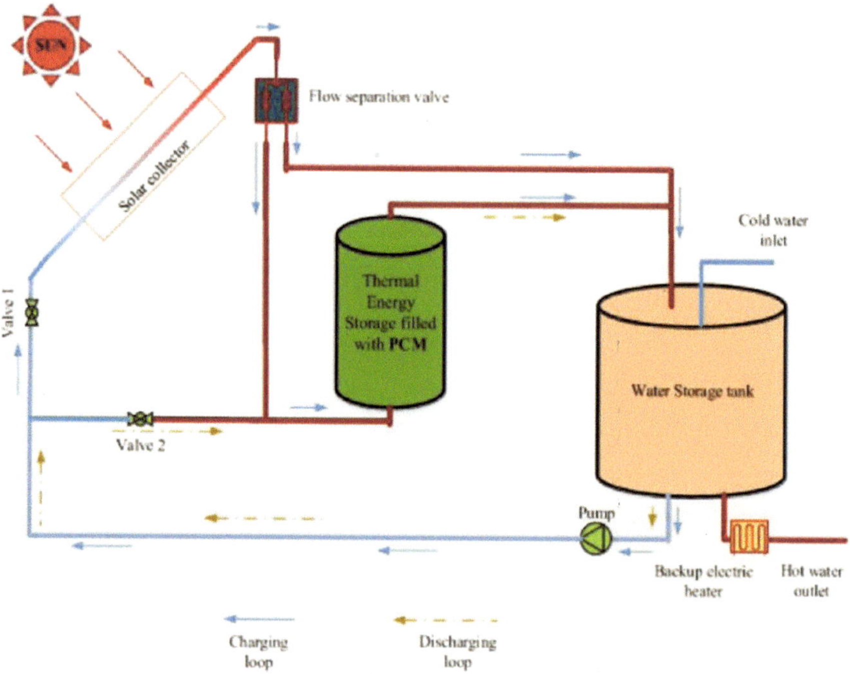

FIGURE 8.14 Solar water heating system with TES [13]. (Permission from Elsevier.com.)

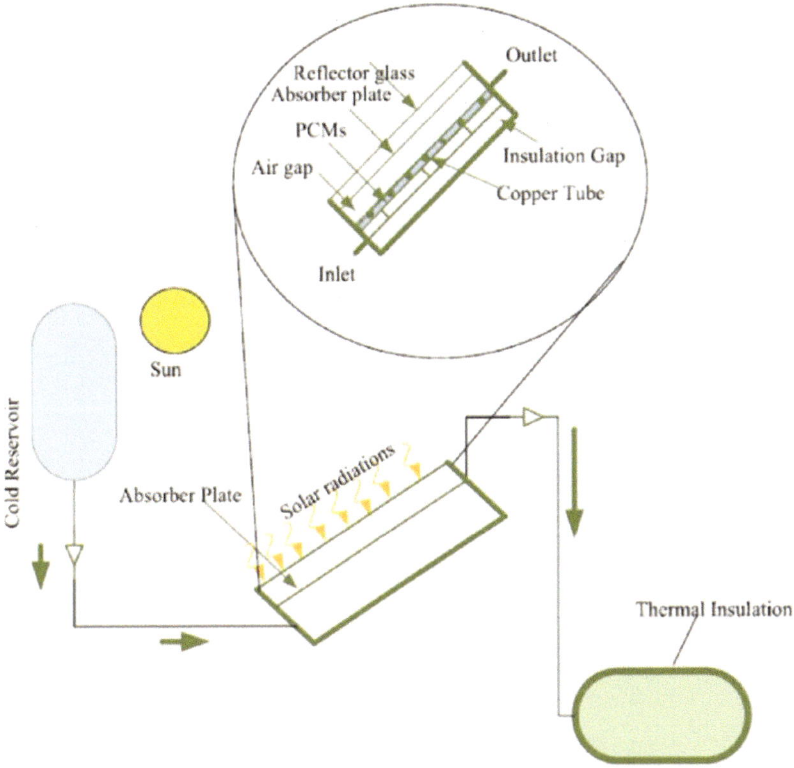

FIGURE 8.15 Experimental setup of solar water heater [14]. (Permission from Elsevier.com.)

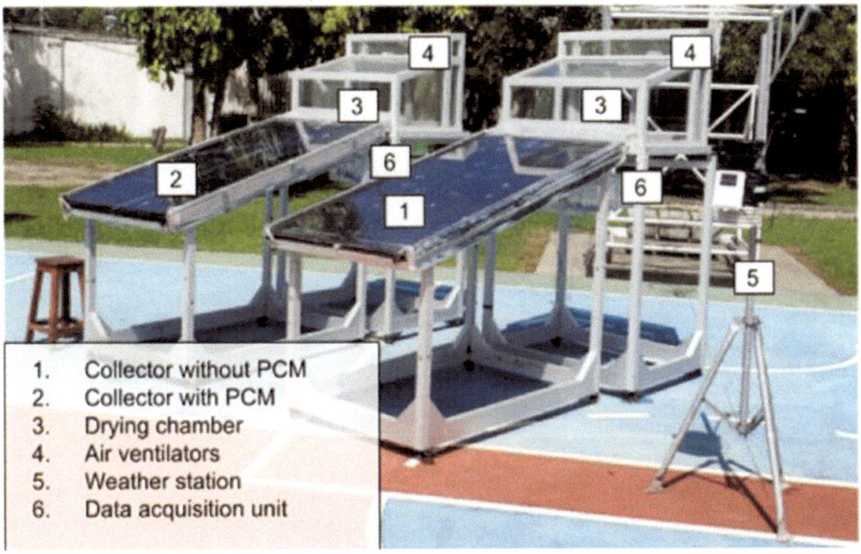

FIGURE 8.16 Experimental setup and its parts [15]. (Permission from Elsevier.com.)

a flat plate solar collector using different thermal conductivity enhancement materials. Three cases of geometries for the solar collector thermal ES such as straight wall (Case A), wavy wall (Case B), and wavy wall with a Y-shaped fin (Case C) have been considered as shown in Figure 8.17.

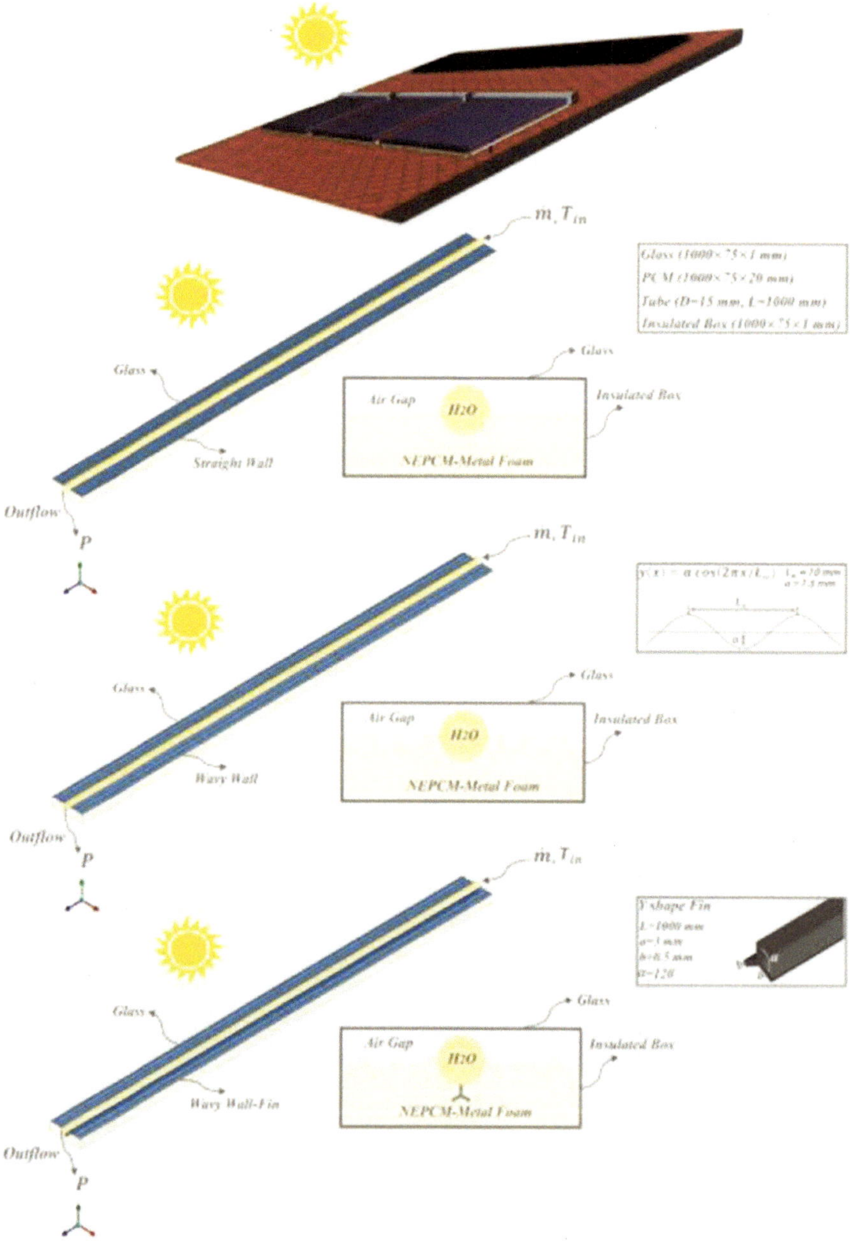

FIGURE 8.17 Schematic of flat plate solar collector with PCM: Case A (top), Case B (middle), and Case C (bottom) [17]. (Open access journal)

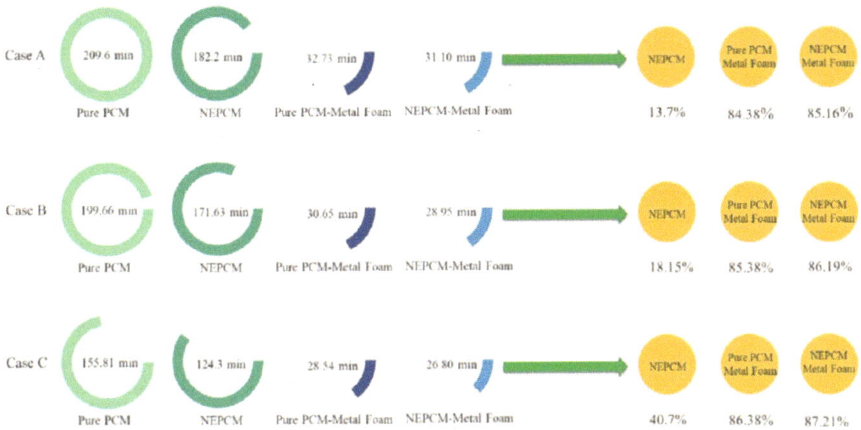

FIGURE 8.18 Time saving of melting process in all cases with different configurations [17]. (Open access journal)

A mathematical model consisted of differential equations for the continuity equation, momentum balance in three directions and energy equation, including PCM and required initial and boundary conditions with some assumptions have been developed and solved using the commercial finite volume code ANSYS software. Melting time and time-saving percentage for different cases of A, B, and C are demonstrated in Figure 8.18 which proves that in all cases with the addition of nanoparticles, the melting time has been sharply decreased due to the heat transfer enhancement. The results of comparison for different cases indicate that compared with pure paraffin, nanoparticles in case B and C have lower melting time by 18.15% and 40.70%, respectively, while for the metal foams and nanoparticles with foams, the melting time has been reduced by 84.49% and 89.18%, respectively.

Their findings prove that metal foams have more impact in cycling times in such applications.

A comprehensive review of the application of PCM as a TES in solar water heaters has been presented by Peng et al. [16] and concluded that several factors such as operation condition and PCM characteristics are effective on the function of the system combined with TES.

Optimization of the efficiency of a hybrid flat plate solar collector integrated with a layer of PEG6000 and magnesium hydroxide ($Mg(OH)_2$) as a composite PCM and heat sink under semi-arid climate conditions in Abha, Saudi Arabia, has been studied by Algarni [18]. The experimental setup includes a glass cover, pipes, absorber plate, and insulating layer for the collector which is consisted of eight straight pipes with dimensions of 1,700 mm in length, 6 mm thick, and heat pipe spacing of 110 mm as shown in Figure 8.19. The experimental data has been collected for three different cases: Case 1, the conventional flat plate, Case 2, flat plate with PCM, and Case 3 flat plate with PCM and heat sink. Composite PCM has been prepared by mixing PEG6000 combined with $Mg(OH)_2$ or magnesium oxide (MgO) in an absolute ethanol solution and the solution underwent for 30 min sonication to remove ethanol form the solution at 80°C.

Thermal efficiency over time for three cases is demonstrated in Figure 8.20 which shows the maximum performance of 66% for the case of flat plate with PCM (Case 2).

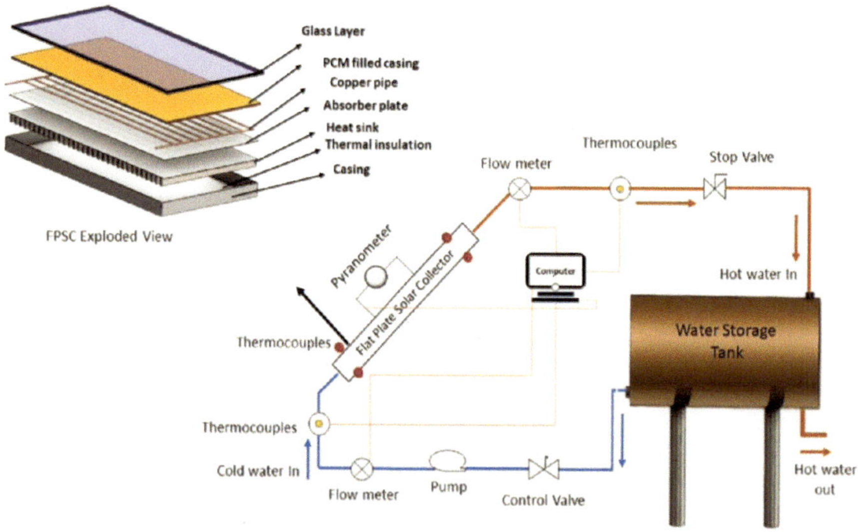

FIGURE 8.19 Experimental setup of solar collector [18]. (Permission from Elsevier.com.)

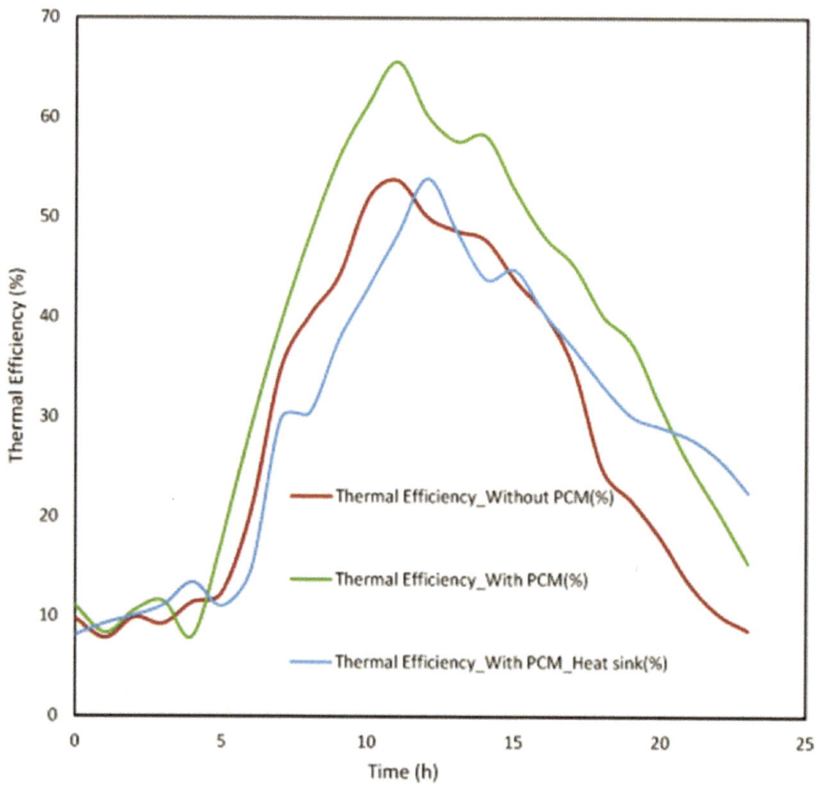

FIGURE 8.20 Thermal efficiency of all cases [18]. (Permission from Elsevier.com.)

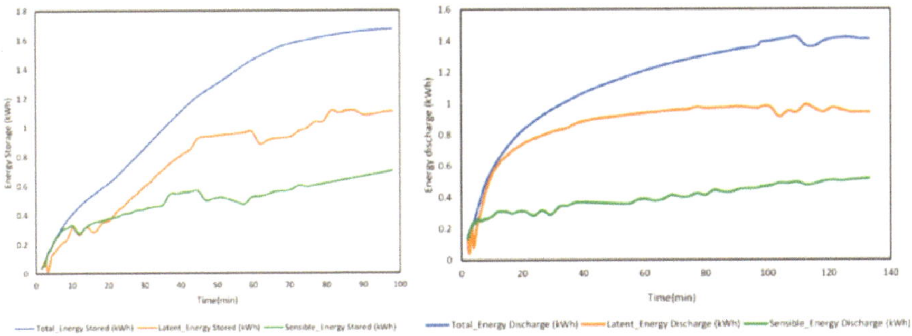

(a) (b)

FIGURE 8.21 PCM energy storage (a) and energy discharge (b) per time [18]. (Permission from Elsevier.com.)

The thermal efficiency increases with time sharply reaching the maximum and then gradually falling.

The distribution of ES and energy discharge in terms of total, latent, and sensible energies during charging and discharging process are illustrated in Figure 8.21.a and 8.21.b, respectively. It can be observed from the curves that during both charging and discharging processes, a temperature difference existed between the PCM layer and working fluid that causes a heat transfer between the layers and led to the charging and discharging processes. It also shows that the uniform constant temperature of $PEG/Mg(OH)_2$ was reached after 95 min, and during the charging process, the total, latent, and sensible energies were 1.65 kWh and 0.687 kWh, respectively. Similar trends will be observed from the discharge energies for the sensible, latent, and total at 125 min will be 0.521 kWh, 0.963 kWh, and 1.45 kWh, respectively.

Similar trends will be observed from the discharge energies for the sensible, latent, and total at 125 min will be 0.521 kWh, 0.963 kWh and 1.45 kWh, respectively.

8.4 SUMMARY

Solar energy from the Sun is one of the most valuable renewable energy sources in our daily life which can be converted to electricity and thermal energy via PV/T solar panels. Our planet receives nearly 1.8×10^{11} MW of total energy from the sun which is much higher than the overall energy consumption of all the combined sources. Solar energy can be harnessed either directly or indirectly for heating and cooling of buildings utilizing different technologies. Solar panel efficiency may fall due to several reasons from which surface temperature increase is the most effective parameter. Panels' surface temperature must be kept at an optimum temperature of 25°C for maximum performance and there are different techniques for cooling solar panel surfaces. In this chapter, different types of solar panels used for electricity generation such as PV panels, solar water heaters, and solar collectors have been discussed and different techniques for surface panel cooling have been described as active and passive cooling methods. The limited nature of solar energy during the daytime obliges the application of thermal ES units to make the solar energy applicable during the nighttime or cloudy days.

One of the most effective cooling techniques for solar panels is the application of PCMs in passive form integrated into the back of the panel as a thermal ES which absorbs extra heat from the panel during the day to provide an optimum surface temperature for the panel and release it to the environment during the nighttime. Experimental techniques have been reviewed and some of the results have been presented for the PVs and thermal panels and it was shown that cooling techniques involving the combination of PCMs and thermal conductivity enhancers such as nanoparticles make the cooling system more effective when compared with the active cooling methods such as air and water cooling.

REFERENCES

1. Dahn J., Ehrlich G., Reddy T. (2011) Linden's handbook of batteries. New York: McGraw Hill.
2. Mousavi Baygi S.R., Sadrameli S.M. (2018) Thermal management of photovoltaic solar cells using polyethylene glycol 1000 (PEG1000) and a PCM, Therm. Sci. Eng. Progr., 5, pp. 405–411.
3. Skoplaki E., Palyvos J.A. (2009) On the temperature dependence of photo-voltaic module electrical performance: a review of efficiency/power correlation, Sol. Energy, 83, pp. 614–624.
4. Wang Q., Yang L., Song J. (2023) Preparation, thermal conductivity, and applications of nano-enhanced phase change materials (NEPCMs) in solar heat collection: a review, J. Energy Storage, 63, p. 107047.
5. Al-Waeli A.H.A., Kazem H.A., Chaichan M.T., Sopian K. (2019) PV/T principles and design. In: Lotsch H.K.V., editor. Photovoltaic/thermal (PV/T) systems principles, design, and applications. Cham, Switzerland: Springer.
6. Goetzberger A., Hoffmann A. (2005) PV systems. In: Letcher T., editor. Photovoltaic solar energy generation. Freiburg: Springer.
7. Belessiotis V.G., Papanicolaou E. (2012) History of solar energy. In: Abou Jieb Y., editor, Comprehensive renewable energy. Athens, Greece: Elsevier, pp. 85–101.
8. Bilen K., Erdogan I. (2023) Effects of cooling on performance of photovoltaic/thermal (PV/T) solar panels: a comprehensive review, Solar Energy, 263, p. 111829.
9. Jieb Y.A., Hossain E. (2022) In photovoltaic systems fundamentals and applications. USA: Springer.
10. Shoaib Sheik M., Kakati P., Dandoriya D., Ravi M.U. (2022) A comprehensive review on various cooling techniques to decrease the operating temperature of solar photovoltaic panels, Energy Nexus, 8, p. 100161.
11. Govindasamy D., Kumar A. (2023) Experimental analysis of solar panel efficiency improvement with composite PCMs, Renew. Energy, 212, pp. 175–184.
12. Maghrabie H., Elsaid K., Sayed E., Radwan A., Abo-Khalil A., Rezk H., Abdelkareem M., Olabi A.G. (2022) J. Energy Storage, 48, 103937.
13. Mahfuz M.H., Anisur M.R., Kibria M.A., Saidur R., Metselaar I.H.S.C. (2014) Performance investigation of thermal energy storage system with PCM for solar water heating application, Int. Commun. Heat Mass Transf., 57, pp. 132–139.

14. Mandal S.K., Kumar S., Singh P.K., Mishra S.K., Singh D.K. (2020) Performance investigation of nanocomposite based solar water heater, Energy, 198, p. 117295.
15. Palacio M., Ramirez C., Carmona M., Cortes C. (2023) Effect of PCMs in the performance of a solar air heater, Sol. Energy, 247, pp. 385–396.
16. Peng L., Salem M., Blazek V., Prokop L., Al-Bahrani M. (2023) TES applications in solar water heaters: an updated review, Sustain. Energy Technol. Assess., 57, p. 103230.
17. NematpourKeshteli A., Iasiello M., Langella G., Bianco N. (2023) Increasing melting and solidification performance of a PCM-based flat plate solar collector equipped with metal foams, nanoparticles, and wavy wall-Y-shaped surface, Energy Convers. Manag., 291, p. 117268.
18. Algarni S. (2023) Evaluation and optimization of the performance and efficiency of a hybrid flat plat solar collector integrated with PCM and heat sink, Case Stud. Therm. Eng., 45, p. 102892.

9 Application of Phase Change Materials in Transportation and Logistics

9.1 INTRODUCTION

Vehicles cabin temperature increases dramatically during the seasons especially summer when parked outdoor under the sun light and even in cloudy days can lead to excessive sweating in humans resulting in imbalance of water and salt metabolism which causes discomfort conditions for the passengers while entering the car. Improvement of thermal comfort of vehicle drivers can be achieved by using air conditioning systems; however, an increase in the air conditioning consumes much energy especially for single-passenger cars as thermal environment improvement of the entire cabin results in waste of resources. Controlling the indoor vehicle temperature at those conditions as both external (sun) and internal (engine) heat sources plays an important role in heat gain. Application of phase change materials as a thermal energy storage medium inside the car near the roof or other suitable places can absorb indoor extra heat, result in lower air and steering wheel surface temperature, and maintain a comfort temperature for the passengers. This chapter explains how PCMs can be applied in the vehicle compartment to achieve comfort conditions. Other applications of PCM as an energy storage system for public transportation, food transportation, and logistics will also be discussed.

9.2 APPLICATION OF PCMS IN A VEHICLE

PCMs can be utilized in the automotive industry as an energy storage system at different parts of the vehicle such as compartment thermal comfort, engine cooling, catalytic converter preheating, seats cooling/heating, engine preheating, pressure regulator, evaporator preheating, and other applications in internal combustion engines. Selection of a suitable PCM is an important task since there are many criteria such as high latent heat, appropriate melting point, high thermal conductivity, high specific heat being a good chemical stability, little volume change during phase transition, non-toxic, non-flammable, high density, and non-polluting for the selection of a suitable PCM that should be taken into consideration for the specific application. Some of the PCMs that are used in vehicle applications as a thermal energy storage are listed in Table 9.1.

DOI: 10.1201/9781003518280-9

TABLE 9.1
Physical Properties of PCMs Used for Vehicles [1]

Compound	Melting Temp. (°C)	Heat of Fusion (kJ/kg)	Thermal Conductivity (W/mK)	Liquid Density (kg/m³)
H_2O	0	333	0.612	998
$LiCO_3\ 3H_2O$	8.1	253	n.a.	1720
$ZnCl_2\ 3H_2O$	10	n.a.	n.a.	n.a.
$K_2HPO_4\ 6H_2O$	13	n,a,	n.a.	n.a.
$Na_2CrO_4\ 10H_2O$	18	n.a.	n.a.	n.a.
$KF\ 4H_2O$	18.5	231	n.a.	1447
$Mn(NO_3)_2\ 6H_2O$	25.8	125.9	n.a.	1738
$Na_2SO_4\ 10H_2O$	32.4	254	0.544	1485
$LiNO_3\ 3H_2O$	30	n.a.	n.a.	296
$Na_2CO_3\ 12H_2O$	34	246.5	n.a.	1442
$(NO_3)_2\ 6H_2O$	36	147	0.464	147
$Na_2HPO_4\ 12H_2O$	36	265	n.a.	1522
$CaCl_2\ 6H_2O$	29	187.5	0.538	1560
$Ba(OH)_2\ 8H_2O$	48	265.7	0.653	1937
$NaOH\ H_2O$	58	n.a.	n.a.	n.a.
$Cd(NO_3)_2\ 4H_2O$	59.5	n.a.	n.a.	n.a.
$Fe(NO_3)\ 6H_2O$	60	n.a.	n.a.	n.a.
$NaOH$	64.3	227	n.a.	1690

n.a.: not available.

9.2.1 Cooling/Heating of a Vehicle Compartment

Almost half of the solar radiation falls under visible light and the remaining being infrared and ultraviolet. When a car is parked under the sun light, the glass-made windshield of the car is transparent to the visible part of the solar radiation allowing it to enter the car cabin and at the same time the glass behaves opaque to the infrared radiation emitted from the interior grey surfaces and prevents it to exit the car cabin. This causes the cabin temperature and humidity to increase dramatically, thereby makes discomfort for the car passengers when they get into the car.

A new analysis based on the mathematical modelling and simulation for the influence of two PCMs with their seven orientations on car cabin temperature reduction based on two-phase simulation for the PCM to reduce the computational time has been performed by Srusti et al. [2]. Available solar radiation models in ANSYS FLUENT 2020R2 have been used to track the solar ray more efficiently. From six available radiation models in ANSYS FLUENT, only two discrete ordinance (DO) and surface to surface (S2S) with very similar physical output can be used. S2S model has been selected due to the less computational time in which only the car cabin has been considered and other parts such as doors, seats, back and front tyres, engine compartment, and boot space were neglected as shown in Figures 9.1.a and 9.1.b.

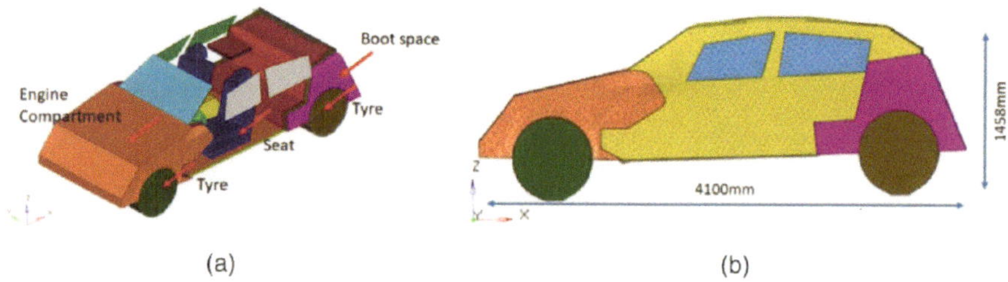

FIGURE 9.1 Volumes neglected in the simulation (a) and car dimensions (b) [2]. (Permission from Elsevier.com.)

Two types of PCMs with the properties listed in Table 9.2 have been selected based on the criteria of human thermal comfort namely as HS24, a salt hydrated salt PCM and OM29, an organic material with suitable melting point temperatures and an acceptable stability for nearly 2,000 cycles.

Seven different orientations have been considered as follows:

1. Flat PCM (OM29) partially covering the roof
2. Flat PCM (OM29) covering the roof
3. 1-curvature of PCM (OM29) covering the roof
4. 1-curvature of PCM (HS24) covering the roof
5. 2-curvature of PCM (OM29) covering the roof
6. 2-curvature of PCM (HS24) covering the roof
7. 3-curvature of PCM (HS24) covering the roof

Temperature values for 2-curvature of OM29 and HS24 covering the roof at 90 min are illustrated in Figures 9.2.a and 9.2.b, respectively. As shown in Figure 9.2.a, the 2-curvature PCM provides a gap through which the cooled air moves down into the cabin from the roof that helps in reducing the driver head temperature and rear passenger head temperature to 38.9°C and 37.3°C, respectively, when compared to

TABLE 9.2
Physical Properties of OM29 and HS24 PCMs [2]

Physical Properties	OM29	HS24
Density (kg/m^3)	923	1565.5
Melting temperature (°C)	28.5 - 29.5°C	24 - 25°C
Solidus pecific heat (kJ/kg °C)	2.32 (20 - 28.4°C)	2.04 (20 - 23.9°C)
Latent heat (kJ/kg) / equivalent specific heat (kJ/kg for 1°C)	194 (28.5 - 29.5°C)	199 (25 - 26°C)
Liquidus specific heat (kJ/kg °C)	2.71 (28.6 - 40°C)	2.42 (25.1 - 40°C)
Thermal conductivity (W/mK)	0.293 (20°C) 0.172 (40°C)	1.05 (20°C) 0.55 (40°C)

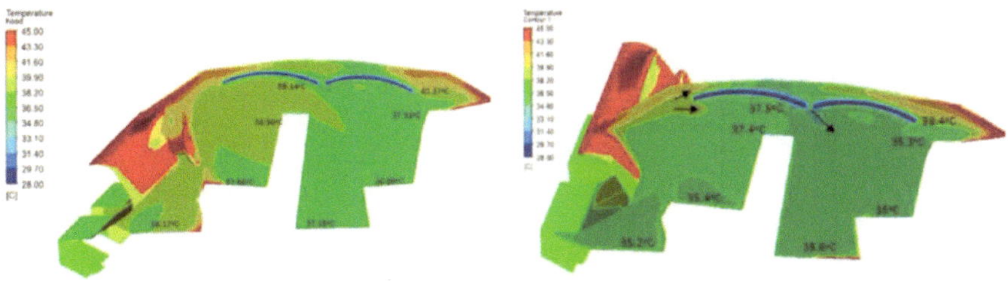

(a) (b)

FIGURE 9.2 Curvature temperature results for OM29 (a) and HS24 (b) [2]. (Permission from Elsevier.com.)

the driver head temperature of 49.5°C without PCM. The average cabin reduction temperature in this case is 7.9°C. When OM29 is changed to HS24 in Figure 9.2.b, the driver temperature head and rear passenger head temperature decreases to 37°C and 34°C when compared to the driver head temperature of 49.5°C without PCM.

Elemental averaged temperature distribution inside the car cabin at the end of 90 min is depicted in Figure 9.3 in which it is found that the temperature of IP panel absorbing solar heat increased close to 85°C, when compared to the other orientation inside the cabin.

A composite PCM consisting of Rubitherm RT55 paraffin wax and five different Al_2O_3 nanoparticle fractions from 5% to 25% has been used for heating of automobile cabin in cold climate condition by Gurbuz et al. [3] to investigate the effect of additive in PCM on the solidification process of PCM. The objective of the study was to recover the exhaust waste heat energy by storing in the PCM container in cold climate conditions for cabin heating under the next first cold start conditions, i.e., one night or a few hours later.

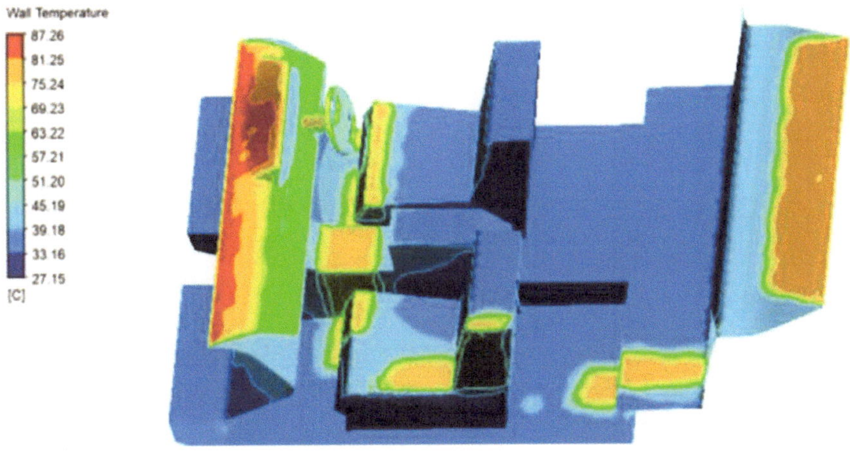

FIGURE 9.3 Temperature distribution due to solar load after 90 min [2]. (Permission from Elsevier.com.)

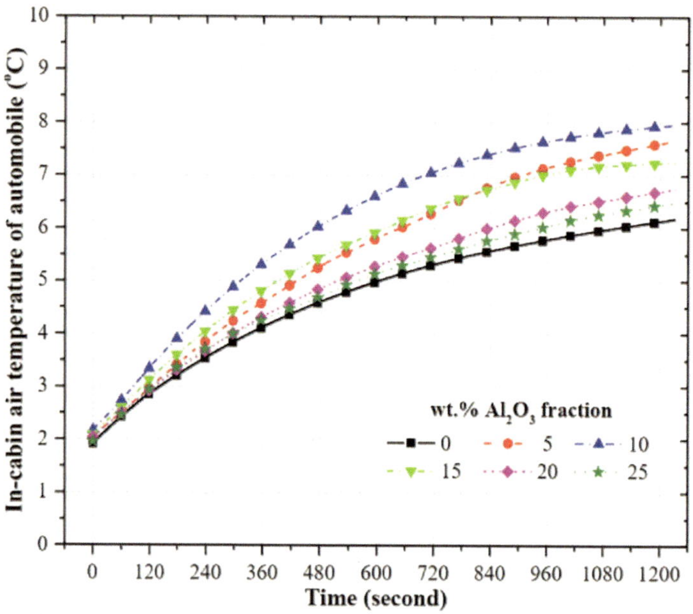

FIGURE 9.4 Interior cabin temperature distribution vs. time [3]. (Open access journal)

Cabin interior temperature distribution for different Al_2O_3 fractions is illustrated in Figure 9.4 which indicates that the cabin temperature increased up with addition of 10 wt.% of Al_2O_3 fraction compared to pure PCM and started to decrease again with further increase of the additives to the PCM. This is due to the thermal conductivity enhancement using the nanoparticles and performing better heat transfer in the system.

They finally concluded that cabin temperature can be increased by 24.2% and 29% when using 5% and 10% of Al_2O_3 in PCM, respectively.

Incorporation of PCM into a 5 door Seat Ibiza 2007 model car has been experimentally analysed by Oro et al. [4]. The car was divided in five areas, front right, front left, back right, back left, and roof and different air temperature measurements were done using some calibrated temperature sensors as demonstrated in Figure 9.5.

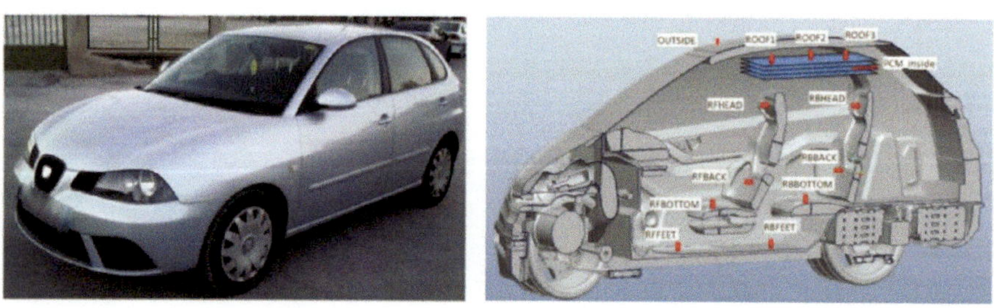

(a) (b)

FIGURE 9.5 Vehicle used in experimentation (a) and location of temperature sensors in the car (b) [4]. (Permission from Elsevier.com.)

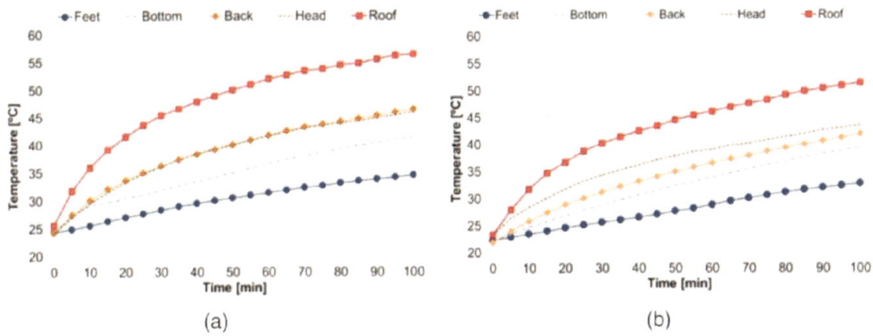

FIGURE 9.6 Air temperature profiles inside the car at different positions without PCM (a) and with PCM (b) [4]. (Permission from Elsevier.com.)

The PCM with weight of 4 kg was fitted under the car roof inside an aluminium structure to have more available area and maximum heat exchange between ambient and the car cabin.

Air temperature distributions at different positions of the car with and without PCM are illustrated in Figures 9.6.a and 9.6.b, respectively. As clearly shown in the figures lower temperatures have been achieved inside the car with PCM. The use of PCM can reduced the temperatures of the roof, head, and back of the car from 56°C, 46°C, and 46°C to 52°C, 44°C, and 42°C, respectively. Temperature reduction of 2–4°C has been obtained using PCM inside the car.

Air temperature distribution of the front and back right side of the car after 100 min are shown in Figure 9.7 which indicates that both systems start at the same

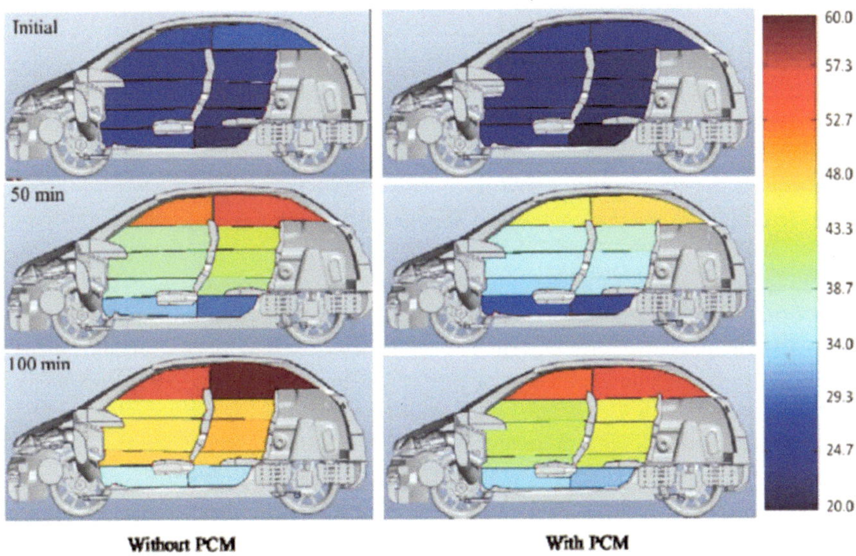

FIGURE 9.7 Air map temperature of the car with and without PCM at different times [4]. (Permission from Elsevier.com.)

temperature of 20°C and then temperature started to increase after leaving the car under the sun but in all times the temperatures inside the car with PCM are lower than the reference car.

Experimental investigation on the control of a public transport on hot sunny days and with external heater of 300 W and internal light bulb of 25 W has been conducted by Nihidul Islam et al. [5] in sodium sulphate dehydrate with a melting point of 32.4°C and latent heat of 252 J/g has been utilized as a PCM placed in between the ceiling and the roof of a public bus model. The effect of PCM on the heat transfer rate has been studied in a bus model with layer of PCM in the ceiling. Their experimental results reveal that when sun light was considered as an external heat source and PCM was not used, the indoor temperature of the bus model increased up to 37°C for 100 min, while the roof surface temperature was 54°C. The presence of PCM could decrease the temperature by less than 2°C in comparison to the conditions without PCM. It was concluded that single-layer PCM is not enough to counter internal heat gain from the internal heat sources and increasing the PCM-layer thickness would increase the delay period of the indoor temperature rising. Therefore, to reduce the effect of the internal heat source, an inner layer of PCM with a moderate melting temperature of 15–25°C along with high latent heat such as manganese nitrate hexahydrate could be utilized to absorb a substantial amount of the internal heat.

9.2.2 PCM in Food Transportation and Logistics

One-third of the total amount of produced food the world annually, around 1.3 billion ton are wasted due to the Food and Agriculture Organization (FAO) report [6] which is mainly depends on the countries. The amount of food loss is related to the lack of required conditions during the first stages of the cold chain, from the producer to the retailer of the customer while the food wastes is defined as the edible food being discarded and rejected at the seller and/or customer level [6].

The cold-chain logistics refers to systematic engineering of food harvesting, preconditioning, transport, bulk storage, retail, domestic, and food services, develops from perishable harvested products to the final consumer consisting of a series of subsequent activities which ensure that fresh food products can maintain at the required temperature throughout the supply chain. As reported by FAO, the food losses and wastage add up to 940 billion dollars financial loss per year and contributes to the annual emissions of 4.4 Gton CO_2 [6]. Optimum ways of foods cold-chain logistics system for low carbon economy requirement, and distributed food chain logistics are one of the main research topics for researchers in the last decade. PCMs utilized in a thermal energy storage system can help to support food cold-chain logistics system which is one of the main current concerns in the world as shown in Figure 9.8.

Expanded graphite (EG) with a particle size of 80 meshes and an expended volume of 250 mL/g, as a thermal conductivity enhancer has been incorporated into a eutectic PCMs of decyl alcohol (DA) and lauric acid (LA) with a mass ratio of 12:1 by vacuum adsorption technique and selected as a novel composite PCM for the vaccine delivery at temperature range of 2–8°C by Liu et al. [8].

The cold box with dimensions of $85 \times 85 \times 80$ cm^3 has been made of polyurethane with vacuumed insulation panel, including the cold plates of DA-LA/EG which have been precooled at −20°C for 24 h to store cold as illustrated in Figure 9.9.

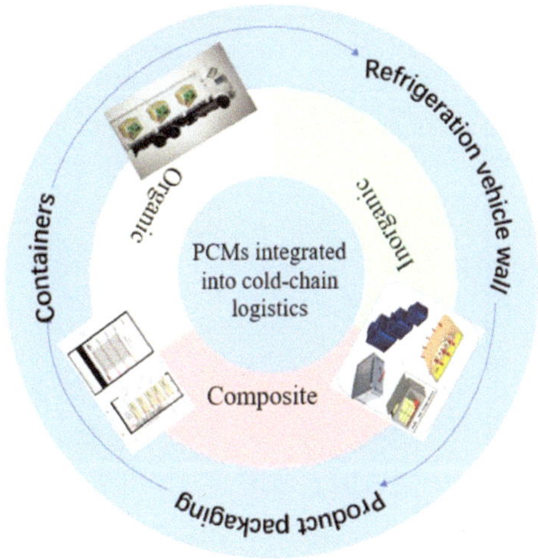

FIGURE 9.8 Application of PCM in cold-chain logistics [7]. (Permission from Elsevier.com.)

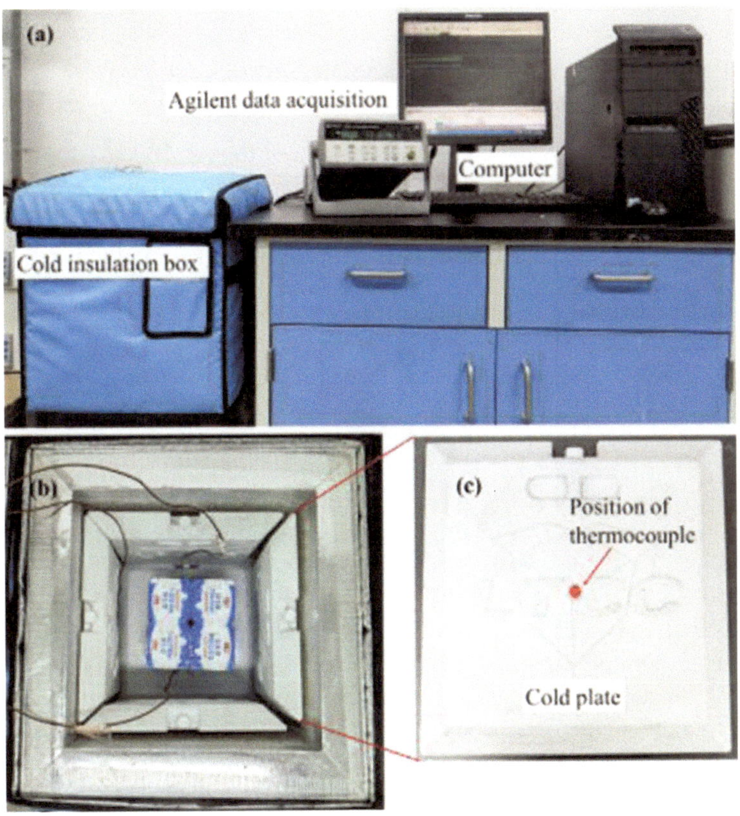

FIGURE 9.9 Cold box insulation apparatus [8]. (Permission from Elsevier.com.)

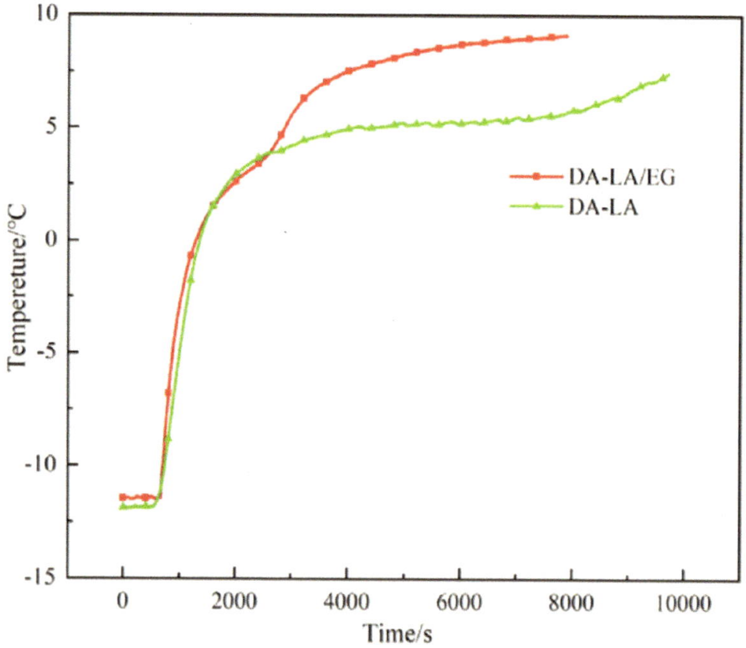

FIGURE 9.10 Cooling curves of composite PCMs [8]. (Permission from Elsevier.com.)

The cooling curves of the composite PCMs over time is depicted in Figure 9.10 which indicates that the melting/solidification time has been decreased by 79.47% after adding EG to the PCM which is because adding EG to the PCM increases thermal conductivity and heat transfer is enhanced by conduction. Addition of EG to the PCM will also increase the contact areas since DA-LA/EG are divided into several small pieces and as the result nucleation rate increases.

The cold insulation experiment has been carried out at ambient temperature of 31°C in daytime, to verify the actual applicability of composite PCM in the cold box. The results are shown in Figure 9.11 which proves that taking 2–8°C as the limitation, the duration of cold insulation for each plate is approximately 10 h.

A heat transfer model for the application of a composite PCM for cold-chain logistics in the range of 0–8°C has been developed by Xu et al. [9]. Caprylic acid (CA) with a melting point of 14.4°C and latent heat of 158.8 J/g and LA with a melting point of 23.4°C and heat density of 200.7 J/g have been selected as a composite PCM and EG has been added to the PCMs as thermal conductivity enhancer. The experimental apparatus used during the study include magnetic stirrer, low and constant temperature tank, Agilent, DSC, hot disk thermal constant tester, and high- and low-temperature alternating box as demonstrated in Figure 9.12.

The cooling step curves of composite PCMs at different CA/LA ratios are shown in Figure 9.13.a which indicates that as by increasing the mass ratio of CA in the composite PCMs the phase change onset point first decreases and then increases with lowest freezing point at the ratio of 40:60. Temperature profiles of composite PCMs with

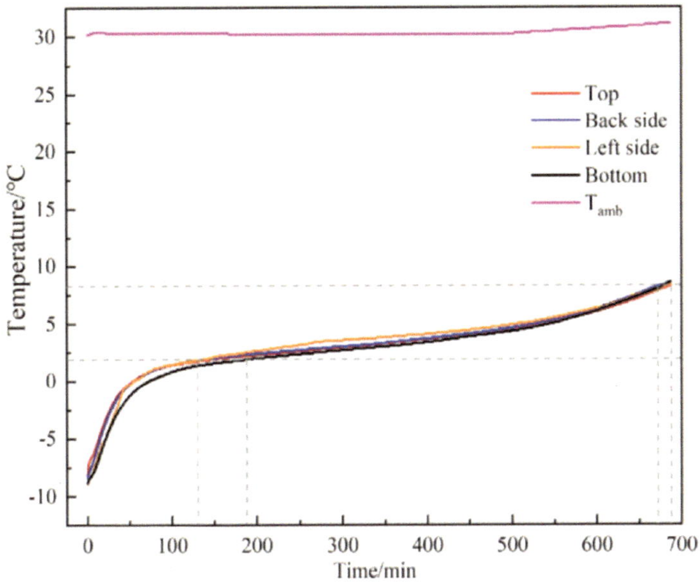

FIGURE 9.11 Cold insulation time with load [8]. (Permission from Elsevier.com.)

and without EG are represented in Figure 9.13.b which shows that as phase transition of CA/LA lasted about 3,400 s, this value for the CA-LA/EG is about 310 s due to the porous structure of EG that enhances the heat transfer performance of the material.

They finally concluded that addition of 8% EG to the composite PCMs can improve the leakage, the mass loss, and after 100 cycle tests the latent heat of was 168.9 J/g which is 1.5 J/g lower than before the cycle.

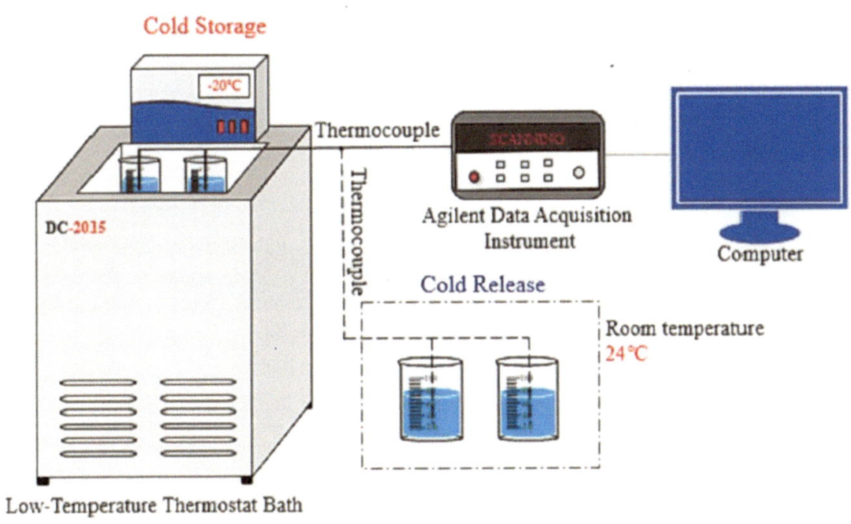

FIGURE 9.12 Experimental setup of cooling system [9]. (Permission from Elsevier.com.)

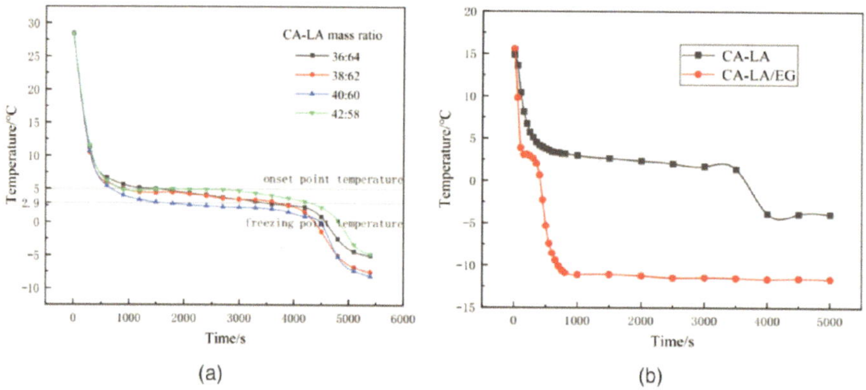

FIGURE 9.13 Step cooling curves of PCMs vs. time (a) step cooling curves with and without EG (b) [9]. (Permission from Elsevier.com.)

One of the most significant tools for ocean observation and exploration is underwater thermal vehicles that uses a thermal engine to harvest ocean thermal energy and whose duration and range of underwater vehicles can be increased by using composite PCMs. The vehicle structure includes hydraulic oil layer, rubber hose layer, PCM layer, and pressure resistance shell layer as illustrated in Figure 9.14. Evaluation of graphene added PCMs for the heat transfer enhancement in underwater thermal vehicles has been performed by Wang et al. [10]. The phase transition temperature of the PCM is selected for the thermal engine which lies between the higher surface temperature (upper warm) on the ocean and the lower temperature of the deep ocean (lower cold). Therefore, when the engine is on the sea level with high

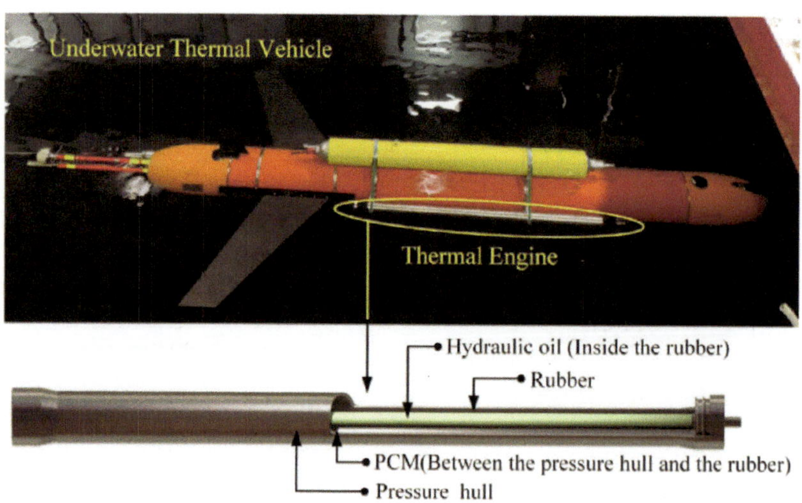

FIGURE 9.14 Schematic diagram of underwater thermal vehicle [11]. (Permission from Elsevier.com.)

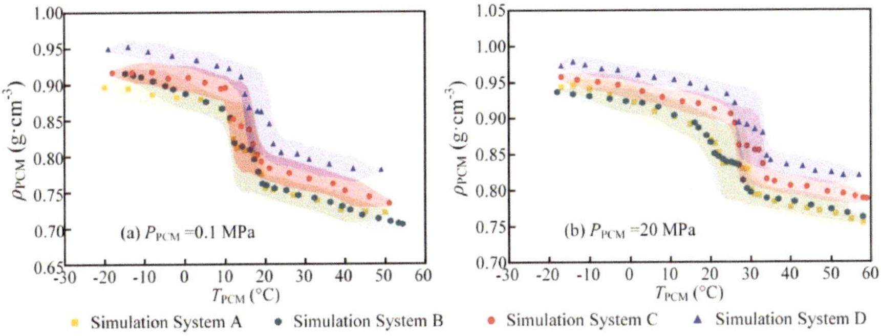

FIGURE 9.15 PCM density vs. temperature for different cases at two pressures [10]. (Permission from Elsevier.com.)

water temperature, extra heat will be absorbed by the PCM during melting, while in the deep sea at low water temperature, the PCM solidifies.

The mathematical model of thermal engine has been developed based on the thermal engine structure and its working principles using some simplifying assumptions. N-hexadecane has been selected as a PCM (case 1) and graphene has been added to the PCM with 1.2%, 4.5%, and 9.0% (cases 2–4) for thermal conductivity enhancement to investigate thermophysical properties of PCM such as melting temperature, thermal conductivity, and specific heat at 0.1 MPa and 20 MPa.

Density error bands of four simulation cases for two pressures vs. temperature are shown in Figure 9.15 which indicates that variation of PCM density is similar for all cases and at two pressures. In both solid and liquid phase regions, the density decreases linearly with temperature increase and vary sharply with temperature around melting temperature indicating phase change in the system.

The variation of PCM thermal conductivity with temperature for all four cases at two pressures of 0.1 and 20 MPa are demonstrated in Figure 9.16 which shows that the thermal conductivity at 0.1 MPa increases in both solid and liquid phases as the addition ratio of graphene increases as expected. When the pressure is increased to

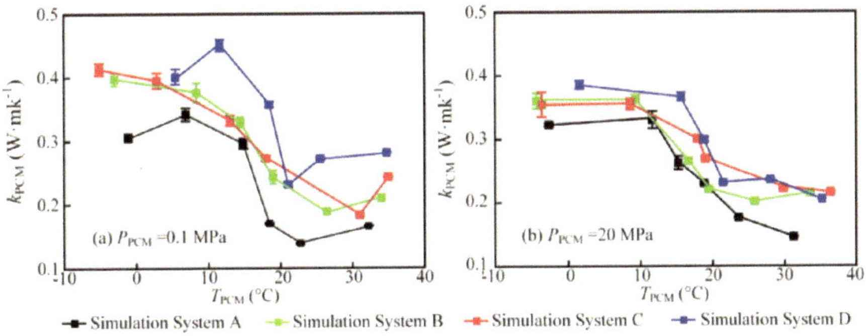

FIGURE 9.16 PCM thermal conductivity vs. temperature for different cases at two pressures [10]. (Permission from Elsevier.com.)

20 MPa, the same pattern is observed which proves that pressure does not have any impact on the PCM thermal conductivity.

Their simulation results show that the thermal engine system with PCM, including 1.2% graphene, has the best performance and the melting time can be decreased by 11.97% compared with the one with PCM without graphene.

Application and property evaluation of EG/ice PCMs formed through vacuum impregnation and in-situ polymerization of sodium polyacrylate in cold-chain logistics has been studied in an 8-L insulation box by Tang et al. [12].

For the preparation of the composite PCMs EG was initially pressed into blocks of varying densities and sodium acrylate was added to the solution and finally, the EG block was immersed in a sodium acrylate solution and the solution was adsorbed at different pressures to enhance the sodium acrylate content with the EG as shown in Figure 9.17.

A mobile thermal storage box composed of three components of six PCMs bags, a portable insulation bag, and a refrigerated box composed of a polyethylene shell with 5 mm thickness and a polyurethane insulation with 30 mm thickness was prepared for medical supplies transportation purposes as illustrated in Figure 9.18.

The results show that the cooling time of the composite PCMs in the cold box has been measured to be 64.7 minutes from 25°C to −10°C with cooling times of 55.7% and 32.8% compared to composite PCMs of tetradecane and water. Also the synthesized PCMs could control the saltwater at 8°C for 1,124.1 minutes.

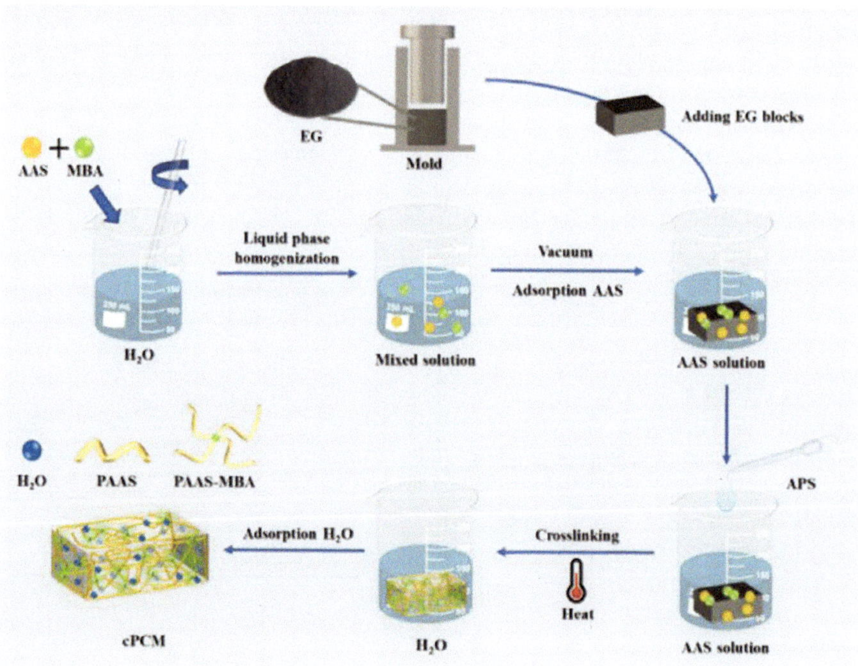

FIGURE 9.17 Preparation steps for composite PCMs [12]. (Permission from Elsevier.com.)

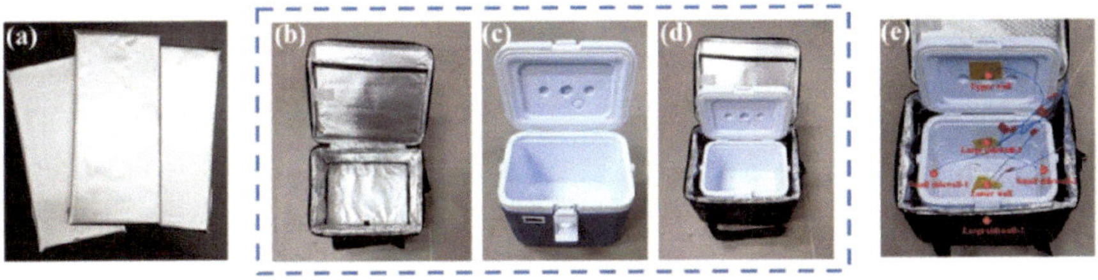

FIGURE 9.18 Cold storage assembly. (a) Packaged PCMs, (b) portable insulated box, (c) cold storage equipment, and (d) refrigerated box wrapped in insulation bags [12]. (Permission from Elsevier.com.)

A composite sheet fabricated from PCM made of O_2 plasma treated activated carbon fibre with porous structure of 2–50 nm used as a matrix to encapsulate polyethylene glycol (PEG) for logistics temperature control has been developed by Zhao et al. [13]. The sheets made of the composite PCM has a latent heat of 146–180 J/g, good stability under high temperature and thermal cycle and adjustable melting temperature by changing PEG molecular weight.

Recent literature reviews on the applications of PCMs in cold-chain logistics for agricultural and aquatic products presented by Syah Mustafa et al. [14] and Li et al. [15], respectively, summarizes the classification, performance optimization technology, and selection methods of PCMs in which the application scenarios and equipment types of phase change cold storage technology in cold-chain logistics has been described.

9.3 SUMMARY

In this chapter, the studies from the literature on the application of PCM for the thermal control of vehicles in cooling and heating environments have been reviewed. The results proves that a suitable selected PCM can decrease or increase the vehicle compartment temperature with the environment temperature a few degrees based on the amount of PCM incorporated in the car and obtain a relative thermal comfort for the passengers.

Low-temperature technologies are used for the cold-chain logistics process of agricultural, aquatic, diary and other food products which are commonly used for low-temperature preservation and safety of the food products. Some prospects on the application of PCMs as a cold energy storage system in transportation and logistics have been presented and discussed. It was shown that organic PCMs are only used in refrigerated, ice temperature, and super chilling temperature processes and are not widely utilized in cold-chain logistics due to the high price while inorganic PCMs with lower prices are few. That is why PCMs are not widely applied in cold-chain logistics of aquatic products. The only challenges for the utilization of PCMs in transportation are their low thermal conductivity that can be treated by adding nanomaterials such graphene, EG, and some of nanometallic powders to the PCM as thermal conductivity enhancement additives.

REFERENCES

1. Ugurlu A., Gokcol C. (2012) A review on thermal energy storage systems with PCMs in vehicles, Electronic J. of Vocational Colleges, May 2012.
2. Srusti B., Kumar M.B. (2022) A novel method to investigate the influence of PCM orientation in reduction of car cabin temperature, J. Energy Storage, 47, p. 103650.
3. Gurbuz H., Aytac H., Hamamcioglu H., Akcay H. (2022) The effects of Al_2O_3 addition on solidification process of PCM: a case study on heating of automobile cabin in cold climate conditions, Int. J. Autom. Sci. Technol., 6(3), pp. 275–283.
4. Oro E., Jong E., Cabeza L.F. (2016) Experimental analysis of a car incorporating PCM, J. Energy Storage, 7, pp. 131–135.
5. Islam Md N., Hossain N., HassnAhmed D. (2022) Investigation of PCMs for interior temperature regulation in public transport, Clean Energy, 6, pp. 178–192.
6. Food and Agriculture Organization of the UN and Food Wastage Footprint (Project), Food wastage footpring full-cost accounting: Final report. Accessed June 2022. www.fao.org/3/i3991e
7. Feng T., Ji J., Zhang X. (2023) Research progress of phase change cold energy storage materials used in cold chain logistics of aquatic products, J. Energy Storage, 60, p. 106568.
8. Liu L., Zhang X., Xu X., Lin X., Zhao Y., Zou L., Wu Y., Zheng H. (2021) Development of low-temperature eutectic PCM with expanded graphite for vaccine cold chain logistics, Renew. Energy, 179, pp. 2348–2358.
9. Xu X., Zhou Y., Zhang S., Zhai X., Zhang X. (2023) Preparation and heat transfer model of stereotyped PCMs suitable for cold chain logistics, J. Energy Storage, 60, p. 106610.
10. Wang G., Yang Y., Wang S. (2022) Thermophysical properties analysis of graphene added PCMs evaluation of enhanced heat transfer effect in underwater thermal vehicles, J. Mol. Liquids, 348, p. 118048.
11. Wang G., Yang Y., Wang S. (2020) Ocean thermal energy application technologies for unmanned underwater vehicles: a comprehensive review, Appl. Energy, 278, p. 115752.
12. Tang L., Ling Z., Zhang Z., Fang X. (2024) Enhanced mechanical and thermal properties of expanded graphite/ice PCMs through in-situ polymerization of sodium polyacrylate in cold chain logistics, Chem. Eng. J., 494, p. 152869.
13. Zhao L., Zhao Y., Wei D., Huang J., Wen B., Ma Y., Deng O., Li Z., Zhang K. (2024) Hierarchical porous carbon fiber felt loaded with PEG as hybrid phase change energy storage sheet for temperature-controlled logistics, J. Energy Storage, 97, p. 112779.
14. Syah Mustafa M.F.M., Namasivayam N., Demirovic A. (2024), Food cold chain logistics and management: A review of current development and emerging trends, Journal of Agriculture and food research, under publication. https://doi.org/10.1016/j.Jafr.2024.101343
15. Li M., Xie B., Li Y., Cao P., Leng G., Li C. (2024) Emerging phase change cold storage technology for fresh products cold chain logistics, J. Energy Storage, 88, p. 111531.

10 Application of Phase Change Materials in Free Cooling/Heating

10.1 INTRODUCTION

Phase change materials are used to decrease building energy consumption of heating/cooling systems which accounts for 40% and even more in the summer in hot climate regions, of the world primary energy production [1]. This has an important impact on the greenhouse environmental gas emissions and requires actions to improve the building energy performance, including their heating and cooling systems. Thermal energy storage (TES) systems utilizing PCMs as a latent heat storage can be applied to increase the energy efficiency of the buildings and meet the peak demand for cooling or heating. Reduction of energy consumption in the building envelope can be achieved by two main techniques; active and passive methods from which the greenhouse gas emission can be reduced. Design of microclimate, shading and thermal capacitance to reduce cooling load, and application of TES in the building envelope are among the passive cooling techniques. In general, cooling and heating techniques can be categorized into two methods:

- Reduction of heat/cool entry to the building via optimum envelope design
- Passive and hybrid cooling/heating systems

This chapter reviews the methods that PCMs can be applied as an energy storage system to capture heating/cooling energies when available and use them when required to reduce building energy consumption and shifting peak load.

10.2 BUILDING ENVELOPS DESIGN

Environmental conditions of a place where the building is located such as outdoor temperature, wind velocity, solar radiation, building structure (wall thickness and area ratio of window to wall), building materials (thermal conductivity and specific heat), and indoor heat source and air change rate have a direct impact on indoor building temperatures. The essential factor on the heat flow rate through the building walls and ceiling is a well-designed insulation that reduces the cooling/heating energy costs and keeps the building in comfort conditions during the winter and summer seasons. Building thermal mass which works as a battery can absorb solar heat from the sun during daytime, while keeping the building with comfortable temperature, and release the same amount of energy during the night to the building if required. Heat flow delay through the building envelope by as much as 10–12 h via using correct thermal mass

DOI: 10.1201/9781003518280-10

can provide warmer house at night during wintertime and a cooler house during the day in summer. Passive cooling techniques have been classified to three methods:

- Ventilation techniques
- Evaporative cooling system
- TES systems

10.2.1 Ventilation Techniques

In this method which is one of the most comfortable and healthy methods for the indoor environments, outdoor environmental air is sent into the building when the ambient temperature is less for cooling purpose to meet the fresh air requirement based on the occupancy. The ventilation techniques for cooling energy building are:

- Natural ventilation
- Advanced natural ventilation
- Mechanical ventilation
- Night ventilation

Natural ventilation: In this method, natural air flows around the building envelope and is controlled by the occupants via windows opening.

Advanced natural ventilation: In this technique, the flow and the direction of the ventilating air are controlled by natural force other than windows like thermal chimneys and windcatchers.

One of the traditional Persian-originated architectural designs for natural ventilation in desert areas that creates natural or passive ventilation in buildings are windcatchers as illustrated in Figure 10.1.

FIGURE 10.1 Windcatchers for as a passive building cooling in Abarkooh, Yazd, Iran [2]. (https://en.wikipedia.org)

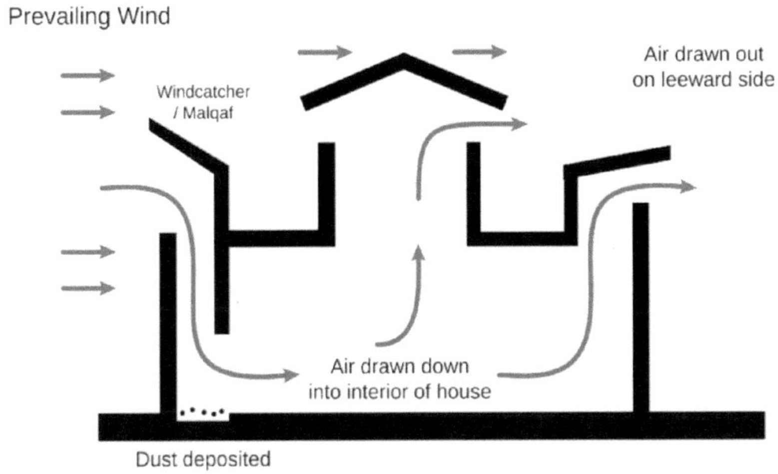

FIGURE 10.2 Concepts of a windcatcher [2]. (https://en.wikipedia.org/wiki/Windcatcher#/media/File:Malqaf.svg)

There are various designs of windcatchers such as unidirectional, bidirectional or multidirectional based on the local weather conditions and wind directions and how they change with altitude, on the daily temperature cycle and how much dust needs to be removed. They also rely on the local weather and microclimate conditions and not all techniques will work everywhere. They are vastly used in Africa, West Asia (Iran, Persian Gulf countries), and India.

In the areas with stronger winds, the windcatchers are taller and have smaller cross-sections and in the areas with very hot wind, some smaller shafts can be built to cool the incoming air. Normally the square horizontal cross-sectional types are more efficient than round ones due to the less laminar flow making better flow separation. Air filtering is required in conditions of dusty and polluted air or in the areas with insect-borne illnesses such as malaria and dengue fever.

In a windcatcher, hot air enters and circulates the building from the openings that are facing the wind and trapping it inside creating a nice breeze inside the building by convective or evaporative principle through the tower and water flowing in qanat for cooling as demonstrated in Figure 10.2.

Mechanical ventilation: Mechanical ventilation and cooling in the buildings are mainly performed by a central or local fans that provide ventilation air.

Natural night ventilation: In the dry areas where there is substantial temperature difference between day and night, night ventilation can be achieved by opening the windows to cool down the indoor temperature and provide comfort conditions during the summer. There is requirement for a fan to enable accelerated night cooling using ambient air for sufficient night cooling. Due to the increase of ambient air temperature at night in the urban areas the efficiency of the night ventilation cooling is low.

Evaporative cooling: In this technique, air entering the building can be cooled by direct evaporating cooling devices in which droplets of water evaporated absorbs the heat of evaporation from air and the air gets cooled as illustrated in Figure 10.3.

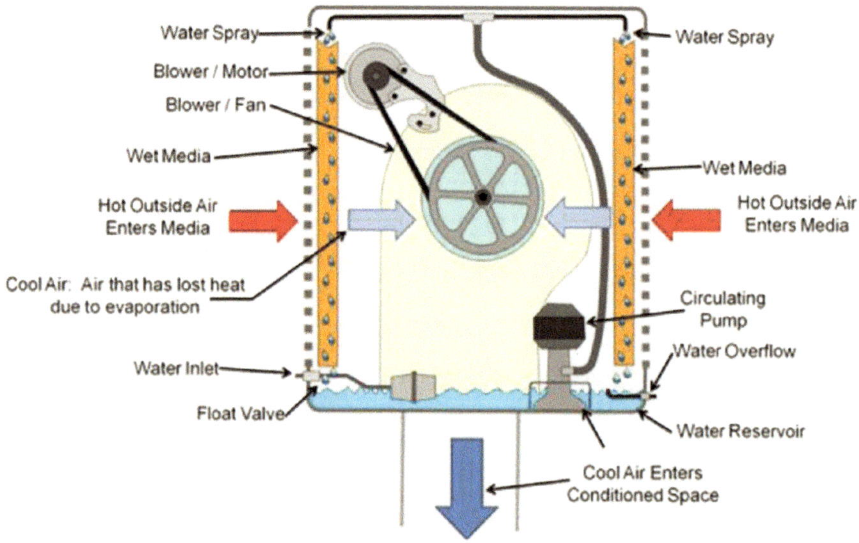

FIGURE 10.3 Direct evaporative cooling [3]. (Permission from https://alopiping.com)

One of the applications of evaporative cooling is in the combination of windcatchers with qanat or underground canal due to the insulation effects and heat capacity of the overlaying earth that maintain lower and stable temperature of the soil and water below ground level during the day and night. In this technique, the directional ports of the tower top in windcatchers may be adjusted with the open side of the tower faces away from the direction of the prevailing wind. This pulls the air into the canal and brought the hot air into contact with the water flow to absorb some of the cool water by evaporation and makes the entering hot air cooler before going to the house as illustrated in Figure 10.4.

10.2.2 Free Cooling/Heating

Integration of latent heat storage systems in building ventilation can store the coldness of the ambient air during the night and release the coldness to the building during the day in the summer. In the winter the heat from the sun can be stored in the TES system during the day and can be used for the building heating during the night. This technique has the best performance for the places where the diurnal temperature range is greater than 15°C. Equal temperature difference is available for charging/discharging if the phase transition temperature of PCMs is at the middle of the diurnal extreme temperatures. Less energy consumption, green house emission reduction, and indoor quality improvement are among the main advantages of free cooling/heating.

Phase change materials have been utilized in a ceiling fan ventilation for indoor thermal comfort assessment by Alizadeh et al. [4] using experimental design and response surface method. Commercial PCM S27 has been used due to the little volume expansion and melting point of 22–24°C. Two identical fully instrumented tests rooms with dimensions of $1.2 \times 1.2 \times 1.5$ m^3 insulated by 0.1 m thick glass wool in the walls and ceiling have been designed and constructed and were elevated 0.5 m above the ground as depicted in Figure 10.5.

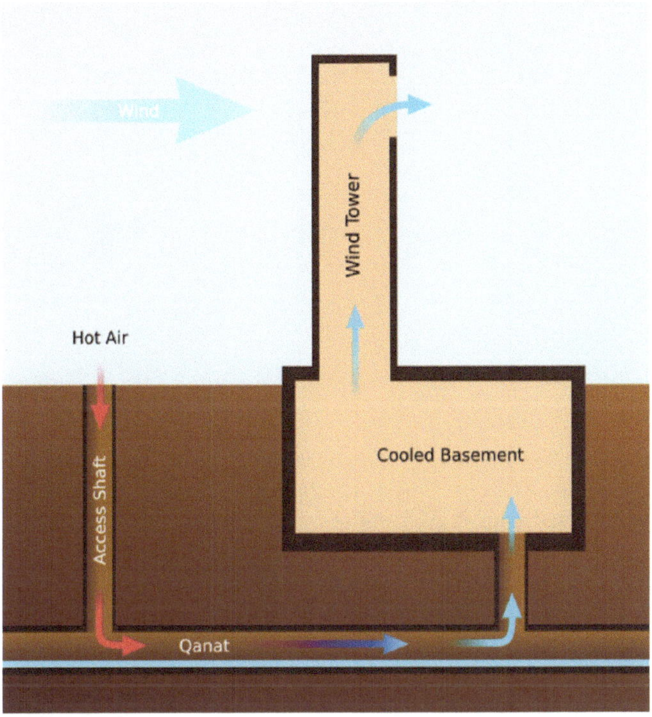

FIGURE 10.4 Combination of windcatchers with qanat for evaporative cooling [2]. (https://en.wikipedia.org/wiki/Windcatcher)

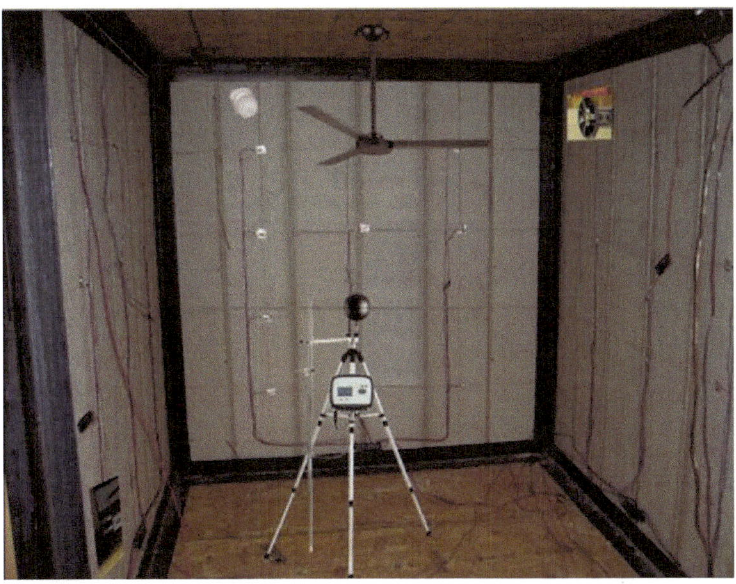

FIGURE 10.5 Schematic of one of the experimental rooms [4]. (Permission from www.Elsevier.com)

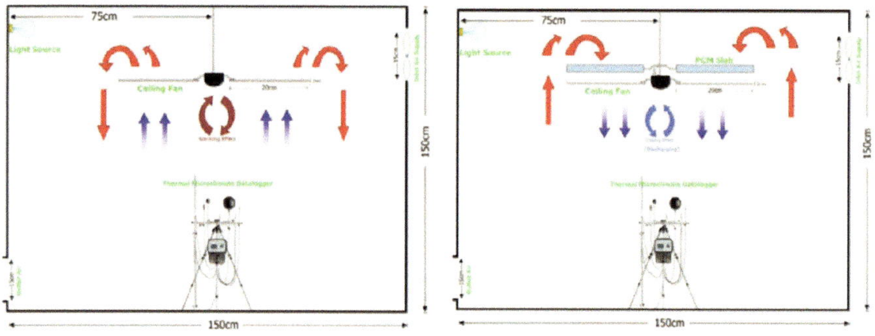

FIGURE 10.6 Concept of ceiling fan utilizing PCM on the top [4]. (Permission from www.Elsevier.com)

The base room which is used for the measurement of all experimental data has an ordinary ceiling fan, an axial three-blade model 3215, while in the second room the ceiling fan is utilized with PCM 2 cm above the fan blades. The airflow by the fan was upward and downward as shown in Figure 10.6.

Temperature distributions for indoor rooms without and with PCM are illustrated in Figure 10.7 which indicates that daytime temperature in room A (without PCM is

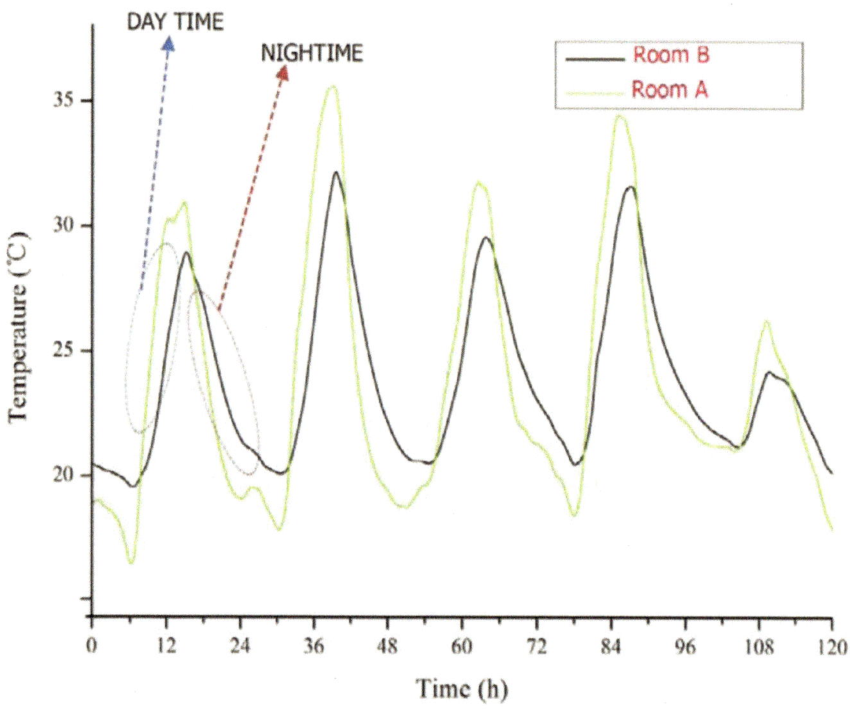

FIGURE 10.7 Temperature profiles of the indoor rooms without (a) and with (b) PCM [4]. (Permission from Elsevier.com)

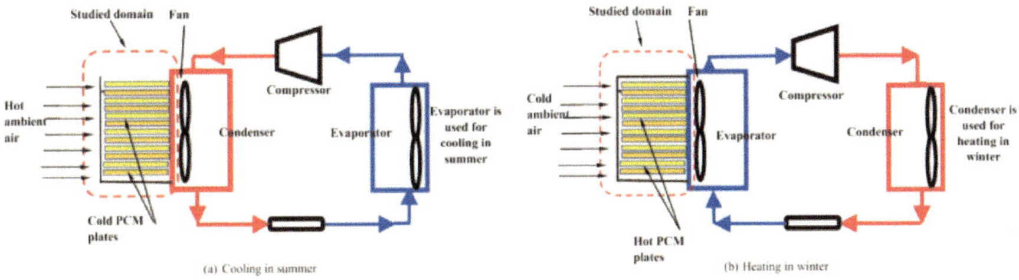

FIGURE 10.8 Hybrid PCMs with A/C unit in the summer (a) and winter (b) [5]. (Permission from Elsevier.com)

higher than room B. During the nighttime, the inverse effect is observed, and room B possesses higher temperature. It is also interesting to mention that temperature fluctuations are smaller for the room with PCM which shows more uniformity of the temperature difference between day and night.

The influence of a hybrid system of two PCMs combined with A/C unit on energy consumption of the building during summer and winter has been studied by Said et al. [5]. Two commercial PCM RT 10 HC with melting temperature of 10–12°C and SP24E with phase transition temperature of 24–25°C have been selected for the TES systems. The physical model includes the cooling impact in the summer and heating in the winter season by coupling the TES with the evaporator and condenser of the A/C unit as illustrated in Figure 10.8.

The experimental unit consisted of a wood made rectangular duct with dimensions of 0.3×0.25 m^2 with 2 kg of PCM24E in six cuboids aluminium made containers with dimensions of $0.3 \times 0.45 \times 0.015$ m^3 as depicted in Figure 10.9.

FIGURE 10.9 Schematic diagram of experimental setup [5]. (Permission from Elsevier.com)

A two-dimensional transient mathematical model has also been developed, including continuity equation, Navier stokes, phase change of PCM, airflow, and heat transfer through the airflow and PCM between the flowing air and PCM and solved numerically using ANSYS-Fluent software. The study has been performed at Mediterranean cities of Egypt with average nighttime and daytime temperatures of 22°C and 40°C during the summer and daytime and nighttime temperatures of 4°C and 16°C at winter season, respectively. The system has been studied for two different configurations of two different spaces between the PCM namely as conf. 1 and conf. 2. Outlet ambient temperature results from the PCM plates in cooling for the hybrid conf. 1 and hybrid conf. 2 and single PCM system during charging and discharging periods are shown in Figures 10.10.a and 10.10.b, respectively. As illustrated in Figure 10.10.a, ambient temperature increases sharply during charging period with time to a maximum value due to the lower temperature of ambient air from PCM temperature while in the discharging as shown in Figure 10.10.b, ambient temperature reduces sharply in a short time since the inlet air is higher than the PCM temperature.

One of the main characteristics of any air conditioning system is Coefficient of Performance (COP). The COP results of the systems vs. time for a single system, hybrid Conf. 1 and hybrid Conf. 2 are presented in Figures 10.11.a and 10.11.b for the summer- and wintertimes, respectively. COP is influenced by the evaporator and condenser temperatures where the lower summer temperature at night and higher winter temperature at daytime increase the COP. In the summer, the COP increases sharply with time to a maximum for a short period of time and then decreases with time until returns to the COP of the unit without PCM. In the wintertime, COP rises suddenly with time and then declines until an enhancement zero.

Application of activated carbon incorporated in n-docosane (n-$C_{22}H_{46}$) with latent heat of 242.7 J/g by vacuum impregnation for the heating efficiency enhancement of the buildings by dry floor heating system has been used by Jeong et al. [6]. The study

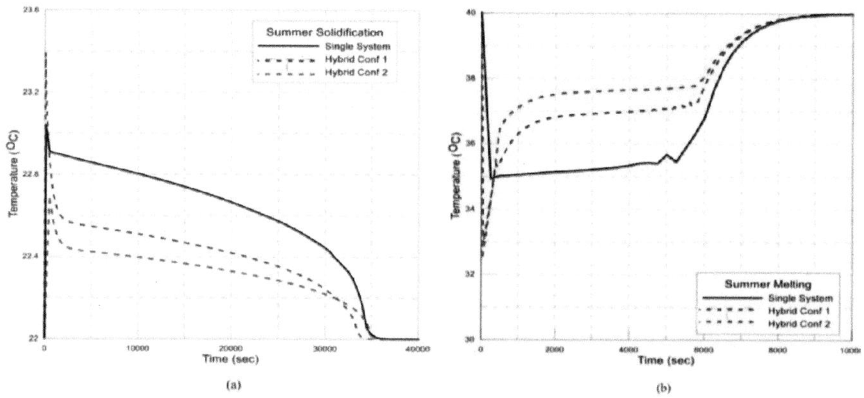

FIGURE 10.10 Temperature distribution during charging and discharging periods vs. time in the summer [5]. (a) Charging (night). (b) Discharging (day time). (Permission from Elsevier.com)

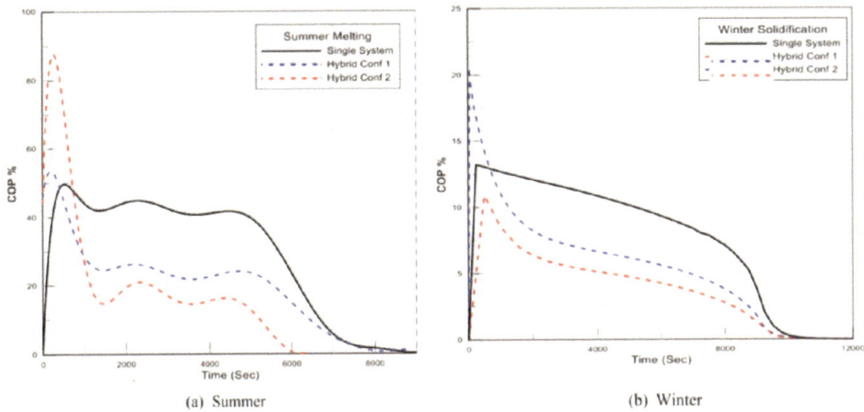

FIGURE 10.11 Variation of COP with time during the summer (a) and winter (b) [5]. (Permission from Elsevier.com)

was conducted in two separate rooms: wet floor heating system vs. PCM dry floor heating systems and dry floor heating system vs. PCM dry floor heating systems as shown in Figure 10.12. Energy savings and temperature reduction results of the PCM dry and general wet floor systems are depicted in Figure 10.13 in which similar time-lag effect is shown in both PCM dry floor heating system and the wet floor heating system.

The temperature difference of 6–7°C between surface temperature of the near and far sides of the pipe of PCM dry floor heating system was also greater than of the wet floor system with maximum 2–3°C. The power consumption during heating and free cooling was measured to be 14 kWh for the wet floor heating system and 7.5 kWh for PCM dry heating system which means that the PCM dry floor heating system utilizes only 53% of the power used by the wet floor heating system.

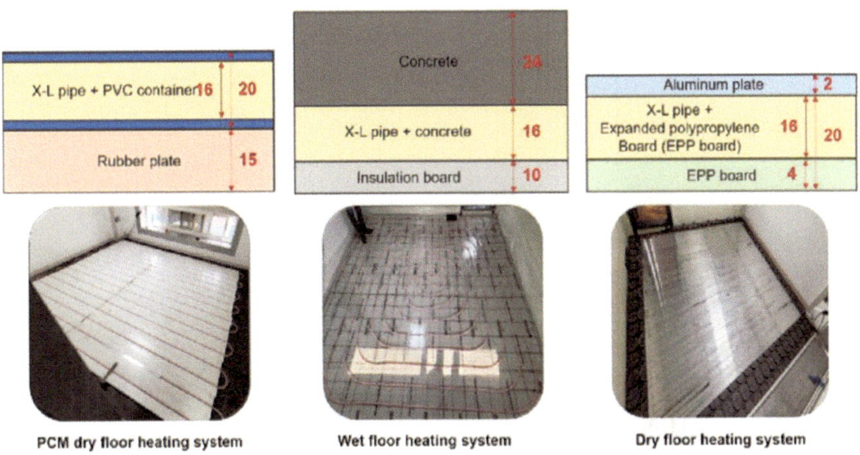

FIGURE 10.12 Wet and dry floor heating systems [6]. (Permission from Elsevier.com)

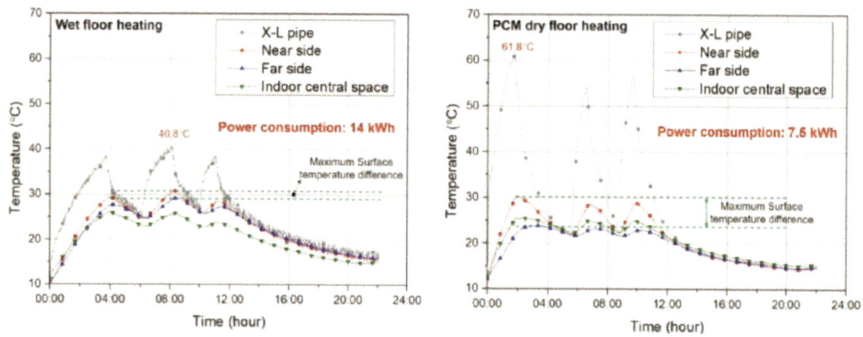

FIGURE 10.13 Evaluation of dynamic heat transfer in dry and wet floor heating systems [6]. (Permission from Elsevier.com)

In general, the wet floor heating systems can have a time-lag effect which is due to the heat storage of the concrete, but can reach to the peak temperature at slower time while for the dry floor heating systems shorter time if required to reach to the peak temperature initially but lack the time-lag effect caused by heat storage system.

A nondimensional analytical model for the application of PCM as heat sinks subject to sinusoidal heating has been studied by Mirza et al. [7] in which two PCM-based heat sinks have been used and expressions for their effectiveness have been derived. The system has been used for the electronic device cooling application which generates a time varying heat load as shown in Figure 10.14.

A eutectic alloy-based PCM (32.5-Bi, 51-In, 16.5-Sn) with melting temperature of 60°C and thickness of 1.0–4.4 mm has been used in the cooling system. Sinusoidal heat input in the range of 1–100 W/cm^2 and periods of 5–640 s has been applied to the system. The external heat transfer coefficients and ambient temperature ranges have been set to be 1,700–18,700 W/m^{2}K and 20–45°C, respectively. The temperature distribution results of the study for air cooled setup at a fan speed of 7,000 RPM with PCM thickness of 1 mm in solid state with a mass of 0.88 g and cross-sectional area of 1.37 cm^2 are presented in Figure 10.15.

The applied heat amplitude of 3–17 W in the period of 5–640 s has been applied for a total combination of 40 at each fan speed. The temperature results of the heated and cooled surfaces are very close together showing that internal thermal gradients

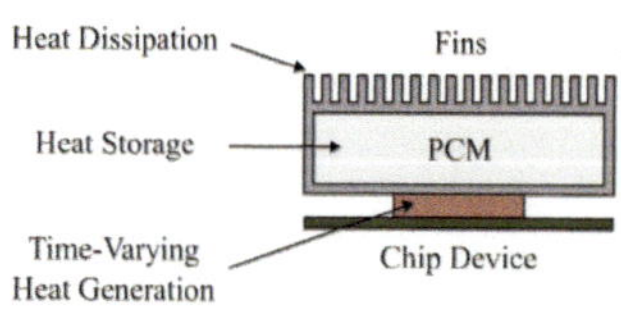

FIGURE 10.14 Schematic of a typical PCM-based heat sink for electronic cooling applications [7]. (Open access journal)

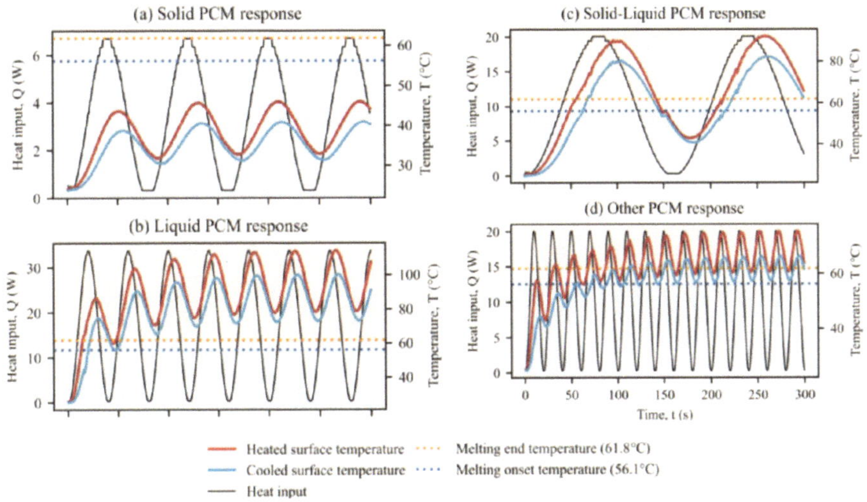

FIGURE 10.15 Experimental results of air-cooled system. (a) Solid response, (b) liquid response, (c) solid-liquid response, and (d) other response [7]. (Open access journal)

for this system is very small and the conduction heat transfer system can be considered as a lump system.

Nanomaterials such as Al_2O_3 and CuO have been added to the PCMs of sp07 and sp11 for thermal analysis of charging/discharging process in different encapsulation geometries for free heating applications by Muzhanje et al. [8]. A mathematical model for the physical system has also been developed and solved using Ansys-Fluent and validated by the experimental data. Their results indicates that rectangular geometry of microcapsules have the shortest phase change transition times and over 41% shorter than circular capsules with the minimum melting times of 79 and 135 minutes using Al_2O_3 with sp07 and sp11, respectively. Thermal performance of the nanoparticles added to the PCMs increases the melting and solidification rates by 4.17–16.90% and the nanoparticles of Al_2O_3 are superior to CuO nanoparticles.

Comprehensive reviews on the development of free cooling-based ventilation technology for building using TES have been presented by Alizadeh et al. [9] and Ghamari et al. [10].

10.3 SUMMARY

One of the major challenges our planet is facing is climate change and countries around the world had been trying to find a solution for the environmental problems and some have taken relevant measures. One of the solutions to the problem is improving energy saving techniques that became an interesting topic, and many developments have already been implemented to save fossil fuel energy. One of most

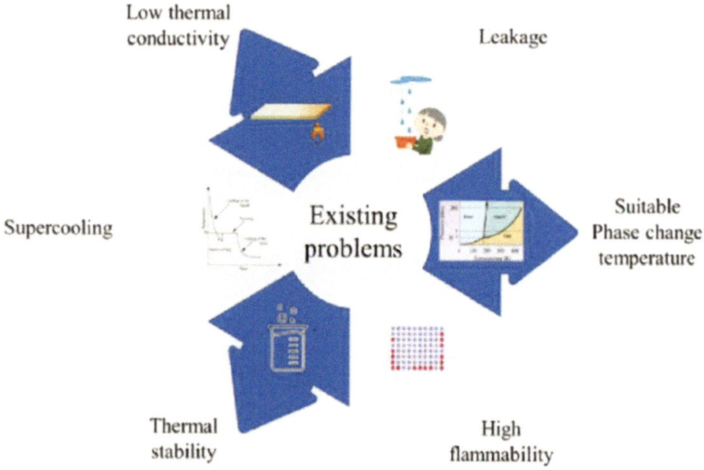

FIGURE 10.16 Existing challenges for application of PCMs [10]. (Permission from Elsevier.com)

attractive solutions to reduce air conditioning energy consumption and shift energy loads from peak time to low consumption period is utilizing PCMs as TES medium that can be used in both active and passive systems. Such systems can store the air coldness at nighttime and release it to the air during the daytime to reduce the air conditioning load and maintenance. The phase transition temperature of the PCMs is driven by ambient temperature variations during day and night. All aspects of application of PCMs in cooling/heating systems have been discussed in this chapter and some of the experimental analyses of using such systems have been explained. There are still some problems with existing PCMs such as low thermal conductivity, leakage, thermal stability, supercooling, and flammability as illustrated in Figure 10.16 that require more research in these areas.

Economic feasibility of application of PCMs in free cooling/heating systems is also one of the important factors that must be taken into consideration for the selection of the system.

REFERENCES

1. Baquera L., Castro J.R., Pisello A.L., Cabeza L.F. (2021) Research progress and trends on the use of concrete as thermal energy storage material through bibliometric analysis, J. Energy Storage, 38, p. 102562.
2. https://en.wikipedia.org/wiki/Windcatcher
3. https://alopiping.com/mag/throwing-water-from-the-evaporative-cooler/
4. Alizadeh M., Sadrameli S.M. (2019) Indoor thermal comfort assessment using PCM based storage system integrated with ceiling fan ventilation: experimental design and response surface approach, Energy Build., 188–189, pp. 297–313.
5. Said M.A., Hassan H. (2021) Impact of energy storage of new hybrid system of PCMs combined with air-conditioner on its heating and cooling performance, J. Energy Storage, 36, p. 102400.

6. Jeong S., Heo J.E., Choi S.Y., Kim S. (2023) Heating efficiency enhanced by combination of PCMs and activated carbon for dry floor heating system, J. Energy Storage, 70, p. 108027.
7. Mirza S., Amon C.H., Chandra S. (2024) Thermal response to periodic heating of a heat sink incorporating a PCM, Int. J. Heat Mass Transf., 229, p. 125761.
8. Muzhanje A.T., Hassan H. (2023) Microcapsule geometry and nanomaterial enhancement of PCMs (sp07 & sp11) for free heating applications, Case Stud. Therm. Eng., 49, p. 103327.
9. Alizadeh M., Sadrameli S.M. (2016) Development of free cooling-based ventilation technology for buildings: thermal energy storage (TES) unit, performance, enhancement techniques and design considerations - a review, Renew. Sustain. Energy Rev., 58, pp. 619–645.
10. Ghamari M., See C., Hughes D., Mallick T., Reddy K.S., Patchigolla K., Sundaram S. (2024) Advancing sustainable building through passive cooling with PCMs, a comprehensive literature review, Energy Build., 312, p. 114164.

11 Medical Applications of Phase Change Materials

11.1 INTRODUCTION

Any application of PCMs in the medical sector or medical equipment, including relevance to the human body, is referred to as a medical application. Heat and cold therapies of human body such as human cushions for physiotherapy or rheumatism, blanket to avoid hypothermia or cooling packs used after surgery or accidents for pain relief, are also among the applications of PCM in medical sector. Wearing comfort enhancement for orthoses and prostheses where PCM is used to reduce perspiration where limb and orthosis or prosthesis are connected. Medicine and vaccines or parts of the body which are used for surgery storage and delivery in cold boxes are among the medical applications of PCMs. Various platforms can be used for the biomedical applications of PCMs such as traditional PCMs (liposome), and their nanoarchitectured composites as thermo-responsive materials, including PCMs as core, shell, and gatekeeper, highlighting the advantages and disadvantages of these architectures for bioactives delivery, imaging anatomical features and engineering tissues as shown in Figure 11.1. This chapter reviews and discussed all the above applications of PCMs in medical sector.

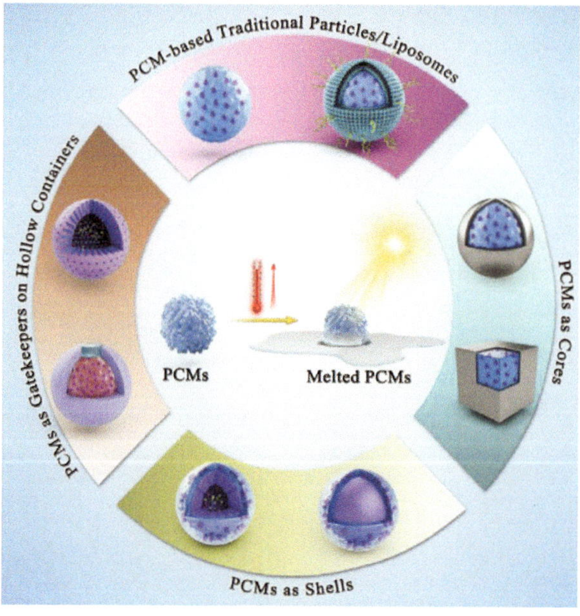

FIGURE 11.1 Schematic of conceptual design of some PCMs [1]. (Open access article.)

DOI: 10.1201/9781003518280-11

11.2 THERMOTHERAPY AND COLD COMPRESS THERAPY

Heating medical applications include heat cushions for physiotherapy or rheumatism, and blankets to prevent hypothermia while cooling packs are used after surgery and accidents or sportive stress for pain relief. Keeping the ambient temperature at a suitable level for the newborn babies who are especially susceptible to an unsuitable ambient temperature and not able to complain using PCM can maintain the temperature at acceptable level for such cases. Specific application for newborns is incubators, warmers, or coolers which are not comparable with regular applications to the human body. Significant cool down of the newborn bodies is called hypothermia which is a serious risk to their health. Utilizing PCMs can address this problem, keeping them either warm or cold based on the requirements. Newborns who suffer from a lack of oxygen during birth, a problem called birth asphyxia. In such cases lowering the body temperature, can reduce the oxygen requirement and therefore avoid brain damage. Cooling the newborn bodies for sometimes called therapeutic hypothermia treatment as illustrated in Figure 11.2 has been developed PLUSS Advanced Technologies Pvt. Ltd. and is the first of its kind in the world which is commercially available since 2014 in more than 165 hospitals around the world [2]. One of the main advantages of Neonate Cooler is its cascade system in which two PCMs with different melting points are employed to ensure stable temperature during its application.

Application of sodium acetate trihydrate as a PCM which is widely used in heat pads, has been utilized by Junghanss et al. [3] for the thermotherapy of Buruli ulcer, a necrotizing disease of skin and soft tissue caused by Mycobacterium ulcerans and can take months to even years. This bacterium grows best at 30–33°C but not above 37°C and this feature of the pathogen can be treated by thermotherapy. The PCM pad can be charged in hot water and then used on the skin. One and half years after thermotherapy all patients were relapse-free and recovered from the disease.

One of the old well-known forms of treatment to relieve pain, inflammation, blood flow, metabolic rate, intramuscular temperature, and nerve conduction velocity is cold compress therapy which can widely applied by sport physicians, physical therapists, or athletic trainers. Cold compress therapy can contract capillaries, decrease local congestion, and relieve pain by decreasing the sensitivity of nerve endings, remove heat excess, and reduce body temperature. Ice, cold water, thermoelectric cooling, cold packs of 15°C filled with PCMs, and chemical sprays can be used in this method.

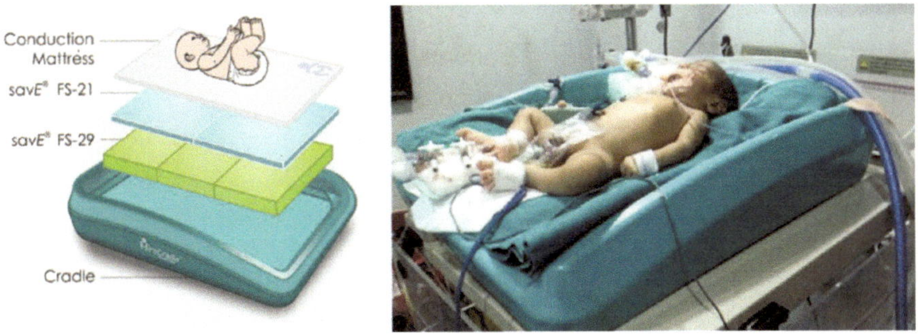

FIGURE 11.2 Application of PCM in neonatal treatment [2]. (www.miracradle.com)

The results of the research conducted by Kwiecien et al. [4] on the efficiency of prolonged PCM-supported cooling on strength loss and pain after eccentric exercise on eight adults' patients, reveals that muscle function was recovered faster and pain associated with muscle damage was reduced faster after PCM cooling.

Carbon nanotube (CNT) bundles assembled PCM composites have been used as thermal energy harvesting and thermotherapy in an advanced portable integrated functional mask for allergic patients by Chen et al. [5]. Polyethylene glycol has been used as an excellent thermal energy guest. The developed CNT sponge served as particulate matter capturer and PEG-infiltrated CNT sponge served as a superior thermal regulator as demonstrated in Figure 11.3.

A diverse range of high intensity movements performed such as accelerating, decelerating, changing directions, impacts and tackles during a typical soccer match, in which the players cover 10–13 km, with 2–3 km cover at high intensities [6]. PCMs have been utilized for the recovery of neuromuscular function following competitive soccer match-play and the effects on 11 male semi-professional soccer players have been reported by Brownstein et al. [7]. The men worn PCM cooled to 15°C (PCM_{cold}) or left 3 hours at environmental condition (PCM_{amb}) after soccer match-play have been tested pre-, and 24, 48, and 72 hours post-match, and participants completed

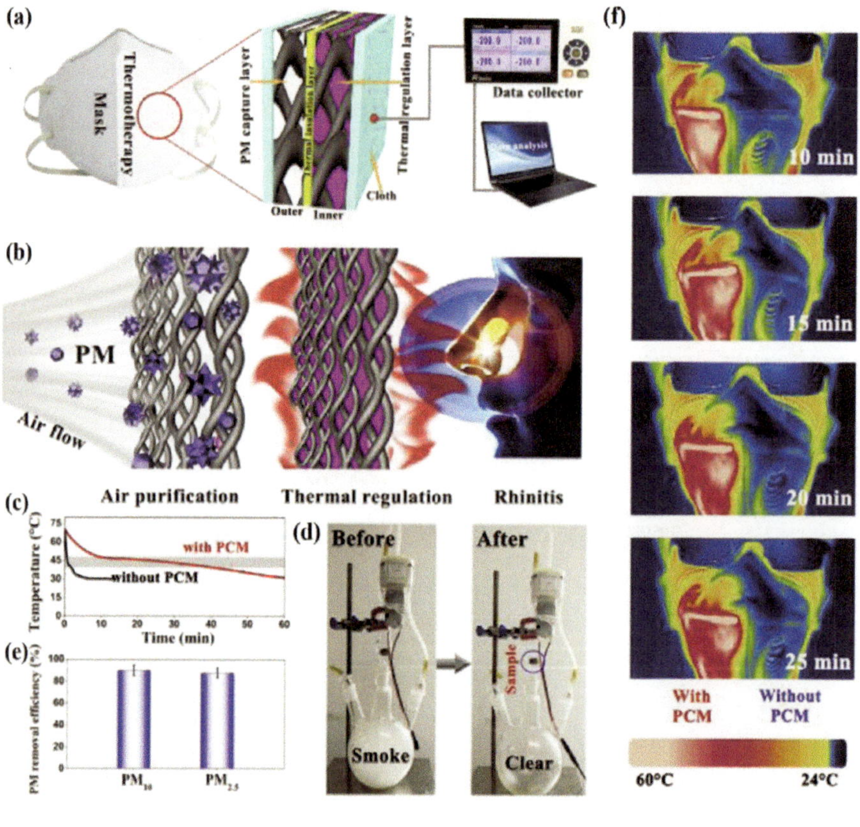

FIGURE 11.3 Schematics of steps for sponge preparation [5]. (Permission from Elsevier.com)

a battery of neuromuscular, physical, and perceptual tests. The results of the study reveal that competitive soccer match-play elicited persistent decrements in maximal voluntary contraction (MVC) force, voluntary activation (VA) as well as strength, fatigue, and muscle soreness. In this study, both MVC and VA were higher than 48 h post-match after wearing PCM_{cold} compared with PCM_{amb}. It has been concluded from the brief questionnaire that wearing PCM_{cold} and PCM_{amb} by soccer players were moderately effective in improving recovery with no difference between the two interventions.

The thermodynamics of soft tissue before, during, and after cold pack therapy using PCM has been reported by Enwemeka et al. [8]. In this experimental study, skin temperatures of 16 healthy male and female volunteers aged 25.4 ± 3.6 year have been monitored and recorded the temperature of the quadriceps muscle at 1, 2 and 3 cm depths below the skin, before, during, and after 20 min of cold pack treatment. A slight increase in temperature has been observed for all four levels during the 5 min pretreatment period however, significant temperature drops at the skin and 1 cm levels beginning from 8 min of treatment. Also, no significant temperature increase has been found at the 2.0 and 3.0 cm depths throughout treatment. However, there has been a sharp temperature increase at 1.0 cm depth just after treatment and return to the baseline levels at variable intersubject times.

They concluded that cold pack therapy can produce significant temperature drop in cutaneous and subcutaneous superficial tissues without directly changing the tissues temperature at or more than 2.0 cm below the skin. Glauber salts ($Na_2SO_4.10H_2O$) solution, including NaCl with melting point of 30°C has been used as a PCM for cooling medical applications of newborn babies by Olson et al. [9]. The PCM solution has been prepared by adding the percentage of *NaCl* between 1.5% and 7% in 0.5% increments to fix the temperature of the solution at 30°C as illustrated in Figure 11.4. Cethyl methyl cellulose (CMC) has been added to solution

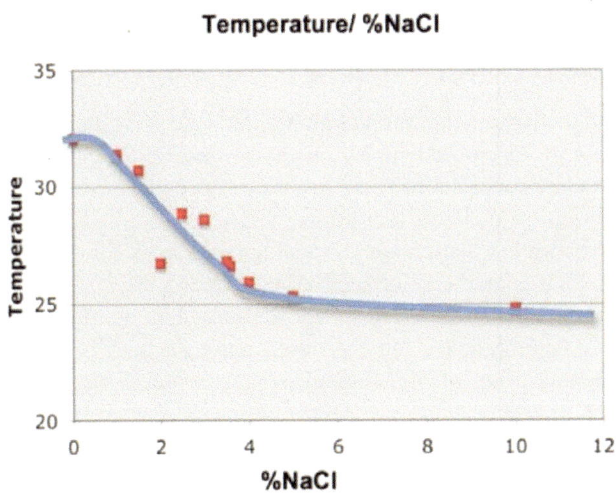

FIGURE 11.4 Effect of adding NaCl to the PCM solution [9]. (Permission from Elsevier.com)

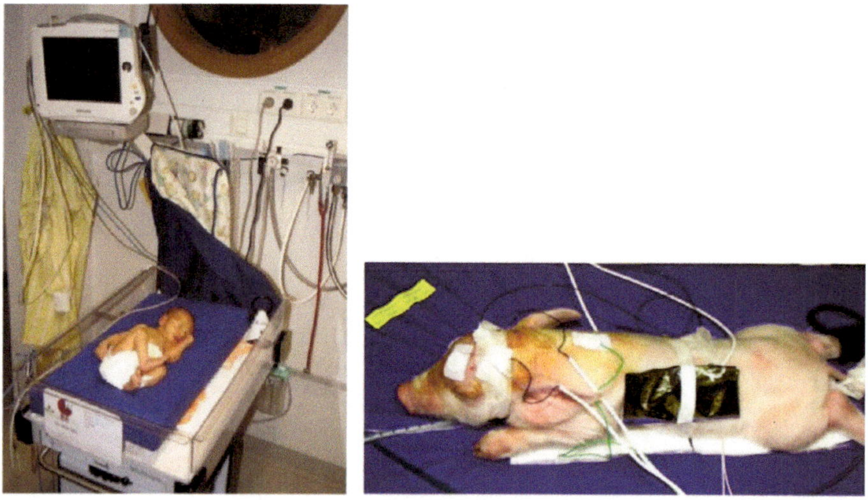

(a) (b)

FIGURE 11.5 Baby cooling by PCM mattress in a neonatal ward (a). An asphyxia piglet on PCM 32 mattress (b) [9]. (Permission from Elsevier.com)

as a gelifier agent with 1–5 g in weight for stabilizing the solution and make it more convenient to handle in the future commercial applications.

The experimental results of using PCM mattresses on babies and animals for cooling purposes as shown in Figures 11.5.a and 11.5.b, respectively, represent that the inability of a temperature undershoot occurring during the cooling time eliminating such problems originating from human error when manually control the cooling temperature.

11.3 PCM AND CANCER THERAPY

One of the severe diseases that threat human life is cancer which can be treated by effective killing of cancer cells using chemotherapeutic drugs. However, chemotherapy has some drawbacks such as low selectivity, drug-resistance, uncontrollability, and serious side effects that limits the application of chemotherapy. PCMs has an ideal substitution that create an ideal nanoplatform for cancer combination therapy due to their characteristics such as loading properties, stable and temperature-responsive phase transition capability, and excellent natural biocompatibility. Among the PCMs, fatty acids, fatty alcohols, and eutectic mixtures of fatty acids with stable melting point of higher than 37°C have received considerable interest in treatment of cancer therapy as thermo-responsive materials due to their low cost, resealable melting point, and excellent biocompatibility and biodegradability. Summary of cancer therapy especially the combination strategies utilizing PCM-based nanoplatforms is shown in Figure 11.6.

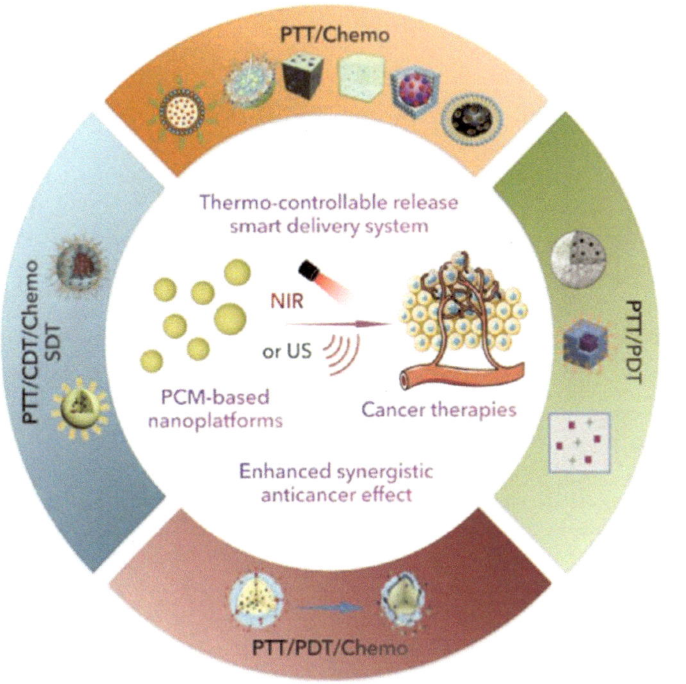

FIGURE 11.6 PCM-based nanoplatforms for cancer therapy [9]. (Permission from Elsevier.com)

11.4 PCMs WITH ORTHOSES AND PROSTHESES

Another medical application of PCMs is in the wearing comfort of orthoses and prostheses in which PCM absorbs and stores heat as it builds up in a residual limb to delay the onset of sweat before starting. The liner stabilizes the skin temperature by releasing stored heat as the body cools, keeping an amputee comfortable [10, 11].

11.5 WOUND AND INJURY HEALING BY PCMs

One of the largest organs of a human body is skin which has a great role in keeping homeostasis and protection of the internal organs from external environment. Some of the skin problems such as injuries, chronic wounds, burns, and skin wound infection which require painstakingly and often long-term treatment can be treated by PCMs.

PCMs can also be considered as additives for variety of skin treatment systems they require thermal protection such as special bandages or dressings for burn wounds or may act as bacterial agents like fatty acids and their derivatives.

Microencapsulated antimicrobial PCMs has been produced by Wang et al. [12] as a promising candidate for wound dressing components. Microcapsules of Ag/SiO_2-n-octadecane-based micro PCM has been produced by emulsion polymerization and used for this purpose as shown in Figure 11.7.

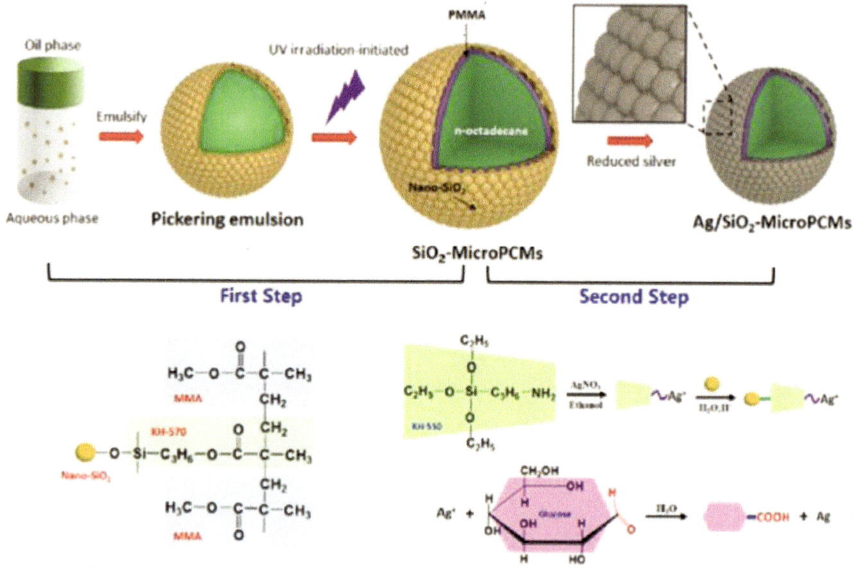

FIGURE 11.7 Schematic diagram of synthesis strategy for micro PCM [12]. (Permission from Elsevier.com)

Lin et al. [8] developed a microparticle system consisted of hydrophobic composites PCMs of 1-tetradecanol and paraffin wax with *NaHS* as an exogenous H_2S depot for patients with diabetes mellitus, prone to formation of refractory wounds. They found that for those patients the treatment not only reduced synthesis and levels of circulating H_2S but also it works as an endogenous gas transmitter with many biological processes. The *NaHs* was also act as an in-situ store for release of H_2S under physiological conditions as illustrated in Figure 11.8.

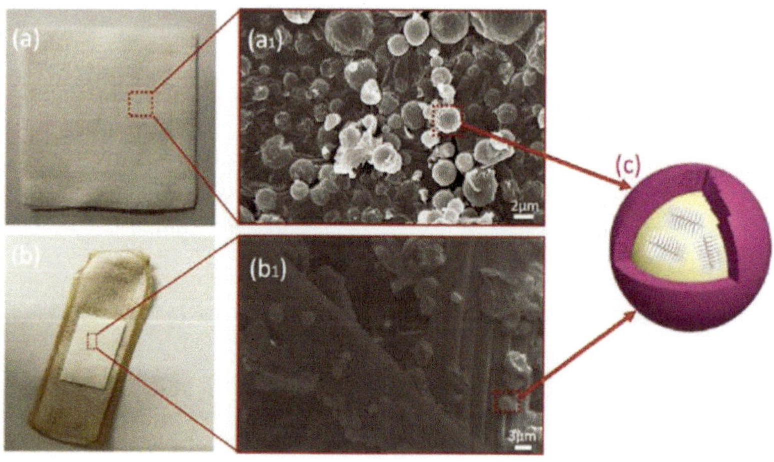

FIGURE 11.8 Medical gauze (a, b), band aid (b1), and model diagram of microcapsule (c) [13]. (Permission from Elsevier.com)

11.6 BONE CEMENT THERMAL MANAGEMENT

Fixing endoprostheses and filling of bone defects from mechanical injuries or diseases of the skeletal system can be treated by using acrylic bone cements which has been used since the 1950s.

This composite cement is composed of two component systems of a liquid (methacrylate monomer or its mixture with butyl methacrylate and a polymerization inhibitor) and a powder (microparticle of poly methyl methacrylate, PMMA, or copolymers of methyl methacrylate with other polymers such as styrene or methyl acrylate).

Due to the exothermic reaction of polymerization, a large amount of heat is released during the acrylate bone cements curing and adversely affects the surrounding tissues and may lead to inferior healing and even tissue necrosis and loosening of the cemented prosthesis [14]. This problem can be solved by using microencapsulated PCM such as paraffin wax into PMMA matrix as an energy storage system to accumulate heat releasing during polymerization of liquid monomer as proposed by De Santis et al. [15]. Negligible effect of PCM on the glass transition temperature and reduction of peak temperature of the PMMA matrix have been observed from thermal and mechanical investigations of the product. Compositions of acrylic bone cement with 10, 20, and 30 wt% of microencapsulated paraffin have been studied by Lv et al. [16] in a similar investigation. They obtained that although the modified product exhibited much higher setting time and significantly lower compression strength, but it possesses lower maximum exotherm temperature and smaller thermal necrosis zone compared with the unmodified cement. The same results have been obtained by Xia et al. [17] in modification of acrylic bone cement with PCM microcapsules containing paraffin with 10 to 30 wt% of PCMs. The results of the studies clearly proves that PCM can improve the properties of the acrylic bone cement however more investigation is required for the optimum setting time and increase of the compression strength.

11.7 SUMMARY

Application PCMs in biomedical, such as thermotherapy, cold compress therapy, drug delivery systems, anticancer therapy, wound healing, and bone cement, became an important group of materials. Constant temperature of PCMs during phase change transition makes them ideal components in thermotherapy processes, and novel developments include interconnected 3-D flexible CNT sponges filled with polyethylene glycol (PEG)-based PCMs. Mattresses filled with PCMs used as cold compress therapies can be applied to relieve pain, inflammation, blood flow, metabolic rate, intramuscular temperature, and nerve conduction velocity. Control release of drugs encapsulated in a solid matrix can also be done by PCMs melting at well-defined temperatures. Application of PCMs as photothermal therapy, photodynamic therapy, chemotherapy, and sonodynamic therapy is shown as a growing trend. Finally, transportation of medicine, vaccine, blood, or organs can also be achieved by delivery in cooling containers incorporated by PCMs.

REFERENCES

1. Chen B., Pan Y., Zhang D., Xia H., Kankala R. (2022) Phase change materials-based platforms for biomedicine, Front. Bioeng. Biotechnol. https://doi.org/10.3389/fbioe.2022.989953
2. Miracradle, MiraCradle website. www.miracradle.com
3. Junghanss T., Um Book A., Vogel M., Schuette D., Weinlaeder H., Pluschke G. (2009), PCM for thermotherapy of Buruli ulcer: a prospective observational single centre proof of principle trial. In: D.J. Diemert, editor. PLoS Neglect. Tropic. Dis., 3(2), p. e380.
4. Kwiecien S.Y., McHugh M.P., Gowatson G. (2018) The efficacy of cooling with PCM for treatment of exercise-induced muscle damage: pilot study, J. Sports Sci., 36(4), pp. 407–413.
5. Chen X., Gao H., Hai G., Jia D., Xing L., Chen S., Cheng P., Han M., Dong W., Wang G. (2020) Carbon nanotube bundles assembled flexible hierarchical framework based PCM composites for thermal energy harvesting and thermotherapy, Energy Storage Mater., 26, pp. 129–137.
6. Mohr M., Krustrup P., Bangsbo J. (2005) Fatigue in soccer: a brief review, J. Sports Sci., 23, pp. 539–599.
7. Brownstein C.G., Ansdell P., Skarabot J., McHugh M.P., Howatson G., Goodall S., Thomas K. (2019) The effects of PCM on recovery of neuromuscular function following competitive soccer match-play, Front. Phycol., 10, p. 647.
8. Enwemeka C.S., Allen C., Avila P., Bina J., Konrade J., Nunns S. (2002) Soft tissue thermodynamics before, during, and after cold pack therapy, Med. Sci. Sports Exerc., 34(1), pp. 45–50.
9. Peilichowska K., Szathowski P., Pielichowski K. (2023) PCM in biomedical applications. Chapter 10. Multifunctional phase change materials. Woodhead Publishing Series, Sawston, Cambridge, U.K. pp. 411–444.
10. Cao C., Yang N., Dai H., Huang H., Song X., Zhang Q., Dong X. (2021) Recent advances in PCM based nanoplatforms for cancer therapy, Nanoscale Adv. R. Soc. Chem., 3, p. 106.
11. Wernke M.M., Schroeder R.M., Kelly C.T., Colvin J.A.D.J.M. (2015) Smart temp prosthetic liner significantly reduces residual limit temperature and perspiration, J. Pros. Orth., 27(4), pp. 134–139.
12. Wang H., Li Y., Zhao L., Shi X., Song G., Tang G. (2018) A facile approach to synthesize microencapsulated PCMs embedded with silver nanoparticle for both thermal energy storage and antimicrobial purpose, Energy, 158, pp. 1052–1059.
13. Lin W.C., Huang C.C., Lin S.J., Li M.J., Chang Y., Lin Y.J. (2017) In situ depot comprising PCMs that can sustainably release a gas transmitter H_2S to treat diabetic wounds, Biomaterials, 145, pp. 1–8.
14. Huo X., Li W., Wang Y., Han N., Wang J., Wang N. (2018) Chitosan composite microencapsulated comb-like polymeric PCM via coacervation microencapsulation, Carbohydr. Polym., 200, pp. 602–610.
15. Santis R., Ambrogi V., Carfagna C., Ambrosio L., Nicolais L., Effect of microencapsulated PCMs on the thermo-mechanical properties of poly(methyl-methacrylate) based biomaterials, J. Mater. Sci. Mater. Med., 17, pp. 1219–1226.
16. Lv Y., Li A., Zhou F., Pan X., Liang F., Qu X. et al. (2015), A novel composite PMMA-based bone cement with reduced potential for thermal necrosis, ACS Appl. Mater. Interfaces, 7(21), pp. 11280–11285.
17. Xia X., Shi R., Huang J., Li Y., Zuo Y., Li J. (2020) Development of a phase change microcapsule to reduce the setting temperature of PMMA bone cement, J. App., Biomater. Funct. Mater., 18, p. 2280800020940279.
18. Dahn J., Ehrlich G., Reddy T. (2011) Linden's handbook of batteries. New York: McGraw Hill.

12 High-Temperature Application of Phase Change Materials

12.1 INTRODUCTION

Sustainable improvements of TES systems can be attained by efficient and resealable cost preparation of high-temperature phase change materials. As discussed in the previous chapters, an energy storage medium can be categorized into three different groups of sensible heat storage, latent heat storage, and thermochemical energy storage. The latent heat storage systems utilizing solid-liquid PCMs due to their merits of isothermal phase transition processes, acceptable energy density, and rational capital investment have tremendous applications for temperature control, peak shifting, and energy savings of thermal systems. One of the main characteristics of PCMs for the specific application is their phase transition temperature during melting/solidification processes. Ranges of melting point and latent heat of different PCMs, including organics, inorganics, and eutectics, are demonstrated in Figure 12.1 which shows that inorganic salts such as nitrates, hydroxides, chlorides, carbonates, and fluorides and some metals like Fe, Cr, Si, Ni, Al, Mn,

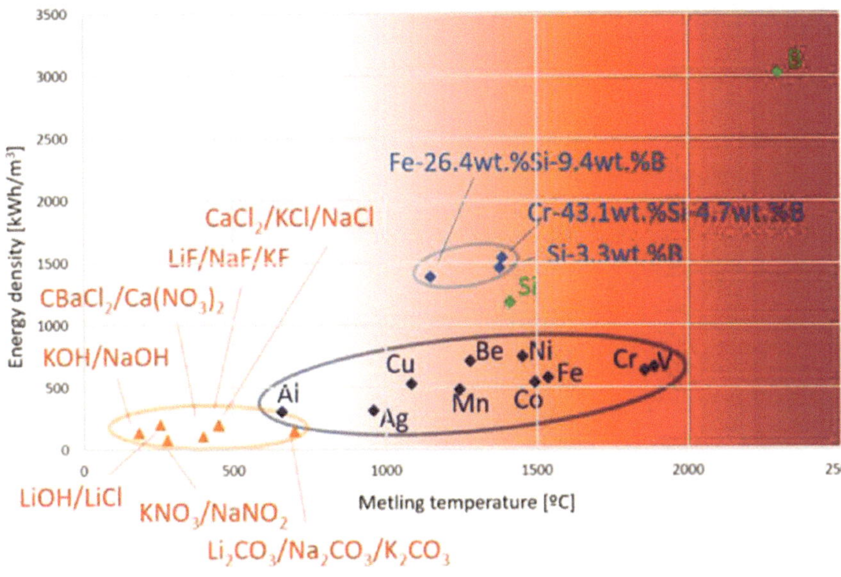

FIGURE 12.1 Energy density vs. melting temperature of different molten salts and metals as PCMs [1]. (Permission from Elsevier.com.)

DOI: 10.1201/9781003518280-12

Ag, and Cu are the most promising PCMs for medium- to high-temperature applications (200–2,000°C) due to their high melting temperature and heat of fusion.

Different applications of PCMs at medium- to high-temperature melting points such as inorganic and metallic-based PCMs will be discussed in this chapter in the following subsections, including the selection principles and criteria for the PCMs, skeleton-supporting materials for encapsulation, and the materials used for thermal conductivity enhancement which are explained in detail.

12.2 APPLICATION OF INORGANIC SALTS

One of the most promising PCM types for high-temperature applications are inorganic salts such as nitrates, carbonates, chlorides, sulphates, fluorides, and their eutectics due to their high thermal properties like specific heat, thermal stability, heat transfer coefficient, and low saturated vapour pressure with low viscosity [2]. They have an excellent heat storage capacity due to their high latent heat through phase transition which is added to their great sensible heat in a large temperature difference. A mixed composite of binary or ternary mixture of inorganic salts has the following advantages in comparison to the single-salt PCMs:

- Desirable melting point achievement
- Higher thermal energy storage capacity
- Reduction in overall costs of the materials

Some of the common high-temperature PCM with their melting points, latent heats, and prices are listed in Table 12.1.

Microcapsules characterization of MgF_2 and LiF as fluorides PCMs with a single shell with volume expansion flexibility as a high temperature has been produced by Jiang et al. [5]. The melting point temperatures and enthalpies of two PCMs have been measured as 1262°C, 718.4 J/g for magnesium fluoride @carbon and 840°C, and 458.2 J/g for lithium

TABLE 12.1
List of Typical Inorganic High-Temperature Salts with Their Prices [3, 4]

Salt roots	Lithium	Sodium	Potassium	Rubidium	Cesium	Magnesium	Calcium	Strontium
Melting temperature, °C								
Nitrate	250	310	337	312	409	426	560	645
Bromide	550	742	734	692	638	711	742	657
Chloride	610	801	771	723	645	714	772	875
Carbonate	732	858	900	837	793	990	1330	1490
Sulphate	858	884	1069	1070	1015	1137	1460	1605
Fluoride	849	996	858	795	703	1263	1418	1477
Latent heat, kJ/kg								
Nitrate	373	177	88	31	71	–	145	231
Bromide	203	255	215	141	111	214	145	41
Chloride	416	482	353	197	121	454	253	103
Carbonate	509	165	202	–	–	698	–	–
Sulphate	84	165	212	145	101	122	203	196
Fluoride	1041	794	507	248	143	938	381	226
Prices, $/ton								
Nitrate	8350–9850	377	508	1000,000	10,000	300	240–255	1300–1600
Bromide	50,000	2800	3000	1600,000	100,000	2200	1500	3000
Chloride	8250	110	580	1580,000	69,800	290	405	800
Carbonate	20,000	405	870	1600,000	10,000	850	150	1500
Sulphate	11,000	145	1200	1650,000	10,000	100	250	1200
Fluoride	200,000	880	1378	100,000	120,000	50,000	500	10,000

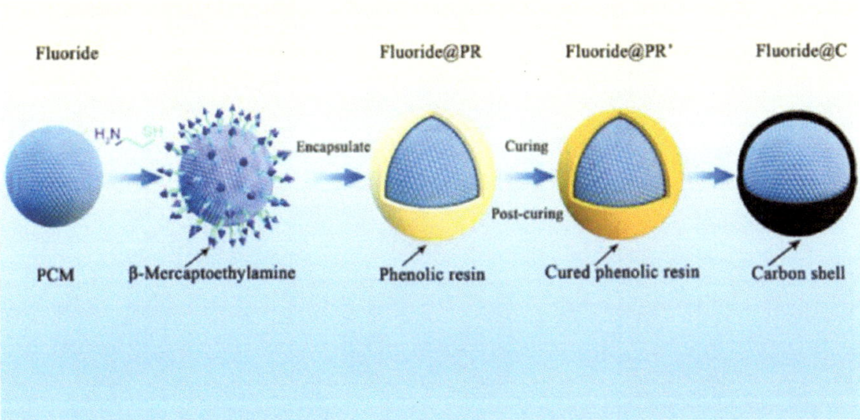

FIGURE 12.2 Microencapsulation of fluoride particles [5]. (Permission from Elsevier.com.)

fluoride @carbon, respectively. Fluoride PCMs particles have been encapsulated on to the surface firstly by a phenolic resin shell using water-induced phase separation process and then crosslinked and carbonized to obtain the fluoride @carbon microcapsule. The dispersion was heated to 95°C for a day to crosslink phenolic resin in situ (fluoride@PR) and had more heat to obtain post-cured microcapsules (fluoride@PR') as illustrated in Figure 12.2.

The prepared microcapsules have been characterized for the enthalpy evaluation, structure and composition using DSC, SEM, FT-IR, EDS, TGA, and XRD analysis. FT-IR and XRD analysis results for different microcapsules of LiF are demonstrated in Figure 12.3. Both microcapsules of LiF show similar structure and content except the surface appearance. The flouride@ Carbon microcapsules remained powder after being heated above the melting point many times while the fluoride powders of LiF and MgF_2 both agglomerated after melting. Also, the hydroxymethyl $C–O$ bond absorption peak in FT-IR results for fluoride@PR' microcapsules at 1,004 cm^{-1}

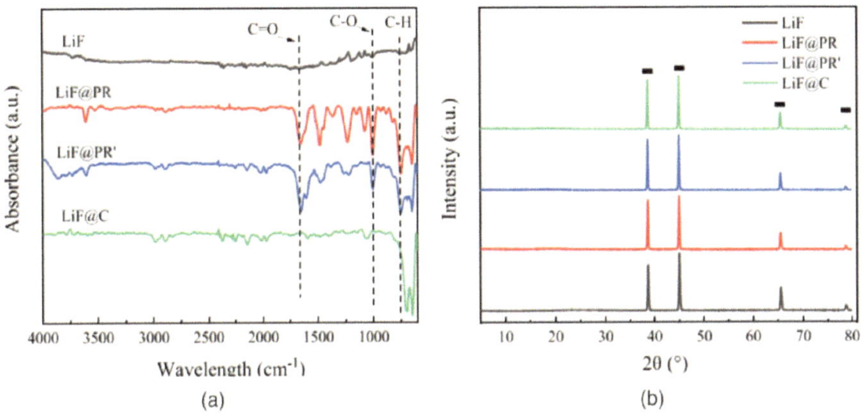

FIGURE 12.3 FT-IR and XRD results of different LiF microcapsules [5]. (Permission from Elsevier.com.)

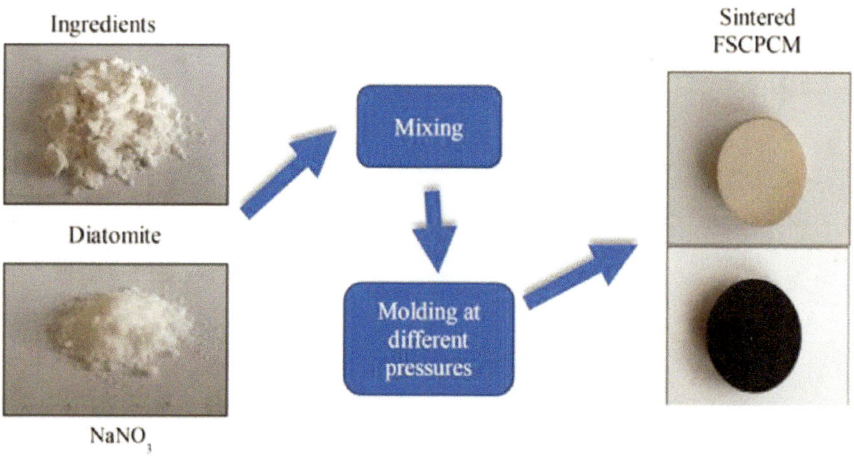

FIGURE 12.4 Preparation steps for shape-stabilized nanoscale PCM [6]. (Permission from Elsevier.com.)

reduced in comparison to fluoride@PR and the identical crystal structure of both fluoride particles of microcapsules have been confirmed by XRD results which indicates that fluoride PCM remained stable during the synthesis process.

Preparation and characterization of high-temperature shape-stabilized sodium nitrate/diatomite PCMs with expanded graphite (EG) nanoparticles as thermal conductivity enhancer for solar energy storage applications have been studied by Soleimanpour et al. [6].

The composite PCMs have been produced by two-step mixing and sintering methods as illustrated in Figure 12.4 for the application of concentrated solar power and solar collectors. The shape-stabilized composite PCMs parameters have been optimized using the design of experiment (DOE) with two responses, i.e., phase change temperature and phase change time stability with and without nanoparticles. The thermal properties of the nanoparticles have been measured by a DSC and an IR camera.

A comparison between the DSC results of shape-stabilized PCM of $NaNO_3$/Diatomite/Nano EG and nano-diamond for two runs is depicted in Figure 12.5. As shown in the figure, the peak temperature for the composite PCM is 306.64°C with the onset temperature of 301.6°C which is acceptable with the result of IR camera

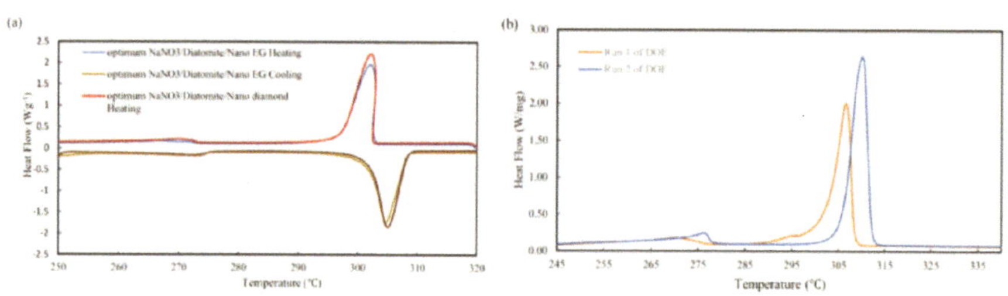

FIGURE 12.5 DSC curves of different composite PCMs [6]. (Permission from Elsevier.com.)

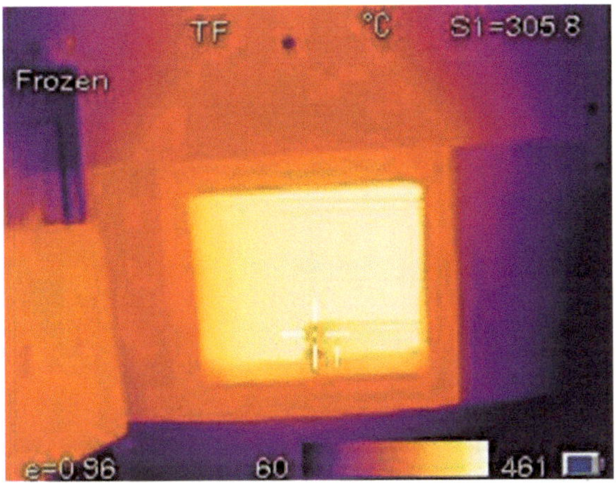

FIGURE 12.6 A photo of PCM by IR camera [6]. (Permission from Elsevier.com.)

(303.3°C) presented in Figure 12.6. There are two endothermic peaks in the composite at 271.87°C which is corresponded to a solid-solid crystal phase change in $NaNO_3$ and 304.64°C which is due to the melting process although both samples have similar exothermic and endothermic peaks and melting points.

Thermal properties of PCMs can be studied by DSC and IR camera although the second option is more economical and quicker. The result of IR camera for $NaNO_3$ composite is shown in Figure 12.6 in which the range of different temperatures is represented in different colours with a provided quid colour bar in the right and bottom corners of the image.

They finally concluded that 62.91 wt% of $NaNO_3$ has an optimum absorption ratio and an exceptional physical property. The nano-EG and nano-diamond added to the composite PCM for thermal property enhancement and the final corresponding melting points of the composite PCMs were 304.64°C and 304.98°C, respectively. Therefore, the optimum composite PCMs are ideal as the TES medium in concentrated solar powers and solar collectors.

Macroencapsulated carbonate eutectic salt PCMs containing $NaCO_3$ (57%) and Li_2CO_3 (43%) with stable inorganic double shells strategy to coat the salt microcapsule with inner EG layer and outer ceramic kaolin layer have been fabricated by Zeng et al. [7] for high-temperature applications as depicted in Figure 12.7.

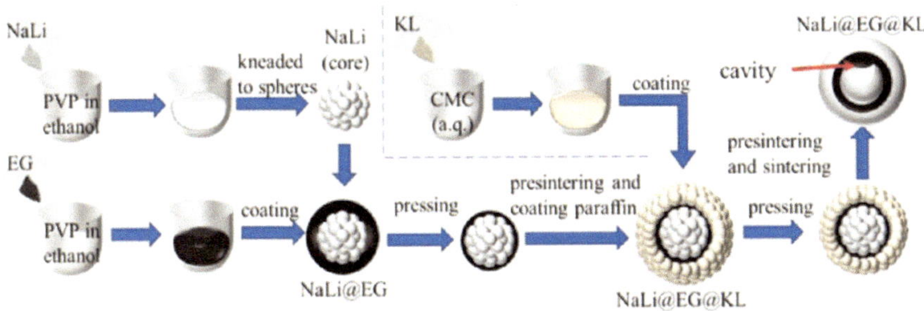

FIGURE 12.7 Preparation steps of eutectic PCM [7]. (Permission from Elsevier.com.)

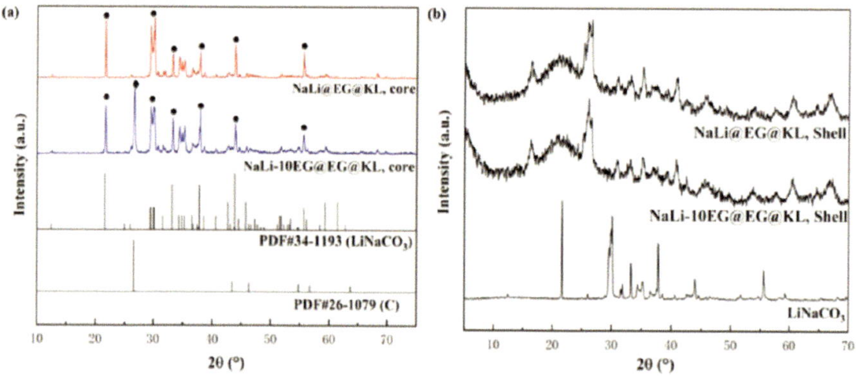

FIGURE 12.8 XRD analysis results for core and shell materials [7]. (Permission from Elsevier.com.)

The core and shell materials have been characterized by XRD, SEM, for the composition and microstructure analysis, DSC for the melting points and latent heat capacity evaluation and a thermal conductivity tester. The XRD results for the cores and shells of *NaLi@EG@KL* and *NaLi@10EG@KL* after cycling test are shown in Figure 12.8, which proves that both capsules are crystal phase of eutectic salts like the raw materials. Both materials show the same pattern for *KL* shell material and the graphite peak was also detected for composite materials due to the composition of EG in the core materials. Leakage prevention can also be proved from the typical main peak results as no indication of eutectic salts was detected in the shells.

DSC analysis curves and thermophysical properties of the salt core PCMs are presented in Figure 12.9 and Table 12.1 for the cores of *NaLi@EG@KL* and *NaLi-10EG@EG@KL* capsules obtained before and after cycling tests which shows similar pattern before and after cycling tests except slight change that may be due the experimental error. The results in Figure 12.9 and Table 12.1 show that the heat

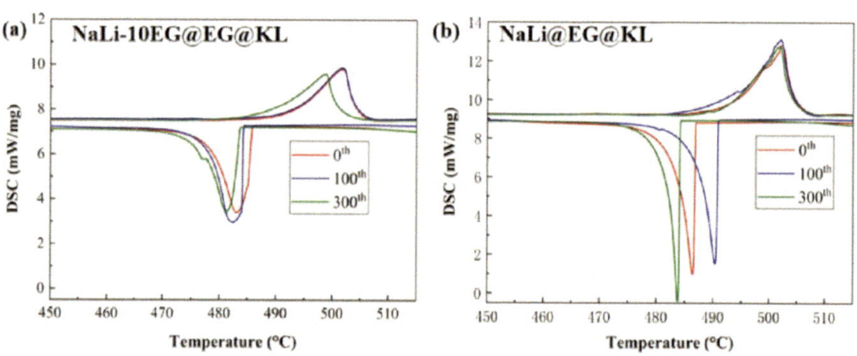

FIGURE 12.9 Melting-solidification curves for two composite PCMs [7]. (Permission from Elsevier.com.)

TABLE 12.2
Summary of Thermophysical Properties of PCM Samples before and after Cycling [7]

Samples	Melting				Freezing			
	T_P (°C)	T_O (°C)	T_E (°C)	Enthalpy(J/g)	T_P (°C)	T_O (°C)	T_E (°C)	Enthalpy(J/g)
NaLi-10EG 0th cycle	501.8	492.7	504.0	284.3	482.4	478.0	484.2	275.4
NaLi-10EG 100th cycle	497.5	488.2	500.2	273.9	483.1	485.7	478	270.6
NaLi-10EG 300th cycle	498.7	489.9	501.0	272.6	481.2	476.9	483.7	269.7
NaLi 0th cycle	502.0	498.6	504.2	313.0	490.3	487.3	490.8	303.8
NaLi 100th cycle	502.2	498.5	504.6	309.5	486.3	483.5	486.7	299.9
NaLi 300th cycle	501.9	495.8	504.0	303.7	483.7	481.9	487.5	303.1

enthalpies of the core *NaLi-10EG@EG@KL* are smaller than other sample due to the addition of EG in this sample. Since the eutectic *NaLi* salt has a phase transition temperature of around 501°C then the adopted temperature range is 450–550°C and 400–600°C for both samples.

The results listed in Table 12.2 also show that the $Na_2CO_3/LiCO_3$ eutectic salt PCM with ($Na_2CO_3/LiCO_3$ = 57/43 wt%) had a high latent heat capacity of 313 J/g and melting point of 498°C. The heat density of the carbonate macrocapsules can also be enhanced by increasing the core to shell ratio, i.e., increasing the capsule size and at the same time reducing the thickness of the shell. Further research is required for the optimum values of capsule size formulation.

Packed bed filled with spherical capsules of composite nanoparticles of $MgCl_2$-KCl-$NaCl/Al_2O_3$ salts has been used as high-temperature energy storage system by Han et al. [8]. They doped the nanocomposite salt with Al_2O_3 nanoparticles in $MgCl_2$-KCl-$NaCl$ with a melting point of 398.9°C.

For the preparation of capsules, the moisture has been removed from the salts and the salts have been mixed with 51:22:27 molar ratio to prepare the base. In the second step the Al_2O_3 nanoparticles and gum Arabic as a dispersant have been added to the deionized water with ultrasonic vibration for 2 hours for uniformity of the mixture. Finally, the suspension has been dried in vacuum at 110°C for six hours as shown in Figure 12.10. Encapsulation of PCM has been

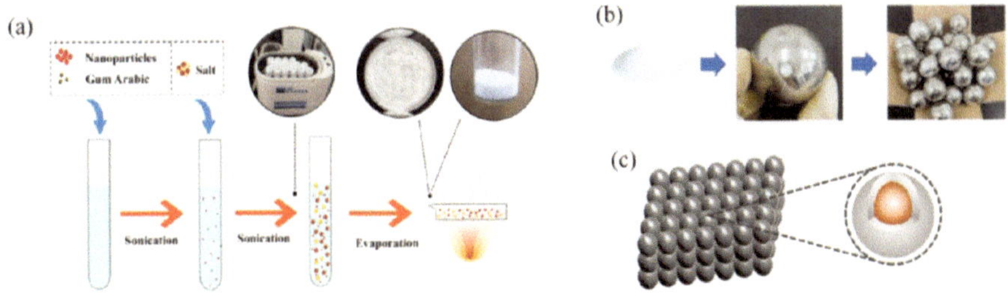

FIGURE 12.10 Preparation of nanoparticle composite PCM for high-temperature energy storage system [8]. (Permission from Elsevier.com.)

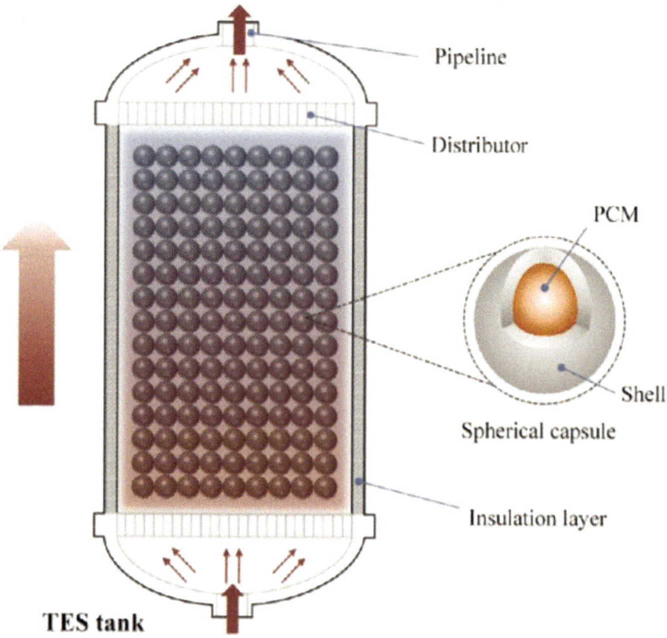

FIGURE 12.11 Schematic diagram of packed bed thermal energy storage system [8]. (Permission from Elsevier.com.)

performed using 316 stainless steel hollow ball capsules with 38 mm diameter and 1.5 mm thickness.

The balls have been used in a cylindrical packed bed thermal energy storage system insulated from outside environment using aluminium silicate fibres to facilitate the experimental analysis as illustrated in Figure 12.11.

The designed and constructed system had an energy storage density of 1,290 MJ/m^3 at thermal efficiency of 86.95%. They also found an increase of 49.55% in average heat storage rate and the heat storage time has been decreased by 29.67% using an increase in blower speed from 800 to 1200 r.min^{-1}.

A composite high-temperature PCM in a shape-stable form has been developed by Jiang et al. [9] using coal fly ash (CFA) as a skeleton material and binary sulphates of Na_2SO_4-K_2SO_4 salt. The effects of CFA on the composite structure are to increase the porous structure of composite PCM as depicted in Figure 12.12. At relatively high content ratio of salts in the composite, the fly ash possesses a porous structure with high specific are and large pore volume while as at low ratio of the salts the skeleton has a poor structure leading to PCM material leakage during the melting process.

The results clearly show that the energy density and thermal conductivity of composite PCM increases with PCM content and excellent chemical compatibility and stability have been obtained from composite of 60 wt% of Na_2SO_4-K_2SO_4 and 40 wt% of coal fly ash. The measured values of composite energy density and thermal

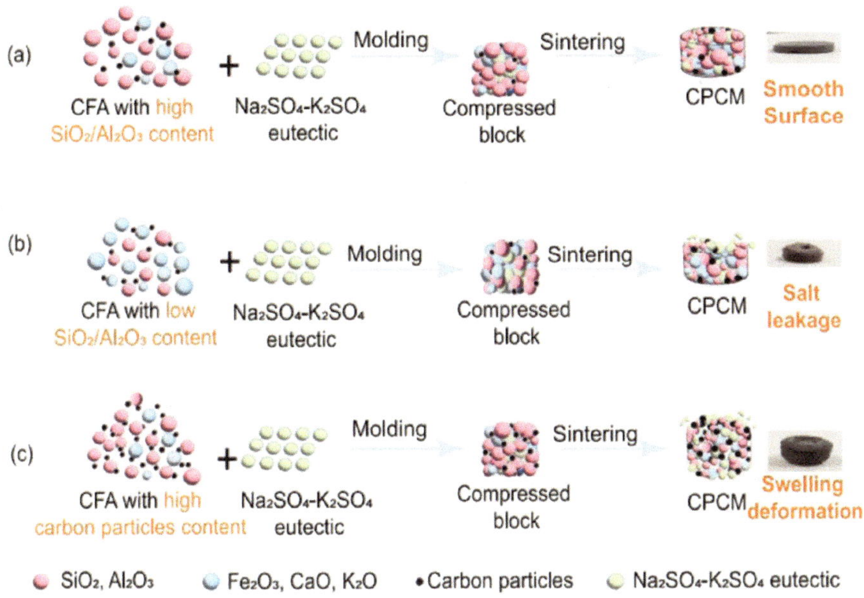

FIGURE 12.12 Influence of coal fly ash on the structure of composite PCM [9]. (Permission from Elsevier.com.)

conductivity have been reported to be 60.43 J/g and 0.61 W/mK, respectively. They finally declared that coal fly ash as a byproduct of coal fire industry can be used a promising skeleton material for the fabrication of high-temperature composite PCM.

An extensive review on the application of K_2CO_3-$LiCO_3$ molten carbonate mixtures and their nano-fluids for the application of thermal energy storage has been presented by Navarrete et al. [4].

12.3 HIGH-TEMPERATURE METALLIC PCMs

Metals can be utilized as solid-liquid PCMs in high-temperature applications such as solar reactors and solar collectors, due to their high-temperature phase transitions, high thermal conductivity relative to common inorganic and organic thermal storage materials, large phase change heat density, and broad applications prospects in various high-temperature industrial waste heat recovery systems however the major challenges with the application of metals is the leakage and corrosion molten metals at high temperature in the process which makes their applications as matrix material to be limited. Therefore, an appropriate encapsulation technique using a suitable material is required to overcome this problem.

One of the byproducts of copper extraction processes is copper slag, impurities that become slag and floats on the molten metal can be cooled down by water to produce angular granules of copper slag. Copper slag has been utilized by Ye et al. [10] to prepare core-shell high-temperature phase change composite materials using solid aluminium ball as a core. A double shell PCM (DSPCM) has been fabricated and

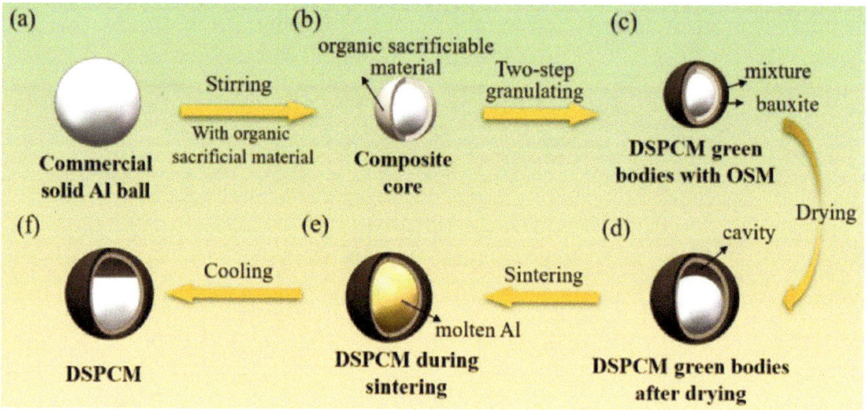

FIGURE 12.13 Preparation steps for high-temperature PCM capsules [10]. (Permission from Elsevier.com.)

characterized using copper slag with phase compositions of fayalite (Fe_2SiO_4) and magnetite (Fe_3O_4) and bauxite as raw materials and alumina-mullite ceramic was coated inside the layer shell to prevent the corrosion of molten metal using bauxite powders as depicted in Figure 12.13. Different copper slag contents in the mixed ceramic powders are numbered DFT0 to DFT3 based on the composition of copper slag in the composites.

Their results show that the prepared DSPCMs after sintering at 1,100°C did not change much when compared to the green bodies with the cavity and the capsules with perfect shell had no crack, breakage or leakage as demonstrated in Figure 12.14.

FIGURE 12.14 Digital images of DSPCM after sintering at 1,100°C [10]. (Permission from Elsevier.com.)

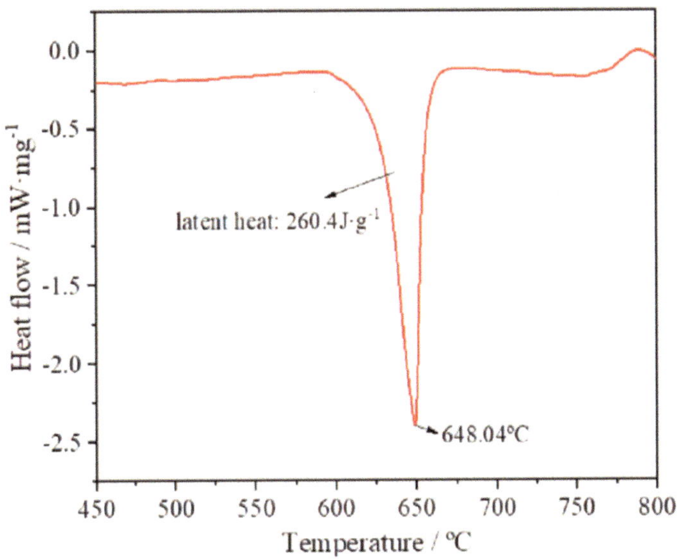

FIGURE 12.15 Latent heat and melting point of the aluminium core [10]. (Permission from Elsevier.com.)

As can be observed from the images, an increase of copper slag in the shell material (DFT0–DFT3) changed the surface colour of the capsules from light to dark brown.

One of the most important characteristics of PCMs is heat storage capacity or latent heat. Melting point and latent heat of DSPCMs which have been measured by DSC are illustrated in Figure 12.15. For the prepared capsules the heat storage capacity included two parts: under 648°C melting point and higher than the melting point. Below 648°C melting points, the heat storage capacity includes the sensible heat of the shell and core as solid, while for higher than melting point ($T > 648°C$), it consisted of sensible heat of the shell and core as liquid and latent heat of the PCM in the core. The DSC results in the figure show the melting point and latent heat of the core as 648°C and 260.4 J/g, respectively.

Application of shape-stabilized Al-25Si alloy powder with size 30–45 μm as a PCM with Al_2O_3 nanoparticles (10 nm) as matrix material to prepare a phase change composite as a high-temperature PCM has been studied by Shi et al. [11]. The composite has been prepared from the mass proportion of Al-25Si with 0–50 wt% in the combination with Al_2O_3 with the steps shown in Figure 12.16. *Al-25Si* and Al_2O_3 as raw materials have been divided into boiling treatment and non-boiling treatment such that the sample 5MPa-BM50 shows it underwent boiling treatment with Al-25Si weight ratio of 50% under a pressure of 5 MPa.

The properties of samples have been determined using XRD, SEM, and DSC for the melting point and latent heat evaluation.

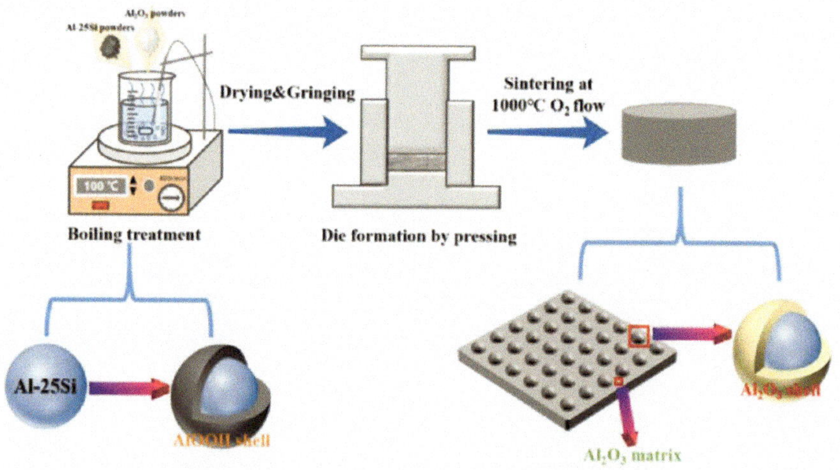

FIGURE 12.16 Preparation steps for shape-stabilized PCM [11]. (Permission from Elsevier.com.)

The results of melting points and heating values for the prepared samples are illustrated in Figure 12.17 for 5MPa-BM100 and 10MPa-BM100 composites PCMs. As can been observed from the profiles in the figure the melting and freezing peak temperatures of 5 MPa are 579.6°C and 562.8°C and for 10 MPa are 580.1°C and 563.3°C, respectively. The corresponding heat values are 220 J/g and 190 J/g for

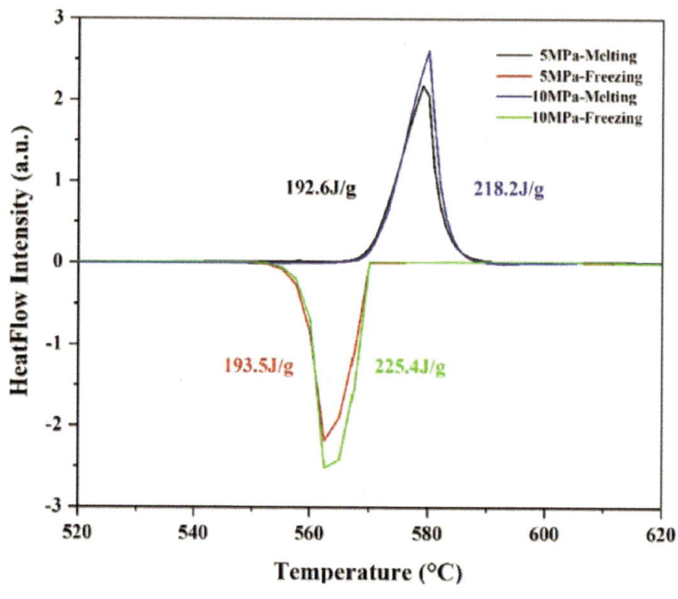

FIGURE 12.17 DSC results for two samples at 5 and 10 MPa [11]. (Permission from Elsevier.com.)

10 MPa and 5 MPa samples, respectively. It can also be seen for the DSC curves that latent heat for height pressure of 10 MPa in compacted tablets is more than 5 MPa since at higher pressure results better sealing of Al-Si that can avoid oxidation during heat treatment however at very high pressure of 15 MPa the compacted sample would leak due to the volume expansion during melting of the alloy.

Macroencapsulation of Cu-Si PCMs by in situ alloying formation for high-temperature thermal energy storage application of 800°C has been conducted by Ge et al. [12]. The fabricated composite balls of $Cu\text{-}Si@Al_2O_3$ macrocapsules have been processed by cladding mixture of Cu and Si powders in spherical shapes with Al_2O_3 shells after two-step sintering treatment as depicted in Figure 12.18.

Properties and thermal characteristics of macrocapsules such as morphology (integrity, size, and colour) have been analysed by a digital camera, crystal phase composition of cores after pre-sintering and sintering by XRD, microscopic morphology and elemental characterization by SEM/EDS and melting point and latent heat evaluations by a DSC. The DSC results of heat density for Cu-10Si, Cu-20Si, and Cu-30Si for the as-prepared and cycled capsules are demonstrated in Figure 12.19 which shows three endothermic and exothermic peaks for heating and cooling periods of Cu-10Si PCM samples (Figure 12.19a) with melting temperatures of 730.4°C, 780.3°C, and 827.8°C corresponding to latent heat of 30.3 J/g, 29.6 J/g, and 89.6 J/g, respectively. For *Cu-20Si* and *Cu-30Si* samples similar profiles are shown in the figure for the melting and solidification of PCMs with

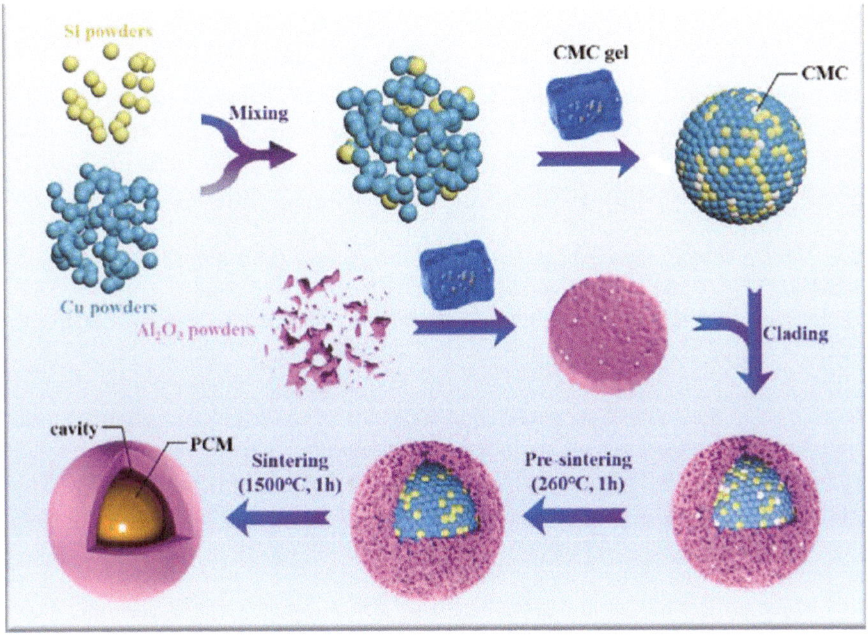

FIGURE 12.18 Preparation steps for $Cu\text{-}Si@Al_2O_3$ macrocapsules [12]. (Permission from Elsevier.com.)

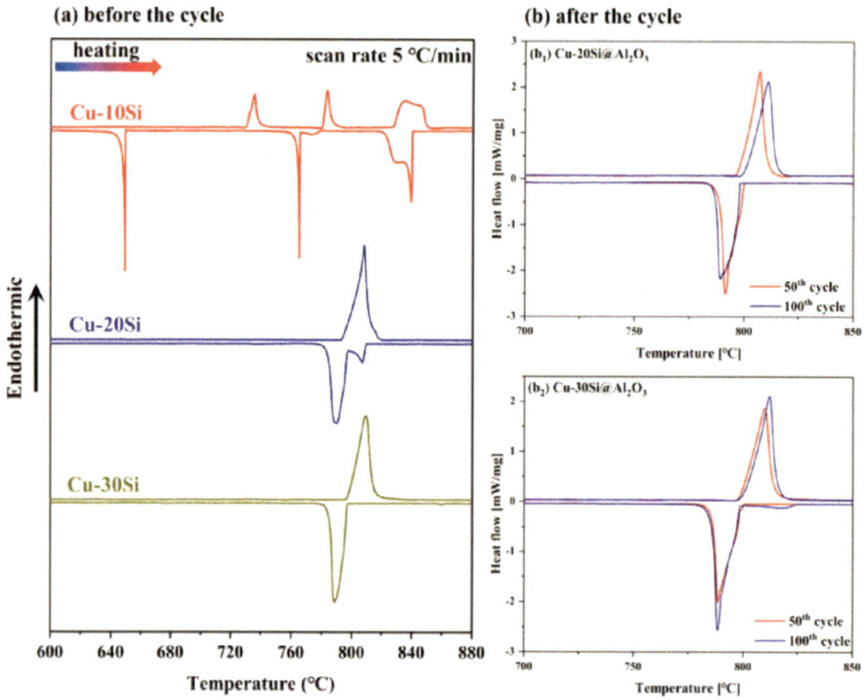

FIGURE 12.19 DSC results of PCM with different mass ratios (a) before and (b) after cycle [12]. (Permission from Elsevier.com.)

both melting temperatures of 801°C due to the identical liquidus temperature for hypereutectic *Cu-Si* alloys and with melting enthalpy of 195 J/g. There is a slightly fluctuations for the melting temperatures of PCMs after 50th and 100th cycles compared to the as-prepared samples. It is also observed from the DSC curves that *Cu-30Si* cycled samples had a slightly higher latent heat (197 J/g) compared with *Cu-20Si* cycled sample (185 J/g) which could be due to the primary Si phase in the samples.

The conclusion from their study shows that the as-prepared *Cu-Si* macrocapsules of PCM can be utilized at high-temperature thermal energy systems.

Metallic PCMs used as a medium for thermal energy storage systems do not require strict mechanical properties and could be made of recycled sources unsuitable for structural significant applications. Aluminium alloys are among the metals with highest recovery rate according to the International Energy Agency (IEA) [13]. The raw material for aluminium is bauxite ore, which is leached to produce alumina and then is converted into pure aluminium in a primary intensive energy consumed aluminium melting furnace as illustrated in Figure 12.20.

Recycling of aluminium scrap into PCMs for high-temperature thermal energy storage applications has been studied by Villada et al. [15]. Three different ingots samples have been synthesized in house based on Al-12.6 wt% Si alloy with different

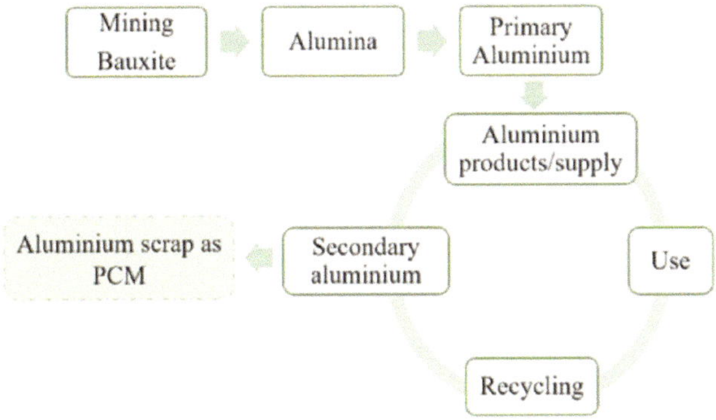

FIGURE 12.20 Life cycle of aluminium production [14]. (Permission from Elsevier.com.)

impurities by casting each ingot from high purity Al-12.6 wt% Si pieces with dimensions of 9.5 mm and purity of 99.9% as depicted in Figure 12.21.

Characterization of the samples have been performed by DSC for the phase transition temperature, enthalpy of fusion and heat capacity, light flash apparatus for thermal diffusivity in solid state, SEM for microstructural analysis and energy dispersive X-ray (EDX), and elemental distribution by Electron Backscatter Diffraction (EBSD) systems, respectively. The results of DSC analysis for three to four samples

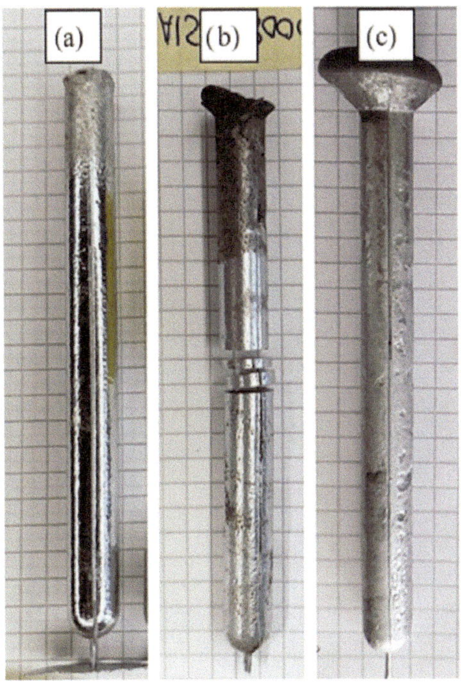

FIGURE 12.21 Al-Si alloy rods after casting [15]. (Permission from Elsevier.com.)

TABLE 12.3
Melting Temperatures and Latent Heats of the Samples [12]

Sample (wt%)	Extrapolated Onset Temp. (°C)	Peak Temp. (°C)	Endset Temp. (°C)	Latent Heat (kJ/kg)
HP-AlSi	579.6 ± 1.0	615.5 ± 2.9	633.3 ± 4.9	487.2 ± 1.5
0.40 Fe-AlSi	572.1 ± 3.5	612.4 ± 2.5	629.2 ± 1.6	481.3 ± 12.7
R3000-AlSi (0.55 Fe-1.15 Mg-1.15 Mn)	556.8 ± 0.3	603.4 ± 1.4	622.1 ± 3.6	496.4 ± 9.2
R2000-AlSi (3.10 Cu-0.45 Fe-1.80 Mg-0.65 Mn-0.70 Zn)	516.2 ± 1.4	590.2 ± 2.9	611.5 ± 4.9	478.9 ± 4.3

of different alloy compositions are listed, and Table 12.3 and the patterns are demonstrated in Figure 12.22. They have also been compared with the commercially prepared high purity Al-12.6 wt% Si (HP-AlSi) to study the effect of the impurities on the thermophysical properties. The latent heat of four alloys has been calculated by integration of area under DSC melting peaks. The heating values for HP-AlSi are found to be 487.2 ± 1.5 J/g (Figure 12.22.a) compared to 504 J/g predicted by the company (ThermoCalc). The melting point for the sample with 0.4% Fe, is 572.1 ± 3.5°C and the enthalpy of fusion is 481.3 ± 12.7 J/g as observed from Figure 12.22.b.

For the R3000-AlSi sample with three additional alloying elements of Fe, Mg, and Mn a two-phase eutectic transition with a reduced melting point of 20°C compared the binary HP-AlSi is shown in Figure 12.22.c. The melting point and latent heat for the last sample, R2000-AlSi, with 5 alloying elements of Cu, Fe, Mg, Mn,

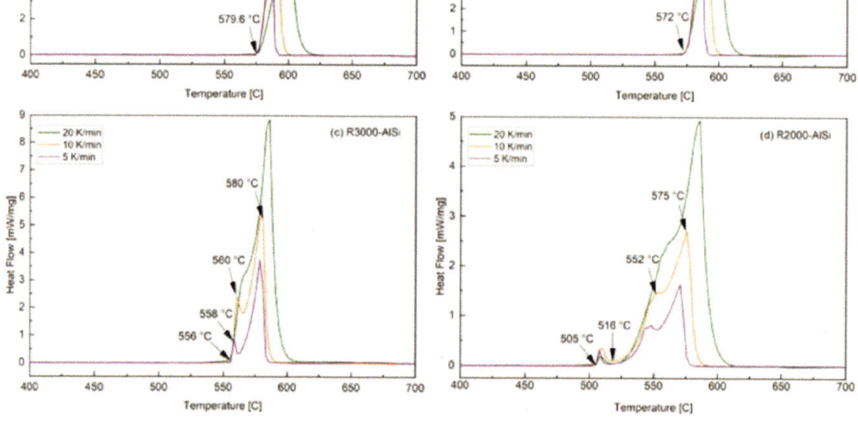

FIGURE 12.22 (a–d) DSC patterns of different PCMs samples [15]. (Permission from Elsevier.com.)

and Zn, shown in Figure 12.22.d are 516.2 ± 1.4 and 478.9 ± 4.3 J/g, respectively. It is concluded from their study that thermal energy storage systems using non-purified materials or additional alloying element has an opportunity to the market for scrap that was previously uneconomical recycle system while significantly decrease the cost of the thermal storage by approximately 12.4 USD/kWh_{latent} when using high purity alloys.

Macroencapsulation of Al-Si alloy has been performed using Al_2O_3 powder shell in direct powder formation form route and in situ powder alloying formation route by Guo et al. [15] for high-temperature energy storage system. Aluminium silicon alloy powder has been prepared using 12% and 25% silicon content as a core PCM capsules coated with Al_2O_3 as a shell and carboxymethyl cellulose sodium with average molecular weight of 250,000 used as an adhesive for the capsule preparation. To optimize the cavity size of the capsules, isostatic pressing at 50 MPa and 200 MPa pressures have been applied to the Si-Al cores. The results indicate that the capsules made from *Al-12Si* with 200 MPa pressure and melting temperature and heating density of 500–700°C and 479 J/g, respectively, exhibit superior thermal storage performance after enduring 1,300 cycles.

Macroencapsulation of Cu-based PCM as a core and alumina-based shell for high-temperature applications exceeding 1,000°C such as solar power stations and industrial waste heat recovery systems has been studied by Zhou et al. [16]. Copper beads and copper-aluminium alloy have been used as core materials coated with alumina made by slip casting process in spherical shell of 25 mm in diameter and 3 mm in thickness. After heating the capsules to 1,200°C in air and cooling in ambient environment, the inlet at the top of the shell was sealed in situ by oxidation reaction as illustrated in Figure 12.23. Finally, the capsules were subjected to ageing tests for 1,000 h and the results prove a little change in compositions and thermal storage properties were recorded.

The capsules were also under tests at 1,100°C for up to 1,000 h and no leakage or damage was found with mass increase of the encapsulate PCM of only 4.5% with the stable thermal energy density of 147 J/g.

A comprehensive review on PCM encapsulation and design methodology for low- to high-temperature thermal energy storage applications has been presented recently by Palacios et al. [17].

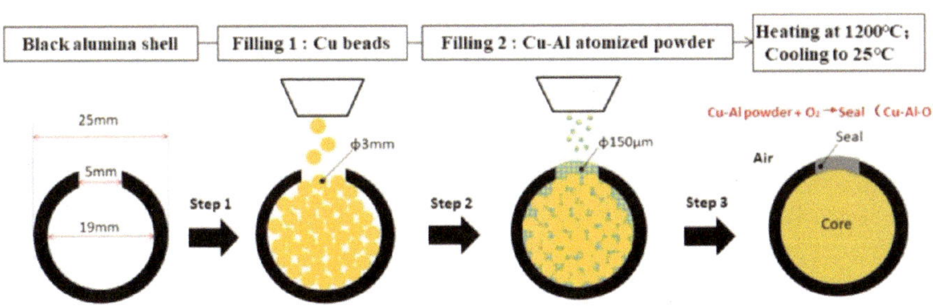

FIGURE 12.23 Preparation steps for macroencapsulation process [16]. (Permission from Elsevier.com.)

12.4 SUMMARY

Metallic and nonorganic-based PCMs have an outstanding latent heat, relatively high thermal conductivity compared to organic materials, high heat storage density and broad application prospectives in various high-temperature industrial waste heat recovery systems. The only drawback with metallic-based PCMs is strong corrosion of molten metal at high temperature that can be solved by an appropriate cladding material as a shell covering the PCM core. Chlorine, fluorine, and carbonate eutectic salts are very promising PCMs as a thermal storage media for high-temperature applications, but like molten metals constrained by their significant corrosiveness and low thermal conductivity. The encapsulation of salt PCMs with a stable inorganic shell and addition of nanoparticles of graphite and metallic powders to the PCMs can overcome these problems. Applications of high-temperature PCMs, including inorganic salts and metallic materials used as a core of the micro and macro capsulation of PCMs have been discussed in this chapter. Some of the previous and recent experimental studies on the preparation, characterization, and applications of both types of nonorganic and metallic-based PCMs have also been described. Finally, the application of scrap aluminium as metallic core for high-temperature PCMs in thermal energy storage has been discussed and found to be economical for the future application at high-temperature processes.

REFERENCES

1. Ramos A., Lopez E., Canizo C., Datas A. (2022) Cost-effective ultra-high temperature latent heat thermal energy storage systems, J. Energy Storage, 49, p. 104131.
2. Li C., Li Q., Lu X., Ge R., Du Y., Xiong Y. (2022) Inorganic salt-based shape stabilized composite PCMs for medium and high temperature thermal energy storage: ingredients selection, fabrication, microstructural characteristics and development, and applications, J. Energy Storage, 55, p. 105252.
3. Zhao X., Zou D., Wang S. (2022) Flexible PCMs: preparation, properties and applications, Chem. Eng. J., 431, p. 134231.
4. Navarrete N., Nithiyanantham U., Hernandez L., Mondragon R. (2022) K_2CO_3-$LiCO_3$ molten carbonate mixtures and their nanofluids for thermal energy storage: an overview of the literature, Sol. Energy Mater. Sol. Cells, 236, p. 111525.
5. Jiang Y., Wang Q., Tian S., Si Y., Bai Y., Li J., Luo Z., Wang D., Zhao T. (2022) Fluoride microcapsules with high phase change temperature, Colloids Surf. A: Physicochem. Eng. Asp., 653, p. 130028.
6. Soleimanpour S., Sadrameli S.M., Seyed Mousavi S.A.H., Jafarpour M. (2022) Preparation and characterization of high temperature shape stable $NaNO_3$ diatomite PCMs with nanoparticles for solar energy storage applications, J. Energy Storage, 45, p. 103735.
7. Zeng L., Zhang L., Zhu H., Sheng N., Zhu C. (2023) Macroencapsulated carbonate eutectic salt PCM with high durability for high temperature heat storage, Sol. Energy Mater. Sol. Cells, 256, p. 112338.
8. Han D., He X., Hou Y., Geng B., Zhang H., Lougou B.G., Shuai Y. (2024) Experimental analysis on improving heat storage efficiency of high-temperature packed bed system using spherical capsules filled with $MgCl_2$-KCl-NaCl/Al_2O_3 nanoparticles composite PCM, Sol. Energy Mater. Sol. Cells, 277, p. 113106.

9. Jiang T., Hu J., Wu S., Shen D., Lu P. (2024) A shape-stable PCM for high-temperature TES based on coal fly ash and Na_2SO_4-K_2SO_4, Sol. Energy, 280, p. 114868.
10. Ye C., Zhang M., Yang S., Mweemba S., Huang A., Liu X., Zhang X. (2023) Application of copper slags in encapsulating high-temperature phase change thermal storage particles, Sol. Energy Mater. Sol. Cells, 254, p. 112257.
11. Shi S., Liu R., Sheng N., Zhu C., Rao Z. (2023) Shape stabilized Al-Si/Al_2O_3 phase change composites for high temperature heat storage, J. Energy Storage, 58, p. 106425.
12. Ge Y., Sheng N., Zhu C. (2023) Macro-encapsulation of CuSi PCM by in situ alloying formation for high temperature thermal energy storage over 800°C, J. Energy Storage, 72, p. 108258.
13. Hong J., Zhou J., Hong J., Xu X. (2012) Environmental and economic life cycle assessment of aluminium-silicon alloys production: a case study in China, J. Clean Prod., 24, pp. 11–19.
14. Villada C., Navarrete N., Rawson A., Kolbe M., Stahl V., Kraft W., Kargl F. (2023) Recycling of aluminium scrap into phase change materials for high-temperature storage applications: thermophysical properties and microstructural characterization, J. Energy Storage, 72, p. 108822.
15. Guo Y., Zhao B., Guo H., Ge Y., Sheng N., Gariboldi E. (2024) Macroencapsulated Al-Si PCMs for high temperature latent TES, Chem. Eng. J., 487, p. 150390.
16. Zhou X., Yamashita S., Kubota M., Zhang C., Hong F., Kita H. (2023) Macro encapsulated Cu-based PCM for high temperature heat storage with characteristic of self-sealing and high density, Appl. Therm. Eng., 229, p. 120491.
17. Palacios A., Navarro-Rivero M.E., Zou B., Jiang Z., Harrison M.T., Ding Y. (2023) A prospective on PCM encapsulation: guidance for encapsulation design methodology from low to high temperature thermal energy storage applications, J. Energy Storage, 72, p. 108597.

13 Application of Phase Change Materials in Electronics

13.1 INTRODUCTION

The advanced portable electronic devices such as compact mobiles and laptop computers which are used in our daily life are designed in more compact size, lightweight, with higher storage capacity and with advancement in wireless technology which also release more heat that must be dissipated to maintain consistent thermal performance. Application of phase change materials (PCMs) in electronics can eliminate this problem and has been shown as a viable option in comparison to active air-cooling methods especially for compact electronics. This problem would cause performance deterioration, component failure, and lifetime deficiency which result discomfort in interaction with the device. This chapter explains and discusses the passive applications of PCMs in thermal management of electronics.

13.2 ENHANCEMENT OF THERMAL CONDUCTIVITY

One of the critical challenges in application of PCMs in electronics is their low thermal conductivity especially in organic type which ranges from 0.1 to 0.4 W/mK and limits the heat transfer rate within the PCM block. It is therefore necessary to boost PCMs thermal conductivity by utilizing some additives as thermal conductivity enhancement such as internal fins, metallic foams, carbon matrix, and nanomaterials to be suitable for electronic cooling applications.

A parametric study of thermal conductivity enhancement for PCMs using carbon foam structured utilized in thermal management of electronics has been performed by Nada et al. [1]. The analysis has been conducted using a validated finite element numerical technique base on volume averaging method and single-domain energy equation. The results of the numerical modelling reveal that reducing carbon foam and PCM thermal conductivities, increasing carbon foam porosity and increasing module height increase the module temperature and delay the approaching steady state temperature of the electronic systems. The calculations also show that using composite PCM with carbon foam structures the controlled temperature could decrease 59°C once the thermal conductivity of the carbon foam was increased by 50%. Thermal conductivity enhancement of a novel composite organic PCM/MXene has been investigated experimentally by Aslfattahi et al. [2]. Synthetized MXene (Ti_3C_2) has been added to PCM Plusice A70 (paraffin wax PW70) with melting point, density, and volumetric heat capacity of 70°C, 890 kg/m^3, and 154 MJ/m^3, respectively, and their thermal conductivity has been measured using KD2 Pro thermal properties analyser (Decagon, USA,

DOI: 10.1201/9781003518280-13

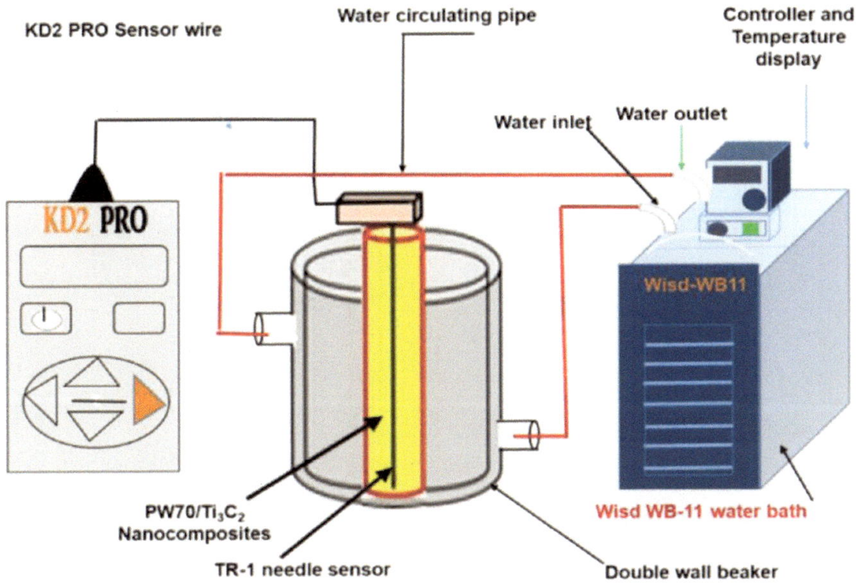

FIGURE 13.1 Thermal conductivity measurement apparatus [2]. (Permission from Elsevier.com.)

version 5) with a probe which works with the principle of a transient hot wire technique as illustrated in Figure 13.1.

Thermal conductivity enhancement of 16% has been achieved by adding 0.3% of nanoparticles to the PCM and the melting point of the composite has been increased slightly from 69.8°C to 71.7°C by adding the same number of nanoparticles to the PCM.

A review on the thermal performance of PCM-based heat sink using thermal conductivity enhancer materials is presented by Maqbool et al. [3]

13.3 THERMAL MANAGEMENT OF ELECTRONICS USING PCMs

A passive thermal management system for cooling of 8 mm diameter, 200 W/240 V cartridge heater as a device using paraffin wax as a PCM with melting temperature of 70°C has been designed and built by Righetti et al. [4] to study the effects of operating conditions on the system performance. The heater has been inserted in an aluminium block of squared area of 42 × 42 mm^2 and 20 mm thick with thermal conductivity of 205 W/mK and was connected to the PCM sample with silicon heat transfer paste and then insulated with rock wool for heat transfer insulation to the surroundings. One of the sample boxes is empty, while the other has been filled with an aluminium 3-D periodic structure as thermal conductivity enhancer for paraffin was, as shown in Figure 13.2. Both boxes have been filled with 50 + 0.1 g of PCM commercialized by Rubitherm as RT70 with phase transition temperature of 70°C. Paraffin wax has been selected due to its non-toxicity, large availability, chemical stability, high latent heat, and low price in comparison to other materials.

Experimental tests have been performed at four ambient temperatures from 10°C to 40°C. The comparison of the temperature fluctuations between two samples

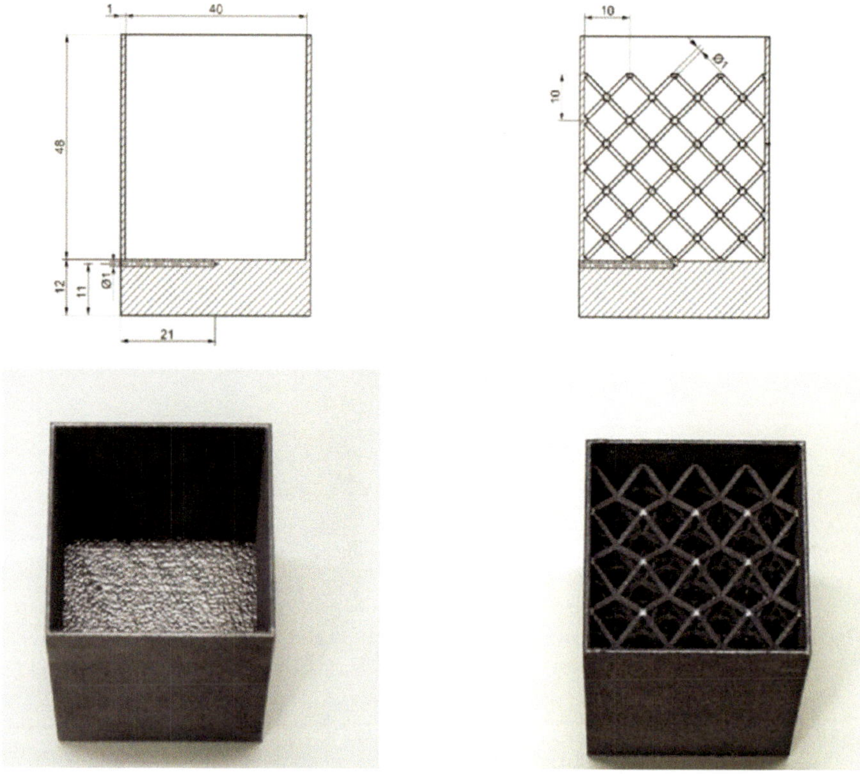

FIGURE 13.2 Schematic and photos of the boxes [4]. (Permission from Elsevier.com.)

(reference and the PCM) during the 10°C ambient temperature charging and discharging tests is depicted in Figure 13.3. As can be seen from the temperature profiles the temperatures are the initially the same for two cases during charging and then deviate during the discharging with higher level for the reference temperature profile. The comparison of the whole cycles reveals that the reference sample took

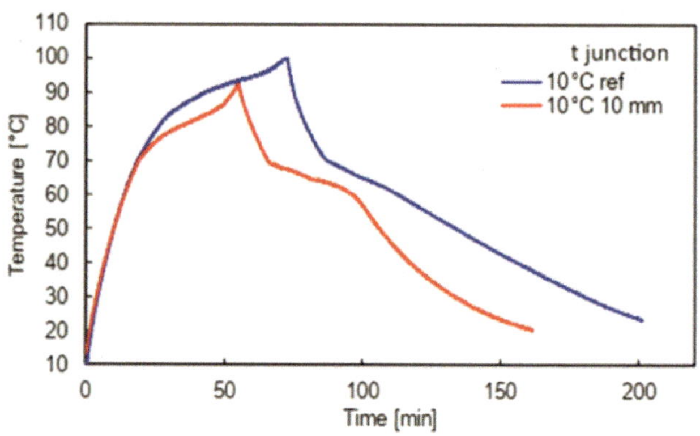

FIGURE 13.3 Comparison of temperature profiles [4]. (Permission from Elsevier.com.)

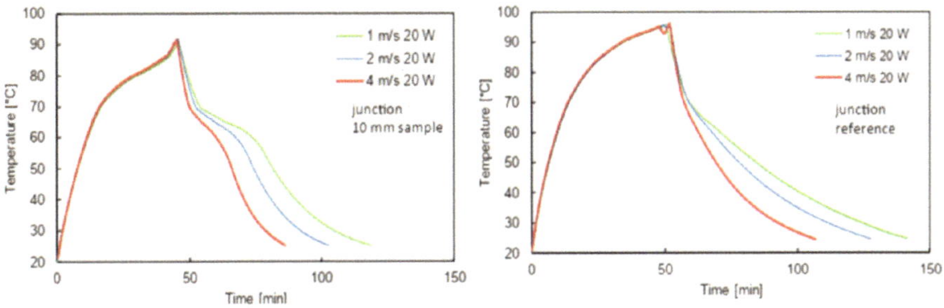

FIGURE 13.4 Junction temperatures in 10 mm sample (a) and reference sample (b) during charging and discharging at different air velocities [4]. (Permission from Elsevier.com.)

about 40 min more time than the 10 mm sample at 10°C ambient temperature which is due to the high conductive structures of aluminium which increase the spreading of the heat and accelerate TMS efficiency.

The junction temperature variations for discharging period of 10 mm and the reference samples at different air velocities of 1–4 m/s are demonstrated in Figure 13.4, respectively, which shows that increasing the air velocity shorten the discharging time while the profiles for all cases during charging are similar.

They finally concluded that more research must be performed to build a robust database that will permit to define clear design for smart, and efficient thermal management system based on PCM.

An experimental and parametric investigations for passive cooling of electronics utilizing PCM incorporated in aluminium fins geometries of circular and square cross sections with two configurations of inline and staggered arrays have been performed by Ashraf et al. [5]. N-eicosane has been selected as a thermal conductivity enhancer to be added to six different commercial PCMs varying melting temperature and heat capacity namely RT-54, RT-44, RT-35HC, SP-31, and paraffin wax with 90% heat sink volume letting a 10% volume for melting volume increase. Power levels of 4–8 W with an increment of 1 W have been applied during the experiment. A real-time assembly has been designed as shown in Figure 13.5 to obtain efficient and more reliable results from the experiments.

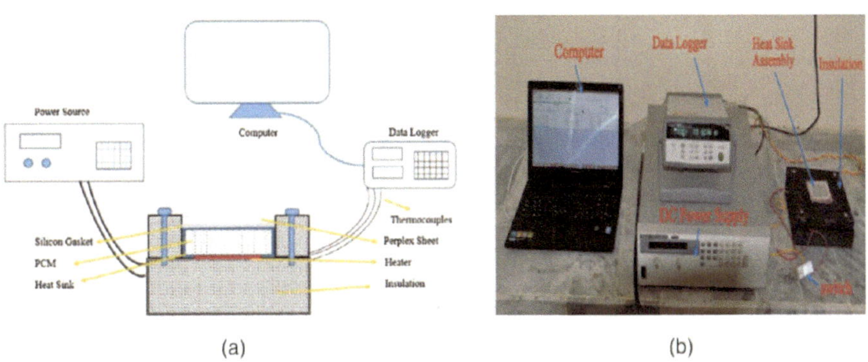

FIGURE 13.5 Schematic diagram (a) and pictorial view (b) of experimental setup [5]. (Permission from Elsevier.com.)

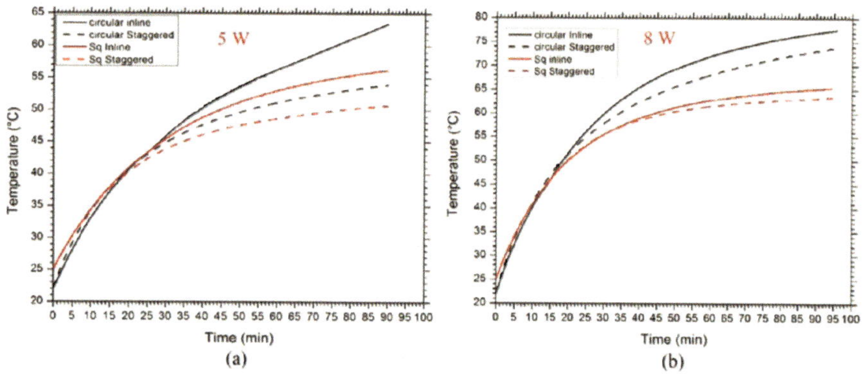

FIGURE 13.6 Temperature profiles of the circular and square-shaped pins for 5 W (a) and 8 W (b) [5]. (Permission from Elsevier.com.)

The first set of experimental tests have been conducted for the comparison of the pins of square and circular shapes. The temperature profiles for the system without PCM are shown in Figures 13.6.a and 13.6.b for 5 W and 8 W, respectively. As expected from the results due to sharp edges of square-shaped fins, there is more turbulence of air in the system helping more heat dissipation.

The temperature distribution of the square-shaped pins with inline and staggered arrangement for 5 W heater utilizing different PCMs are depicted in Figure 13.7 in which the graphical fluctuations predicts that inline arrangement has superior to the staggered. It is interesting to see from the results that for paraffin wax square staggered is initially bellow the inline temperature profile but after one hour show a higher temperature gain. Temperature gain for inline formation is higher than staggered for circular pin-fin configuration.

They finally concluded that for 8 W power level systems paraffin wax is the most efficient PCM with highest enhancement ratio.

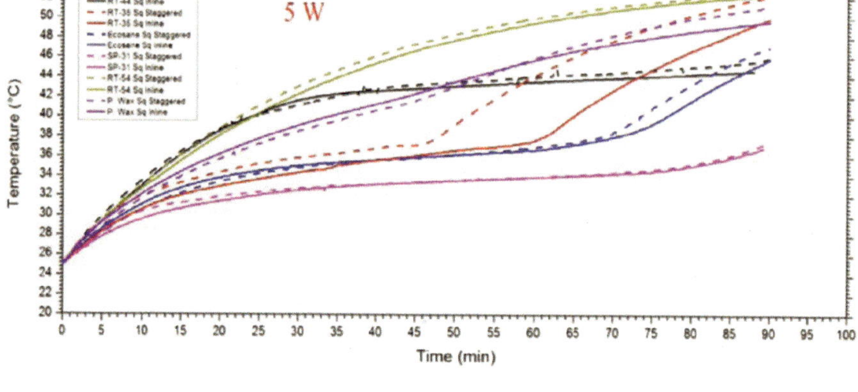

FIGURE 13.7 Temperature profiles of the system with square pins using different PCMs [5]. (Permission from Elsevier.com.)

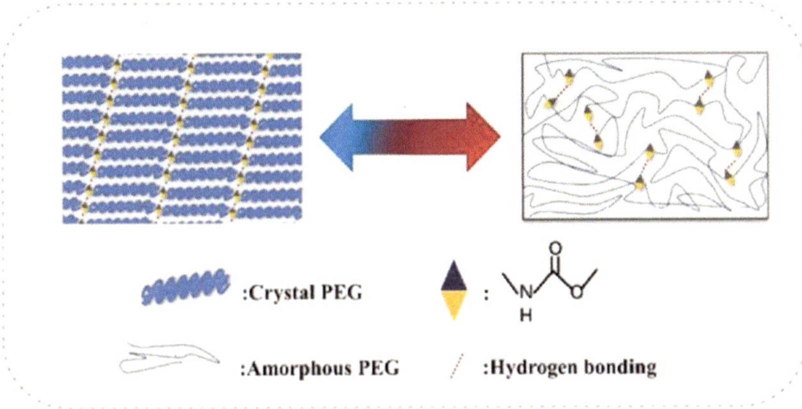

FIGURE 13.8 Working principles of polyurethane [6]. (Permission from Elsevier.com.)

Super elastic and shape stabilized solid-solid PCMs has been used for the thermal management of electronics by Lio et al. [6]. This has been achieved through molecular weight adjustment of polyethylene glycol (PEG) to change the melting temperature and enthalpy of polyurethane (PU) which is also equipped with long-term cycling stability of 700 thermal cycles.

The synthesis of PU has been done using PEG with isophorone diisocyanate (IPDI), and butanediol (BDO) in an oil bath by a two-step polymerization process under nitrogen as inert gas. Many hydrogen bonds can help to cross link the functional groups of PU molecular chain as a PU structure as illustrated in Figure 13.8, which not only prohibit the PCM leakage during melting but also enhance the mechanical properties of PU.

Two step polymerization synthesis of PU with BDO and IPDI and FT-IR spectra of PEG mixture, including PEG4000 and PEG 2000 are illustrated in Figure 13.9

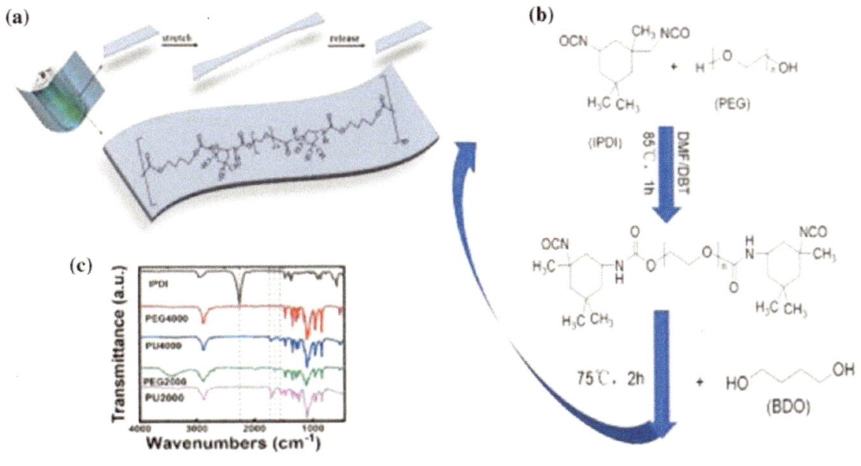

FIGURE 13.9 Schematics of PU film (a), PU synthesis (b), and FT-IR spectra of materials [6]. (Permission from Elsevier.com.)

which shows that the stretching vibration absorption peak spectrum of hydroxyl carbon oxygen (C-O) is around 1,096.24 cm^{-1}, while this value for hydroxyl (OH^-) exists approximately at 3,458 cm^{-1}.

Their experimental observations concluded that the PU mechanical behaviour is mainly based on the number of hydrogen bonds than van der Waals and by increasing the number of hydrogen bonds the tensile property of PU is enhanced and the final PU developed composites has a great potential for the energy management of 5G generation electronics.

Thermal management of downhole electronics cooling has been performed by Peng et al. [7] using a hybrid system of deionized water cooling and PCM in which a spiral and a S-shaped pipes incorporated into PCM and the cold plates to strengthen the heat exchange as depicted in Figure 13.10. Low melting point alloy with thermal conductivity of 18.79 W/mK and melting and solidifying temperature ranges of 77.20°C to 77.73°C and 59.54°C to 60.73°C, respectively. The electronics have been installed on the circuit board and the gaps between the board and the cold plate has been filled with silicon pads composed of silica gel with thermal conductivity, density, and specific heat capacity of 1 W/mK, 1,810 kg/m^3, and 923 J/kgK, respectively. Simulation of the heat source has been conducted using ceramic heat plates with thermal conductivity, density, and specific heat of 30 W/mK, 3,960 kg/m^3, and 850 J/kg K, respectively. A simultaneous simulation model, including the transient flow and heat transfer have been implemented using finite element method by software COMSOL Multiphysics, to study the performance of the system.

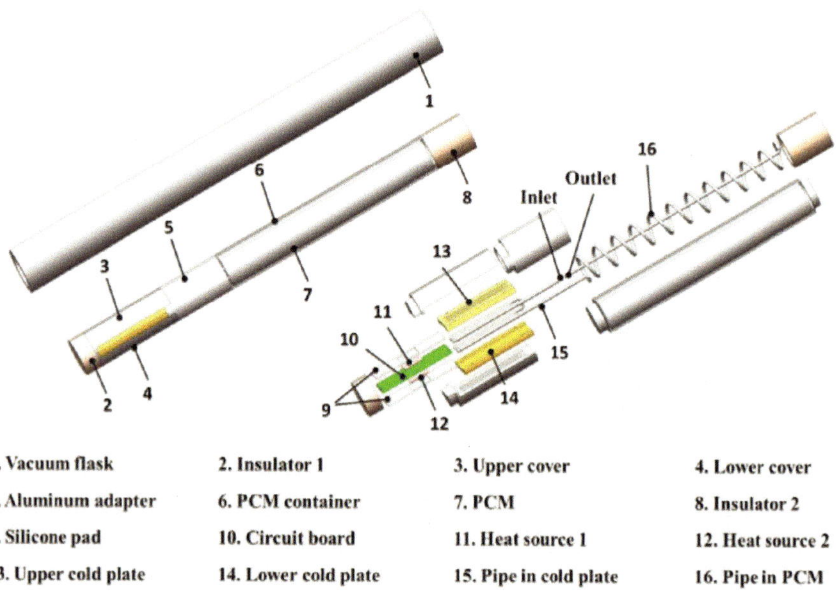

FIGURE 13.10 Experimental setup for downhole electronics cooling [7]. (Permission from Elsevier.com.)

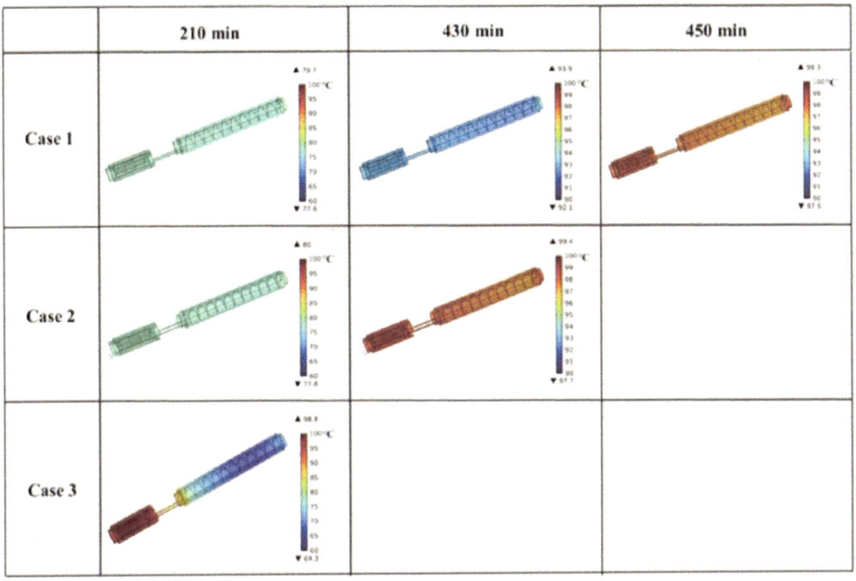

FIGURE 13.11 Temperature distributions of three heat transfer modes at different times [7]. (Permission from Elsevier.com.)

The dominant mechanism of heat transfer through the channels is by conduction. Temperature distributions for different mechanisms of heat transfer in three cases are demonstrated in Figure 13.11 in which case 1 includes forced convection through pipe and heat conduction through aluminium adaptor, and cases 2 and 3 have forced convection through the pipes and heat conduction through the aluminium adaptor. The parameters of the system are inlet flow rate of liquid cooling of 0.5 m/s, and heating power of 30 W for the with ambient temperature of 205°C. The results show that the temperature distributions of cases 1 and 2 are more uniform than case 3 and the maximum temperature of the heat sources approaches the temperature limit at 210 min, 430 min, and 450 min for in cases 1, 2, and 3, respectively.

They finally concluded that the hybrid system water cooling and PCM can reduce the thermal resistance from the heat source to the PCM and extend the worktime of the electronics.

Application of pin-finned heat sinks packed with solid-solid PCM (SS-PCM) and solid-liquid PCM (SL-PCM) as a cooling system for avionics electronics has been investigated by Mayank et al. [8] with a schematic shown in Figure 13.12. PEG-based form-stable PCM and a commercial organic based PCM named as A24 with different thermophysical properties have been used in the cooling system. A mathematical model coupled with the enthalpy method has been developed to study the transient heat transfer of two PCMs under same heat input rates and charging/discharging time.

Their experimental results indicate that at lower heat input the solid-solid PCM perform better in comparison to solid liquid PCM after the heating cycle however,

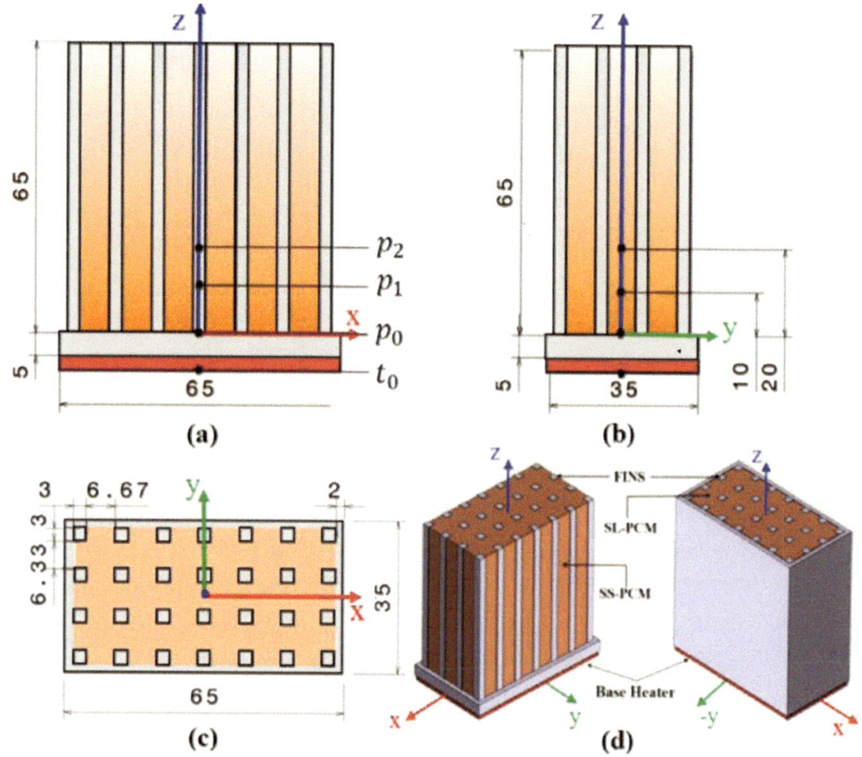

FIGURE 13.12 Schematic of a pin-finned heat sink for electronics cooling [8]. (Permission from Elsevier.com.)

solid liquid PCM shows better performance, larger safe operational period, high energy storage potential, and efficiency compared to solid-solid PCM. The safe operational time for solid-solid and solid-liquid PCMs is increased by an average time interval of 668 s and 788 s, respectively.

Dragon peel wastes have been utilized as a PCM for optimizing heat transfer efficiency of electronics operating under low temperature operation by Arulprakasajothi et al. [9]. Two heat sink configurations of hollow structure without fins and triangular fins with subjected heat input levels of 35 W to 45 W. Thermal conductivity of PCM has been enhanced with additives such as graphene nanoparticles and nanoparticles coated with dragon fruit extract. The results indicate that thermal performance of the heat sink system increased by 23%, 11%, and 6.8%, respectively, when using dragon fruit extract-coated nanoparticles in conjunction with PCM. For the heating load of 40 W the heating and cooling cycles of the heat sink containing PCM and without fins was decreased by 18% while coating the nanoparticles with dragon fruit extract reduced the heating and cooling cycles by 37% compared to heat sink without fins and PCM included.

In general, cooling electronic devices using PCMs is an effective thermal management technique to handle the heat generated by modern electronic systems. However to enhance the thermal conductivity of the PCM some additives must be added to the PCM for better heat transfer in the system.

13.4 APPLICATION OF PCM IN PORTABLE ELECTRONICS

Portable electronics such as laptop computers, tablets, and mobiles are developed with light, thin, and small shape factors to achieve an eye appealing design during extreme mobility. This leads to exceptionally high operating temperatures. Overheating of the components of such systems may affect their performance and stability.

A thermal energy storage system utilizing a mixture of hydrated salts as PCM has been used by Soni et al. [10] for the cooling of a laptop computer model Dell Vostro dual core processor to protect user's lap from overheating. The TES container is made of 1 mm thick stainless-steel sheet with dimensions of 32 × 22 × 1 cm^3 filled with HS-207 PCM with melting point of 30°C and latent heat 238 kJ/kg. The TES performance has been analysed by fitting the TES at the bottom of the laptop and temperature of the laptop component such as cores 1 and 2 of the processor and hard disk have been measured by using CPUID HW monitoring software. Extra heat from the computer components has been absorbed by PCM inside the TES, during charging period, causing the PCM to change phase from solid to liquid. During discharging period when the temperature of the system fell below PCM melting point the PCM changes its phase back from liquid to solid. Temperature profiles of core 1 processor and hard disk, with and without PCM are illustrated in Figure 13.13, in which the maximum temperature of core 1 processor reaches to the maximum of 41°C in 60 minutes. Using TES for cooling this temperature dropped to 37°C after 115 minutes which shows 4°C decrease in temperature and 55 minutes increase in

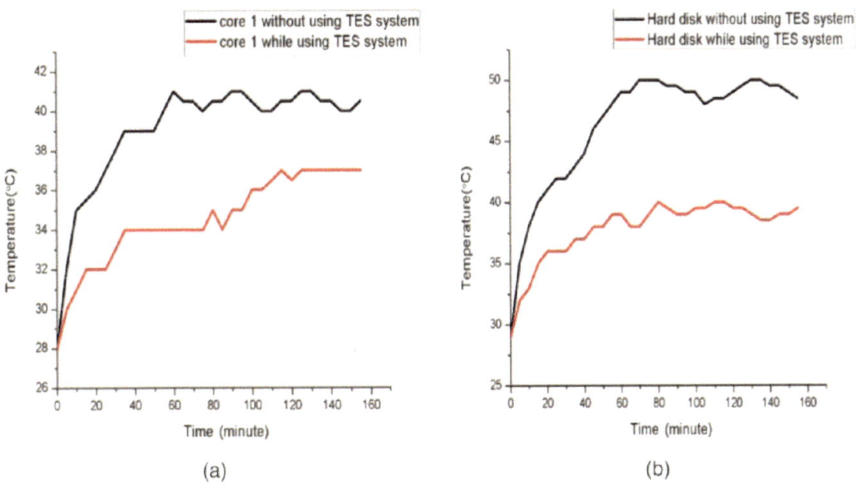

FIGURE 13.13 Temperature profiles of core 1 (a) and hard disk (b) vs. time with and without PCM [10]. (Open Access Journal)

operating time. Similar results can be observed for hard disk where the maximum temperature has been reached to 50°C after 70 minutes with TES while this has been reduced to 40°C after 80 minutes. Their experimental results shows that TES system can decrease the laptop computer components by maximum 65.44% which proves that the performance and stability of the components have been improved by using TES system.

Thermal management of tablet PCs has been investigated by Ahmed et al. [11] utilizing encapsulated PCM in a thin aluminized laminated film for continuous operation. Two types of PCM have been investigated in TES systems as n-eicosane and commercial PT-37 as a safe and inexpensive with suitable melting points 35.6°C and 36.3°C, respectively. Both PCMs have been tested for phase transition cycles under 10,000 and 4,000 phase change cycles and showed very good chemical stability. A tablet has been fixed on the test stand using four 3-D printed clamps with rotation capability and for isolation from external thermal environmental conditions as depicted in Figure 13.14. Encapsulation of both n-eicosane and PT-37 have been done with different base areas in three packs with dimensions of 152.4 × 66 mm^2 base area, 177.8 × 38 mm^2 and 177.8 × 86 mm^2 base area with the average thickness of 2 mm.

Temperatures have been recorded using for thermocouples P1 to P4. Temperature variations of the heater (Point 1), Point 1B on the back cover and Point 1F on the front display have been considered as performance parameters. The experiments have also been performed in three different inclinations of 0°, 45°, and 90° with horizontal. Temperature profiles measured by four thermocouples for size 1 PCM and TES unit for two PCM of n-eicosane and PT-37 are illustrated in Figure 13.15. The results reveal that although eicosane has 16% higher energy density than PT-37 but this does not affect the temperature during unsteady state heating of the laptop. This may be due to the small amount of PCM and smaller melt amount during one

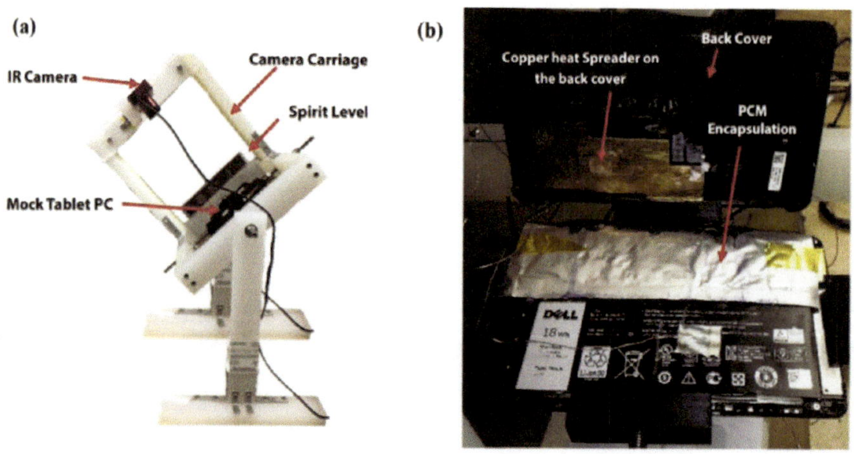

FIGURE 13.14 An inclined test stand (a) and photograph of experimental setup (b) [11]. (Permission from Elsevier.com.)

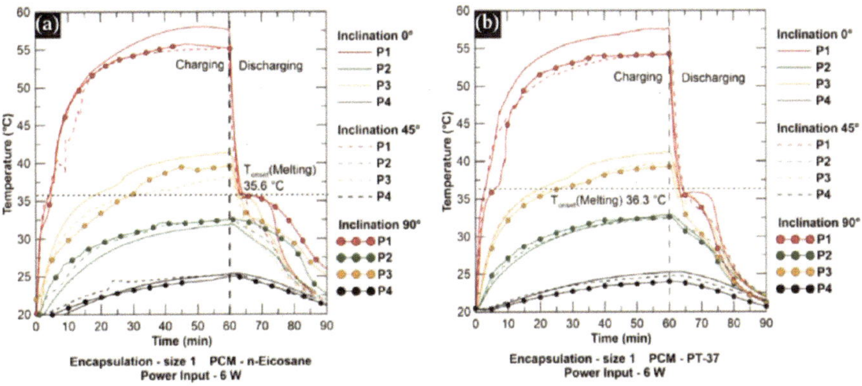

FIGURE 13.15 Temperature distributions of pack size 1 with n-eicosane and PT-37 PCMs with 6 W power [11]. (Permission from Elsevier.com.)

hour of operation. It can also be seen from the temperature profiles that for the horizontal condition (0°) the entire heat transfer is dominated by conduction and the temperature fluctuations for 45° inclination and 90° inclination are very similar together except for some variations during transient heating which may be caused by three dimensional and unstable flow structures in the liquid PCM.

The influence of different power input levels on the phase transitions of two PCM TES for size 1 unit and 0° inclination for the most experienced melting points, i.e., P1 and P3 are depicted in Figure 13.16 which clearly shows that the rate of PCM melting is directly depended on the power output.

Faster PCM phase transition and higher steady state temperature profile can be achieved with increasing heat input and there was no PCM melting at 2 W power input. As a result, no latent heat release is observed during the discharging process

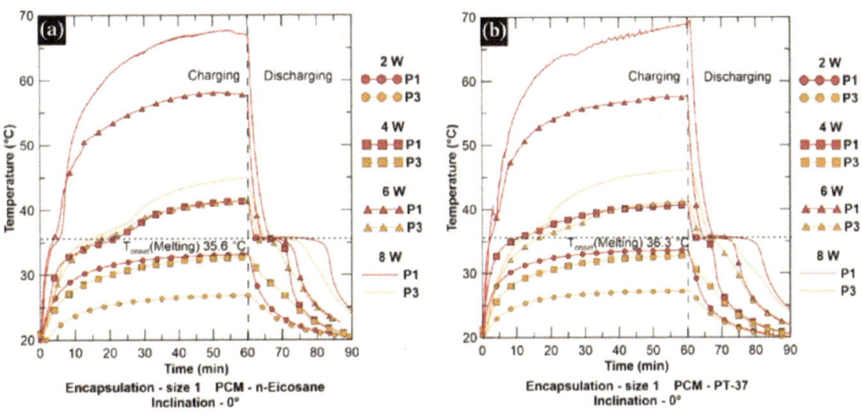

FIGURE 13.16 Influence of power output for the melting temperature of n-eicosane and PT-37 PCMs [11]. (Permission from Elsevier.com.)

with 2 W input power. It is also observed that the time required for PCM solidification at points P1 and P3 during which the absorbed latent heat was released is much longer than that for the melting phase.

The results of the above studies clearly prove that PCMs can significantly improve the heat transfer on the mobile electronics for cooling process especially in the small-scale processes; however, more research is required for the improvement of encapsulation process and thermal conductivity of PCMs.

13.5 SUMMARY

As the electronic devices continue to be designed in smaller sizes and higher storage capacity, the power densities persist in being on rising trend. The range of heat fluxes that are currently generated by microprocessors are over 100 W/cm^2 and hot spots in chips are caused by local heat fluxes of 1 kW/cm^2 or more that leads to excessive local temperatures. Operation of such devices at high temperature besides other failure modes of mechanical, electrical, and corrosion affects the system performance and reliability and as a result their failure through the operation period.

Cooling techniques under steady state operation cannot be applied for the system at transient condition since the primary objective of steady state cooling systems is to reduce the overall thermal resistance of the package so that the maximum power can be released to the environment without exceeding the maximum temperature limit of the system.

PCMs can be utilized as TES besides the electronics to suppress device temperature increase during power surges. The only challenges with using PCMs as a TES is their low thermal conductivity that can be solved by adding nanomaterials such graphite and nanofibre to the PCM before encapsulation. Mettalic and non-metallic PCMs are found to be suitable for cooling of the electronic device for applications involving short duration high power pulses. The effect of ambient temperature on the starting temperature of the electronic devices at rest during the cooling process and the possible effect of the ambient temperature on the PCM selection need to be taken into account in the future research in this area.

REFERENCES

1. Nada S.A., Alshaer W.G. (2015) Comprehensive parametric study of using carbon foam structured with PCMs in thermal management of electronic systems, Energy Convers. Manag., 105, pp. 93–102.
2. Aslfattahi N., Saidur R., Arifutzzaman A., Sadri R., Bimbo N., Sabri M., Maughan P., Bouscarrat L., Dawson R.J., Said S., Goh B., Che Sidik N. (2020) Experimental investigation of energy storage properties and thermal conductivity of a novel organic PCM/MXene as a new class of nanocomposites, J. Energy Storage, 27, p. 101115.
3. Maqbool Z., Hanief M., Parveez M. (2023) Review on performance enhancement of PCM based heat sinks in conjunction with thermal conductivity enhancers for electronic cooling, J. Energy Storage, 60, p. 106591.
4. Righetti G., Zilio C., Doretti L., Longo G.A., Mancin S. (2021) On the design of PCMs based thermal management systems for electronics cooling, Appl. Therm. Eng., 196, p. 117276.

5. Ashraf M.J., Ali H.M., Usman H., Arshad A. (2017) Experimental passive electronics cooling: parametric investigation of pin-fin geometries and efficient PCMs, Int. J. Heat Mass Transf., 115, pp. 251–263.
6. Liao Y., Li J., Li Sh., Yang X. (2022) Super-elastic and shape-stable solid-solid PCMs for thermal management of electronics, J. Energy Storage, 52, p. 104751.
7. Peng J., Deng C., Wei F., Ding S., Hu R., Luo X. (2023) A hybrid thermal management system combining liquid cooling and PCM for downhole electronics, J. Energy Storage, 72, p. 108610.
8. Mayank M., Midhun V.C., Sandip K.S. (2024) Evaluation of solid-solid and solid-liquid PCMs in pin-finned heat sinks for cooling of avionics electronics, Therm. Sci. Eng. Progr., 53, p. 102781.
9. Arulprakasajothi M., Raja N.D., Saranya A., Elangovan K., Murugapoopathi S., Poyyanozhi N., Mesho K.T.T. (2024) Optimizing heat transfer efficiency in electronics component cooling through fruit waste-derived PCM, J. Energy Storage, 80, p. 110238.
10. Soni P., Murty V.V.S., Gupta A., Purohit K. (2019) A passive thermal energy storage system for laptop cooling using PCM, Int. J. Appl. Eng. Res., 14(13), pp. 131–136.
11. Ahmed T., Bhouri M., Groulx D., White M. (2018) Passive thermal management of tablet PCs using PCMs: continuous operation, Int. J. Therm. Sci., 134, pp. 101–115.

14 Application of Phase Change Materials in Greenhouses

14.1 INTRODUCTION

World population growth has increased the food demand that requires more advanced technologies. Newly designed greenhouses have been developed through the years for the safe production of food sources with higher efficiency. Although solar radiation is the main source of energy for the greenhouse but in developing countries, the rate of fuel consumption in the greenhouse especially in cold season is high and there is an urgent requirement for improving heating system. The optimization of thermal energy can impact higher yields per cultivation area and harvest time in a protected environment especially in the countries that fossil fuel sources are hardly accessible. Phase change materials (PCMs) can store extra heat from heating systems (active or passive) in greenhouse, including gas and liquid heaters, solar collectors, and release it to the greenhouse when required. This chapter discusses the latest technological advancements in the development of greenhouse heating/cooling systems incorporated with PCMs.

14.2 GREENHOUSES

An enclosure with a controlled microclimate, sufficient heat, light, and CO_2 provided by either natural or artificial techniques and can be utilized for farming and production of fruits, crops, and vegetables is called a greenhouse. This enclosure can protect the food production from unexpected climate conditions such as storms, heavy rain, drought, temperature fluctuations, etc. The greenhouse envelope is typically made of transparent materials such as glass or plastic to transmit only shortwave radiation from sun while reflecting the longwave radiation. However, the floor and walls re-radiate the amount of absorbed radiation inside the envelope causing the indoor air temperature to increase which is favourable for plant growth on sunny winter days but during the nighttime due to the low outside temperature and the inferior greenhouse thermal insulation, the indoor greenhouse temperature is reduced which leads to plant death. Similarly, an increase in the temperature during the hot summer days affects the plants growth in the system. Therefore, a suitable heating/cooling control system is required to optimize the indoor greenhouse temperature throughout the year. The greenhouse microclimate and the solar radiation captured by the system have been affected by multiple design parameters such as size, shape, and orientation which can control the indoor temperature apart from outside ambient temperature. Basically, there are three types of greenhouses based on the material used for the roof: plastic greenhouse, glass greenhouse, and solar

DOI: 10.1201/9781003518280-14

green house. A plastic greenhouse is made of bamboo and steel and covered by light-transmitting plastic material making an arch-shaped shed for cultivating fruits and vegetables.

Glass greenhouses are manufactured by steel or aluminium alloy, is covered with glass, and is characterized by large lighting area, uniform light, long service life, and good lighting transmission rate, while they are complex in structure and high in cost, so is rarely used. A solar greenhouse as illustrated in Figure 14.1 is common in northern cold winter and hot summer areas with a single-slope plastic greenhouse covered

FIGURE 14.1 General indoor and outdoor of a solar greenhouse [1]. (Permission from Elsevier.com.)

with insulation at night on the front slope and enclosing walls on the west, east, and north sides. They have a good heat preservation, low investment, and energy saving in which PCM can be utilized.

Crop yield and quality enhancement, long harvest period, protection from pets, and disease and local production under extreme environment can be achieved by adding some heating or cooling to greenhouse for critical conditions to keep a suitable sustainable environment and control the microclimate inside the greenhouse. A suitable heating and cooling technique which varies in different regions with different techniques can be utilized at each stage of the crop cultivation to obtain an optimal temperature inside the greenhouse.

14.2.1 Energy Consumption in a Greenhouse

The amount of fossil fuels that will be required to maintain the greenhouse indoor temperature at approximately 15–20°C, has been estimated to be 5–6 kg/m^2 which mainly utilized for heating and cooling purposes [2]. The conventional heating/cooling installation system may have a low cost but the greater operational costs and environmental impacts lead us to utilize renewable sources like solar and geothermal, wind and biomass which can be used directly or integrated with existing conventional systems. One of the drawbacks of renewable energy sources is their performance fluctuations that necessitate coupling the system with a relevant backup such as a thermal energy storage system (TES). Three classifications of TES systems are sensible TES, latent TES, and thermochemical energy storage systems which have been discussed in detail in Chapter 3 of this book.

14.3 PCM APPLICATIONS IN GREENHOUSES

The higher energy density per unit mass of PCMs and storage at constant temperature makes them ideal and more attractive for greenhouse applications. PCM can store excess thermal energy from solar system during the day and release them during the night at constant temperature. Active or passive heating/cooling systems can be adopted by PCMs and can be utilized with local or renewable energy sources to reduce heating and cooling loads. PCMs can be installed in active systems where thermal energy is transmitted to the greenhouse by air and water through a pipe or a duct from the heating source. They can store the excess or wasted heat and release them when the system is off, thereby reducing the fossil fuel consumption and operating costs of the greenhouse. In passive system, PCM with specified phase transition temperature and heat density can be integrated with wallboards or individual TES system and is utilized to control temperature of the greenhouse by the absorption of thermal energy during melting and release them throughout the freezing.

Energy cost and efficiency analysis of utilizing PCM in a greenhouse heating system have been studied by Yan et al. [3] where four PCM plate types with a thickness of 2.5 cm are fitted inside a heat exchanger with dimensions of 20 × 50 × 140 cm^3

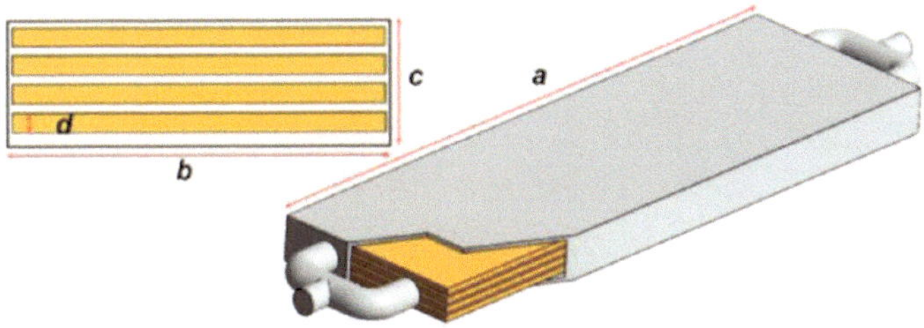

FIGURE 14.2 Schematic diagram of heat recovery system [3]. (Permission from Elsevier.com.)

and have been fitted inside a plate heat exchanger as a PCM heat recovery system as illustrated in Figure 14.2.

During nighttimes when the required thermal energy for the greenhouse is provided by a gas heater equipped with gas burner with power of 220 kcal/h, the flue gas from the heater is utilized by PCM exchanger for melting PCM and during the daytime the stored heat in PCM is released to the air for heating the greenhouse as shown in Figure 14.3. The greenhouse temperature has been controlled to 20°C for the cultivation of Lisianthus flower.

They concluded that the application of PCM heat recovery system can save fossil fuel energy as much as 22.7% and the working period of heater burner has been reduced by 27%.

Analysis of the technical, environmental, and economic potential of PCM for root zone heating in Mediterranean greenhouses situated in Cabrils, north Barcelona with dimension of 19.2 m wide, and 12 m long with a 3 m high gutter and a 5.5 m high ridge covered by a single layer of polyethylene have been performed by Llorach-Messana et al. [4].

Root zone heating application for tomato crop (Solanum lycopersicum Arawak) with optimum root zone temperatures of 15°C and 25°C has been selected that could help crop production continuity during the cold season or extend crop production during November and allow earlier crop initiation in March (Figure 14.4).

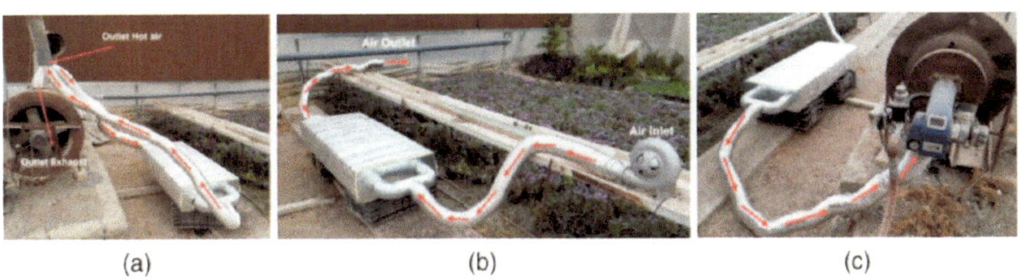

FIGURE 14.3 Charging (a) and discharging (b) modes of TES and preheating air (c) for the greenhouse [3]. (Permission from Elsevier.com.)

FIGURE 14.4 Root zone heating for tomato crop growth [5]. (Permission from Elsevier.com.)

Their investigation reveals that an optimized phase transition temperature for a root zone passive heating system utilizing PCM in Mediterranean greenhouses seems to be 15°C and PCM with melting/freezing temperature of 12°C does not ensure the freezing of the PCM if temperature does not fall under 10°C. Also based on one-night results obtained by the authors, PCM can offer potential to reduce the carbon footprint of gas and oil conventional root zone heating systems.

Thermal performance of passive greenhouses utilizing PCMs in cold climate using simulation by COMSOL Multiphysics has been studied by Ismail et al. [6]. The simulation model focused on the wall materials in which the north wall of the greenhouse is made up of three layers of lime silica brick, PCM layer, and an outer layer of cellulose batt insulation. The single layer south wall was constructed with polycarbonate sheets for the transparency, low thermal conductivity, flexibility, and affordability. Both the east and west walls were made of lime silica brick and insulated with cellulose batt outer layers. The dimensions of the greenhouse were $5 \times 4.5 \times 5$ m^3 and three mechanisms of heat transfer have been considered in the model as illustrated in Figure 14.5.

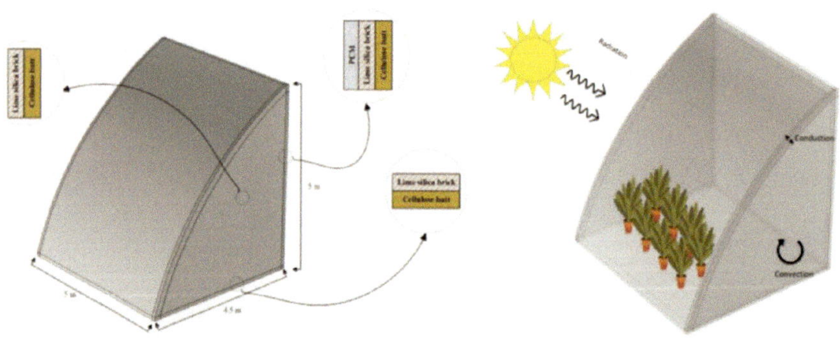

FIGURE 14.5 Greenhouse dimension (a) and heat transfer modes (b) [6]. (Open Access Journal)

TABLE 14.1
Properties of the Walls and PCMs

Material	Melting Point (°C)	Thermal Conductivity (W/m.K)	Latent Heat (kJ/kg)	Density (kg/m³)
n-Tetradecane	4–6	0.2	231	771
n-Pentadecane	10–12	0.17	207	768
Lime silica brick	335	0.9–1.8	–	1800
Cellulose batt	230	0.05	–	50
Polycarbonate	105	0.2	–	1.2

The greenhouse has been simulated using COMSOL Multiphysics to determine the temperature profiles inside the greenhouse at different seasons, based on the ambient properties such as geographic location, and solar radiation in clear sky conditions obtained from the COMSOL database. The properties of the walls and two PCMs used in the simulation are listed in Table 14.1.

The north wall temperature profiles of the greenhouse without and with PCMs are depicted in Figures 14.6.a and b for four different times throughout the day: 8 am, 2 pm, 8 pm, and 2 am of the following day. As can be seen from the results, the temperature profiles closely follow the solar irradiation cycle as is expected which means that the temperature increases during the day and decreases sharply during the night.

The temperature profiles of the greenhouse with PCMs in Figure 14.6.b clearly show a significant temperature decrease when compared with the results without PCM.

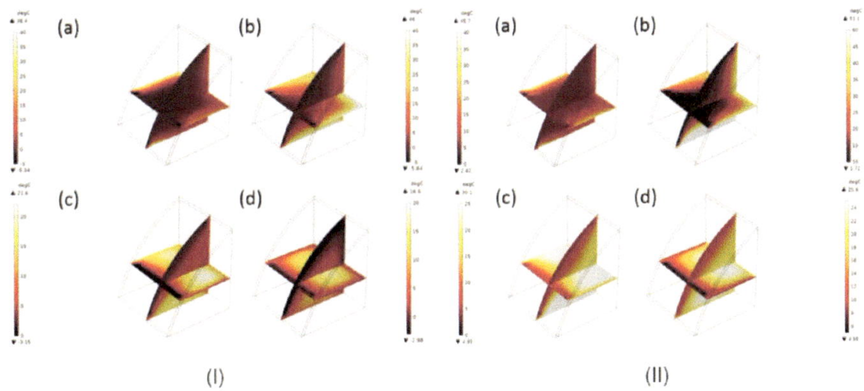

FIGURE 14.6 Simulation results at north wall without (I) and with (II) PCMs for 8 am (a), 2 pm (b), 8 pm (c), and 2 am (d) [6]. (Open Access Journal)

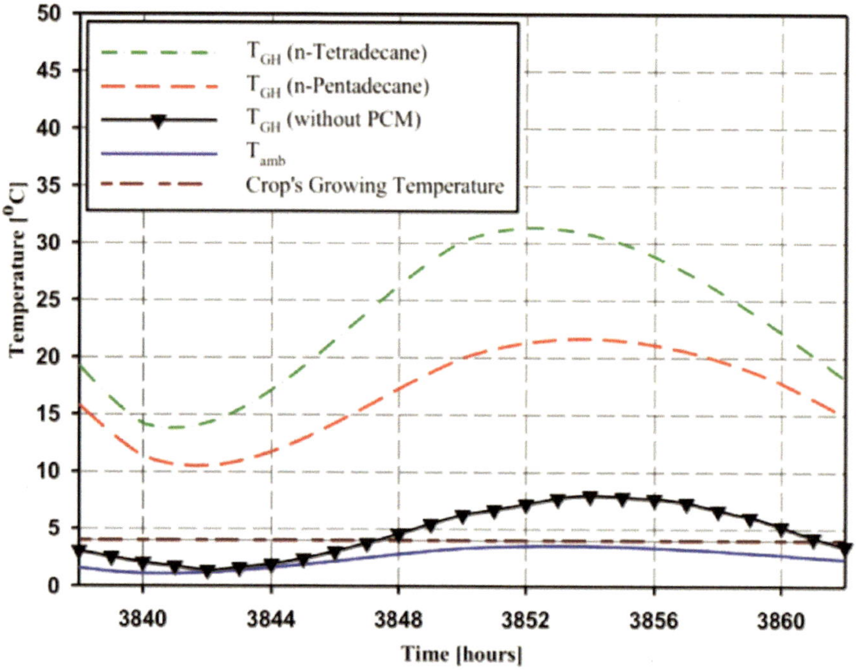

FIGURE 14.7 Temperature profiles of greenhouse for different PCMs [6]. (Open Access Journal)

Temperature profiles of the greenhouse utilizing different PCMs are demonstrated in Figure 14.7 which clearly shows that the type of PCM plays a significant role in determining the thermal efficiency of the greenhouse. As can be seen from the temperature profiles, on average n-Tetradecane has a better performance with an average increment temperature increase of 3–10°C, while the lower performed PCM (n-Pentadecane) had lower performance increase. This proves that the type of PCM can significantly affect the stability and efficiency of temperature regulation in a greenhouse; the best results can be achieved by the selection of a PCM with melting point closer to the minimum required temperature.

An experimental assessment, including design, fabrication, and experimental investigation of a greenhouse utilizing PCM for prediction of thermal behaviour using machine learning to forecast greenhouse air temperature, has been developed by Badji et al. [7]. The case study has been performed in a prototype single-tunnel greenhouse with a 1/10 scale located in Ghardaia Province in southern Algeria with the lowest wintertime temperature of 20°C. Hydrated calcium chloride with a melting point of 29°C, heat of fusion of 246 J/g, and specific heat capacity of 200 J/kgK has been selected as a PCM for the greenhouse. The PCM has been encapsulated in 40 aluminium soda black painted cans with a diameter of 6 cm and height of 12 cm filled with 18 kg of PCM (each can weighs 0.45 kg)

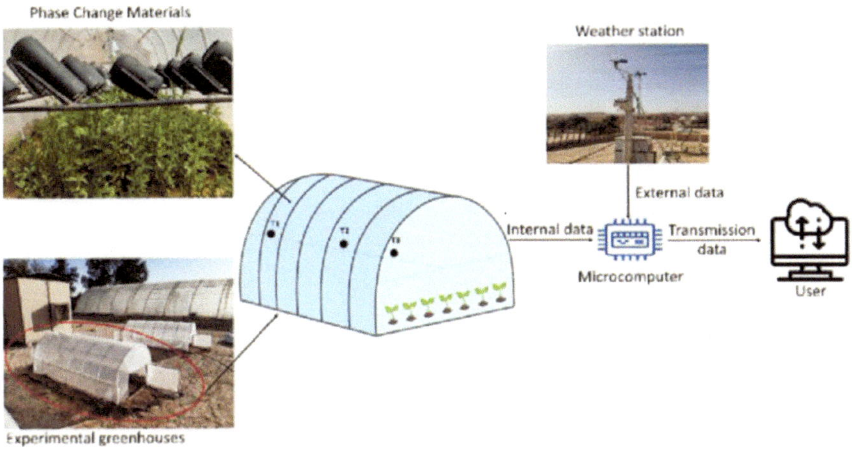

FIGURE 14.8 Experimental setup for the greenhouse [7]. (Permission from Elsevier.com)

with a 30° inclination to the south to absorb the most significant solar radiation as illustrated in Figure 14.8.

The temperature profiles of indoor greenhouse for the month of January with minimum ambient temperature, with and without PCM are illustrated in Figure 14.9 which shows the minimum values of the temperature is at midnight but the internal temperature with PCM is 8°C greater than the same case without PCM.

As can be seen from the temperature profiles in Figure 14.10, the greenhouse without PCM could not keep the indoor temperature over 10°C at night. This shows that the PCMs considerably improved the thermal environment near the ground as depicted in Figure 14.10.

A comparative theoretical and experimental study on the feasibility of utilizing polyethylene 1000 as a PCM in greenhouses has been performed by Mirahmad et al. [8] in which a one-dimensional mathematical model is introduced for a heat

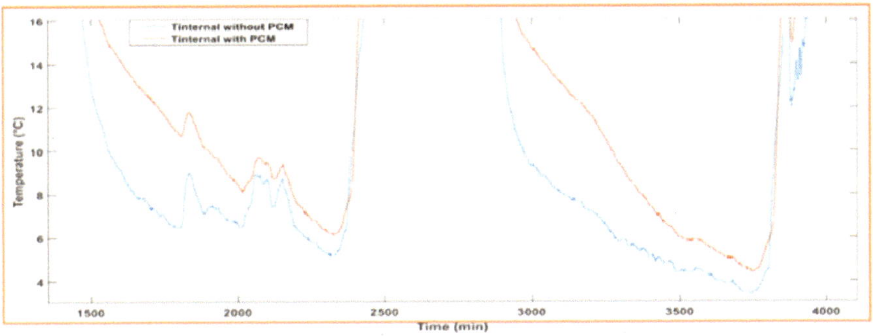

FIGURE 14.9 Internal greenhouse temperatures with and without PCM [7]. (Permission from Elsevier.com.)

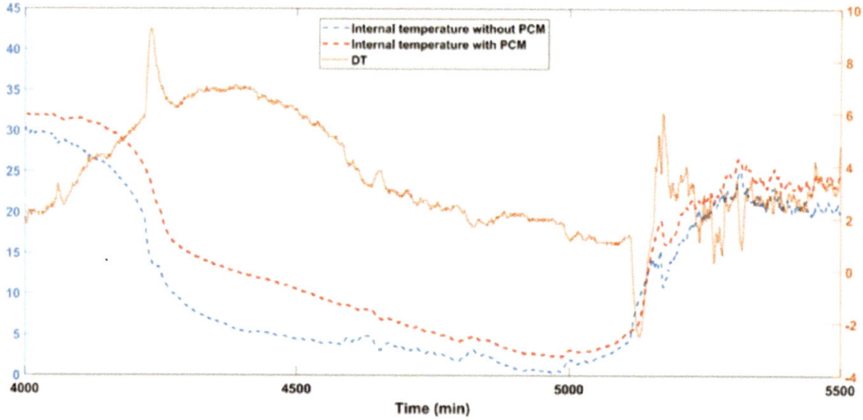

FIGURE 14.10 Indoor temperature profiles with and without PCM [7]. (Permission from Elsevier.com.)

exchange containing flat plats of PCM. Effects of inlet air velocity and temperature on the outlet stream have been investigated using factorial method. Temperature distributions for the cold and hot cycles vs. time at different air flowrates are illustrated in Figure 14.11.a and 14.11.b, respectively. Higher velocities for the inlet air require shorter time for the process termination but at higher velocities this effect becomes weaker.

Experimental analysis of application of 670 kg of calcium chloride hexahydrate based PCM in a solar greenhouse with dimensions of $50 \times 7.6 \times 2\ m^3$ in severe cold regions of Northern China (Fuxin city) has been performed by Meng et al. [9]. The north wall is 2 m high and the roof ridge is 3.4 m high with the slope to 45° to the north. The greenhouse has been divided into four greenhouses of which the east and the west greenhouse length is 15 m; the middle two greenhouses were the comparison greenhouses used in this experiment (east side: PCM

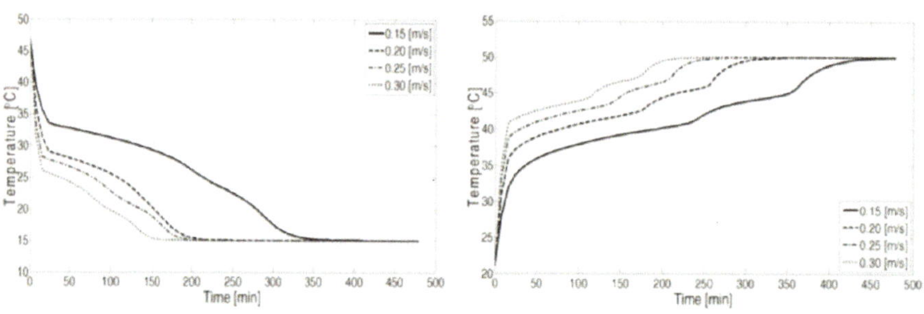

(a) (b)

FIGURE 14.11 Temperature profiles of cold (a) and hot (b) cycles vs. time at different air flowrates [8]. (Open Access Journal)

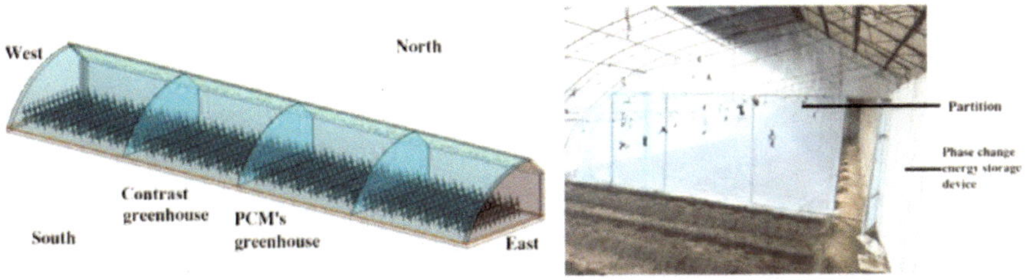

(a) Greenhouse segmentation schematic diagram (b) Greenhouse physical diagram

FIGURE 14.12 Greenhouse schematic diagram (a) and greenhouse physical diagram (b) [9]. (Permission from Elsevier.com.)

greenhouse, west side: contrast greenhouse), both of which are 10 m in length as demonstrated in Figure 14.12.

The PCM has been encapsulated in 150 PVC-U tubes of length 2 m and thickness 2 mm with a diameter of 50 mm which were evenly arranged inside the north wall of the greenhouse with phase change. Schematic diagram of the measuring points in each part of the greenhouse is depicted in Figure 14.13 in which the temperature and humidity loggers were arranged in the middle of the two contrasting greenhouses, 1.7 m above the ground and 3.05 m form the inner surface of the north wall to record the changes in air temperature and humidity with interval of 60 s.

The results of the study have been presented in Figure 14.14 to compare the impact of PCMs on greenhouse during a sunny, cloudy, and snowy days.

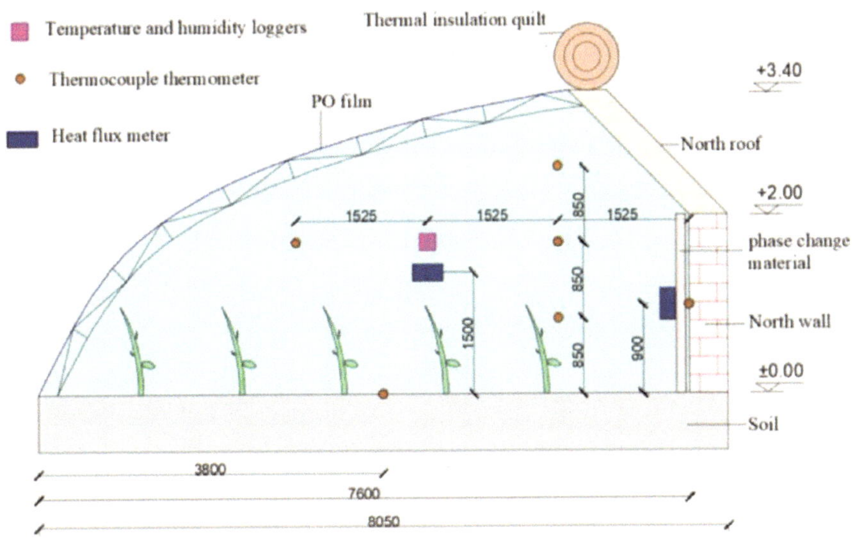

FIGURE 14.13 Schematic of the measuring points [9]. (Permission from Elsevier.com.)

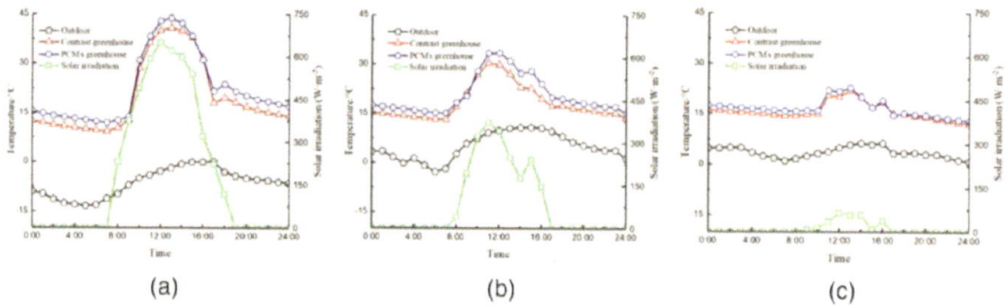

FIGURE 14.14 Temperature profiles of the greenhouses for a sunny (a), cloudly (b), and snowy (c) day [9]. (Permission from Elsevier.com.)

As can be seen from the temperature profile in the figure at night, the maximum temperature differences of the contrast greenhouse were 4.1°C, 2.2°C, and 1.6°C for sunny, cloudy, and snowy conditions, respectively, and the minimum differences were 2.6°C, 1.7°C, and 0.3°C with the average temperature difference of 3.1°C, 1.9°C, and 0.9°C, respectively. It is also revealed from the results that the heat storage of PCM is the best in a sunny day and the lowest in the snowy day.

The internal surface temperature of the north walls for two comparative greenhouses in three different weather conditions are demonstrated in Figure 14.15 which shows the same trends for three different climate conditions and the wall temperatures continued to fall during the night. The maximum differences were 1.9°C and 1.7°C for the contrast and PCM greenhouses, respectively.

Optimization and thermal evaluation of a plastic greenhouse utilizing PCM in southern China have been investigated by Li et al. [10]. A single plastic greenhouse was selected in Chengdu, Sichuan Province in which the average temperature in winter is 5.6°C with extreme temperature of −3.9°C. The greenhouse was built with steel skeletons covered with ethylene-vinyl acetate film as illustrated in Figure 14.16.

An energy balance model, including the heat transfer between the indoor air and the outdoor environment, and the ground was developed to study the influence of the ground heat transfer on the indoor air temperature variations of the plastic greenhouse as shown in Figure 14.17.

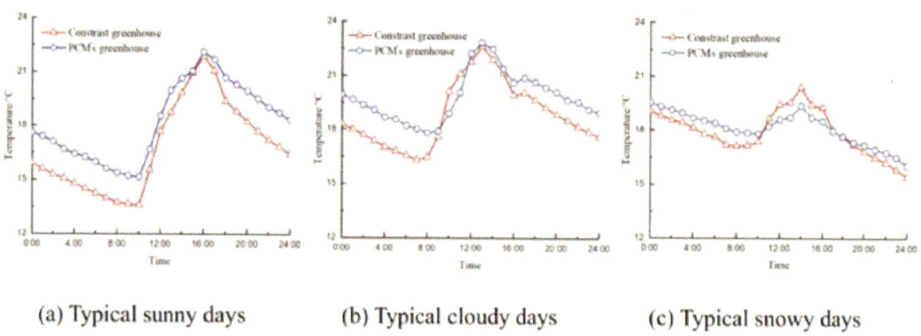

FIGURE 14.15 Temperature profiles of the inner surface of the north wall [9]. (Permission from Elsevier.com.)

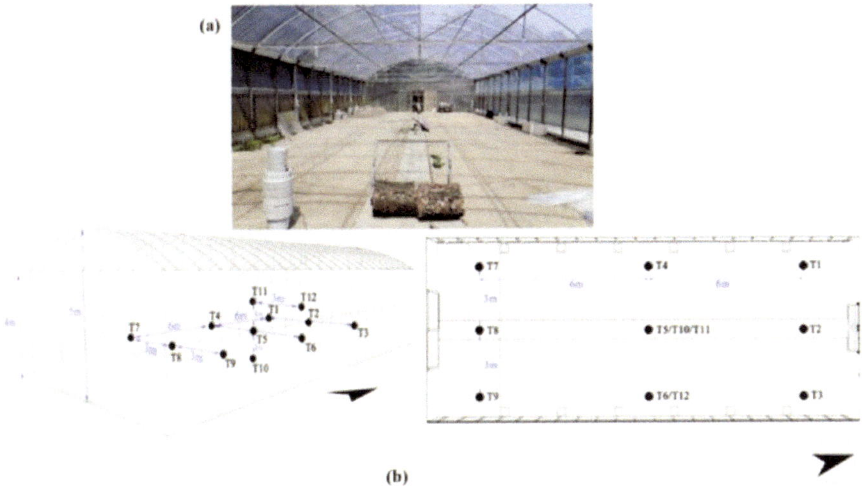

FIGURE 14.16 Schematic diagram of the experimental greenhouse [10]. (Permission from Elsevier.com.)

Temperature profiles of indoor air (T_a) and ground (T_s) inside the greenhouse with and without PCMs are demonstrated in Figure 14.18 which clearly shows that PCMs could reduce the daytime indoor temperature and raise the nighttime indoor temperature, thus reducing indoor temperature variations. The highest achieved ground temperature with PCMs was 13.79°C which was 2.08°C lower the case without PCMs. The variation range of the ground surface temperature with PCMs was 9.36–12.95°C which was higher than the case without PCMs (7.87–11.37°C).

They concluded that application of PCMs in the greenhouse could effectively improve the thermal environment of the plastic greenhouse at night and enhance the thermal storage performance of the ground.

A novel composite PCM consists of $Na_2SO_4.10H_2O\text{-}Al_2O_3$ (NAPCM) incorporated in paints and hollow bricks as illustrated in Figure 14.19 for energy storage

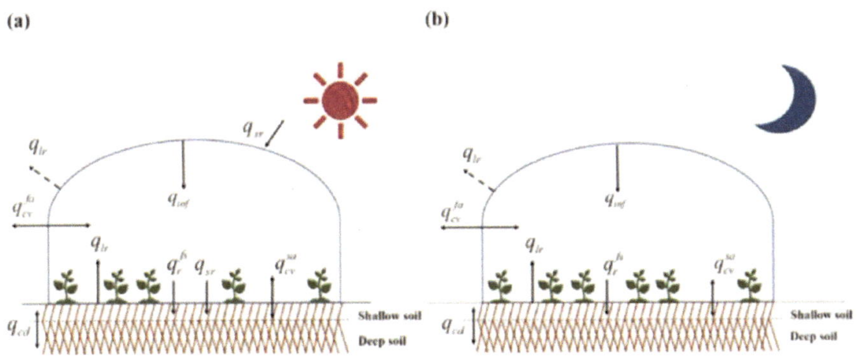

FIGURE 14.17 Heat transfer mechanisms in the greenhouse during the day (a) and night (b) [10]. (Permission from Elsevier.com.)

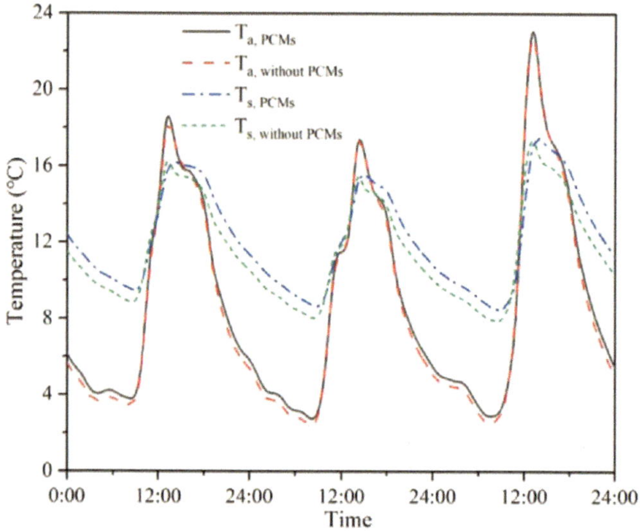

FIGURE 14.18 Comparison of the temperature profiles in the greenhouse [10]. (Permission from Elsevier.com.)

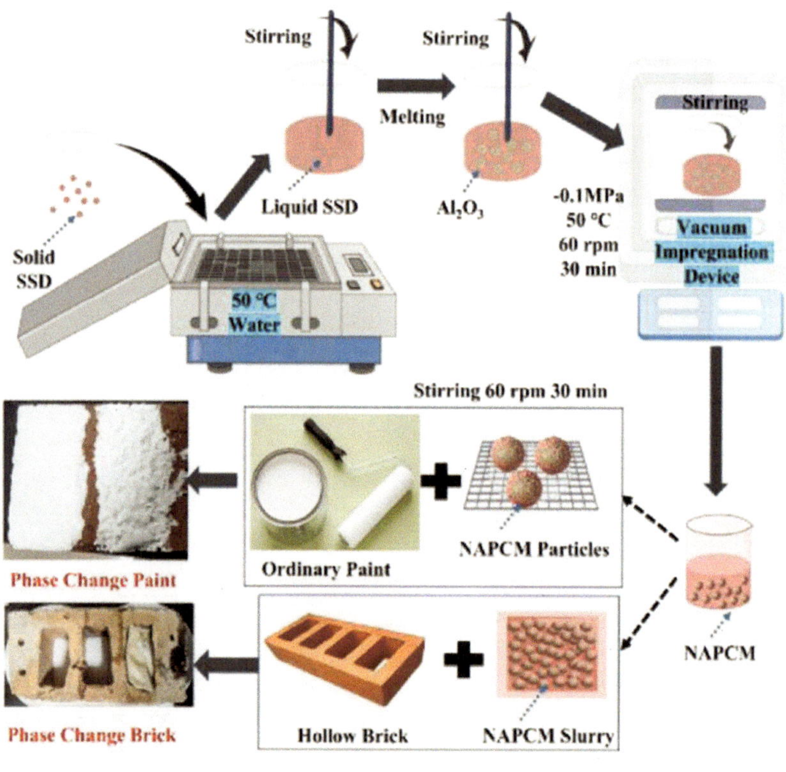

FIGURE 14.19 Preparation steps of composite PCM with bricks and paints [11]. (Permission from Elsevier.com.)

applications of agricultural greenhouses has been developed using vacuum impregnation by Huang et al. [11].

The optimum mass ration of the paint and PCM particles has been taken as 9:1 ratio to ensure the uniformity of of the PCM particles within the paint matrix. Infrared thermal imaging results illustrated circular areas with temperatures much lower than their surroundings indicating the PCM has a good heat absorption during the heating process thereby reducing the rate of temperature change. In comparison to ordinary bricks, the phase change bricks showed superior thermal insulation in the greenhouse. They finally concluded that the phase change greenhouse demonstrated suprior insulation effects, creating a warm environment conductive to plant growth in comparison to the ordinary greenhouse.

Integration of PCM on solar greenhouse back wall and its effects on indoor thermal environment and cucumber production in winter has been developed by Bi et al. [12]. Paraffin-like substance has been used as a PCM mixed with expanded graphite with energy density of 120 J/g to offer substantial heat capacity within a specific temperature range as shown in Figure 14.20.

Their results reveal that the warming effect of the PCM-based bricks positively affects the cucumber growth and the composite wall exhibits an average heat release

FIGURE 14.20 Schematic diagram of the composite PCM and greenhouse wall [12]. (Permission from Elsevier.com.)

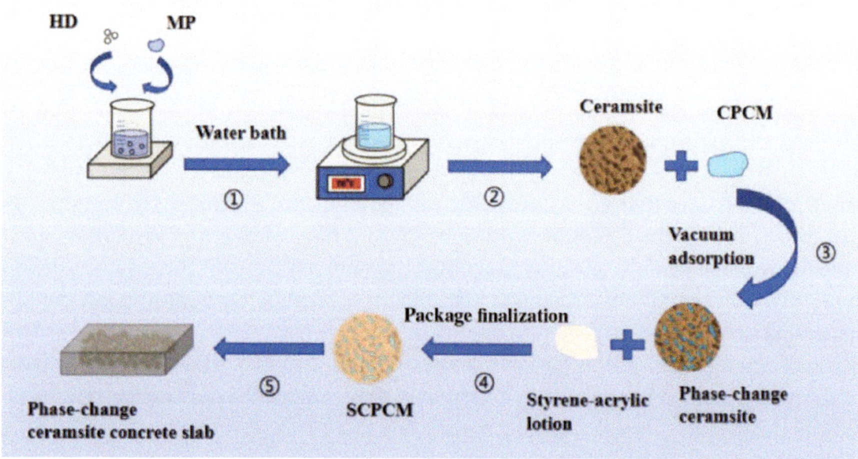

FIGURE 14.21 Preparation steps of composite PCM encapsulation [13]. (Permission from Elsevier.com.)

of 60.47 W/m^2 with average temperature of 21.09°C which shows an increase of 14.03 W/m^2 for the traditional wall of the greenhouse.

Composite PCM made from melt blending of methyl palmitate and hexadecanol as raw materials has been used as a heat storage medium for a solar assisted greenhouse by Yang et al. [13]. The composite PCM then has been encapsulated with syrene-acrylic emulsion to form a shape composite PCM (SCPCM) as depicted in Figure 14.21.

The results show the heating density and melting temperature of the composite PCM to be 221 J/g and 23.48°C, respectively. Concrete slabs made from composite PCM and concrete slabs without PCM have been used for the construction of the greenhouses as shown in Figure 14.22.

Indoor and outdoor temperature profiles of the greenhouses with and without PCM on 27th of January 2024 are illustrated in Figure 14.23. As seen from the temperature profiles, all temperatures increase from 9.00 am to 2.00 pm due to the solar radiation with the minimum and maximum outdoor temperatures of –7.4°C and 4°C, respectively.

FIGURE 14.22 Picture of greenhouses with and without PCM [13]. (Permission from Elsevier.com.)

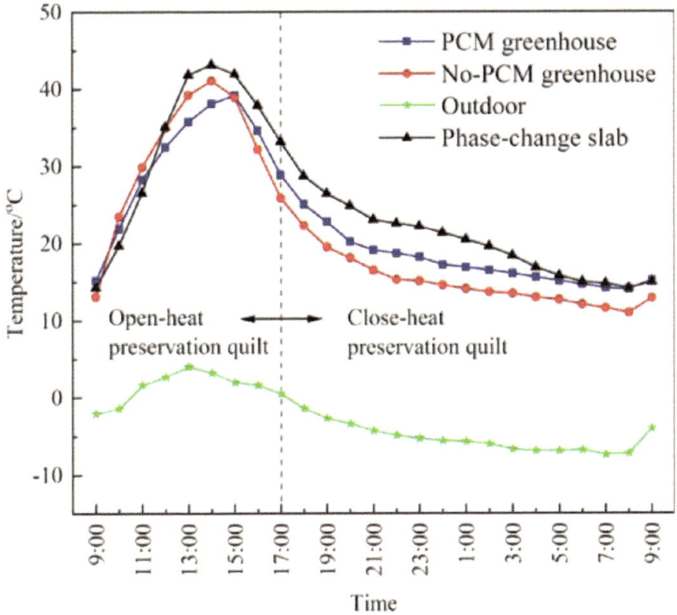

FIGURE 14.23 Temperature profiles of the glasshouse [13]. (Permission from Elsevier.com.)

The research studies clearly prove the significant importance of PCMs on energy conservation of glasshouses especially the solar assisted type for storing the solar energy during the day in the energy storage system and release them to the environment during the nighttime.

Recently published review papers by Nishad et al. [14], Zhu et al. [15], and Nasimi et al. [16] present more information regarding the application of PCMs in agricultural greenhouses.

14.4 SUMMARY

The largest portion of the energy consumption in agricultural sector is for greenhouses. PCMs can provide a feasible and effective solution for decreasing greenhouse energy consumption and environmental pollutions. They can also realize the peak load shifting of solar energy, reduce the indoor temperature fluctuations, heat damage and frost damage prevention, and therefore the building energy demand. Previous studies discussed in this chapter have demonstrated that the application of PCMs in greenhouses can significantly reduce the indoor temperature fluctuations and reduce the energy consumption of the greenhouse building. They can also improve the quality and yield of planted crops and enhance quality of agricultural crops specially when utilized for root zone heating in the wintertime. However, this application is still under development and there are many research gaps and many unsolved problems that are required to be solved for practical applications.

REFERENCES

1. Cao K., Xu H., Zhang R., Xu D., Yan L., Sun Y., Xia L., Zhao J., Zou Z., Bao E. (2019) Renewable and sustainable strategies for improving the thermal environment of Chinese solar greenhouses, Energy Build., 202, p. 109414.
2. Bibbiani C., Fantozzi F., Gargari C., Campiotti C.A., Schettini E., Vox G. (2016) Wood biomass as sustainable energy for greenhouses heating in Italy, Agric. Agric. Sci. Procedia, 8, pp. 637–645.
3. Yan S., Fazilati M.A., Toghraie D., Khalili M., Karimpour A. (2021) Energy cost and efficiency analysis of greenhouse heating system enhancement using PCM: an experimental study, Renew. Energy, 170, pp. 133–140.
4. Llorach-Massana P., Pena J., Rieradevall J., Montero J.I. (2017) Analysis of technical, environmental, and economic potential of PCM for root zone heating in Mediterranean greenhouses, Renew. Energy, 103, pp. 570–581.
5. Mu M., Zhang S., Yang S., Wang Y. (2022) Phase change materials applied in agricultural greenhouses, J. Energy Storage, 49, p. 104100.
6. Ismail M.M., Dincer I., Bicer Y., Saghir M.Z. (2023) Effect of using PCMs on thermal performance of passive solar greenhouses in cold climate, Int. J. Thermofluids, 19, p. 100380.
7. Badji A., Benseddik A., Bensaha H., Boukhelifa A., Bouhoun S., Nettari C., Kherrafi M.A., Lalmi D. (2023), Experimental assessment of a greenhouse with and without PCM thermal storage energy and prediction their thermal behaviour using machine learning.
8. Mirahmad A., Sadrameli S.M., Jamekhorshid A. (2016) A comprehensive study on a latent heat thermal energy storage system and its feasible applications in greenhouses, Iranian J. Chem. Eng., 13(2), pp. 61–73.
9. Meng F., Chu Q. (2023) Heat storage and release performance experiment of externally hung phase change solar greenhouse in severe cold regions of Northern China – taking Fuxin City as an example, J. Energy Storage, 58, p. 106411.
10. Li Y., Liu X., Li W., Jian Y., Arici M., Chen Y., Shen Q. (2022) Thermal environment evaluation of plastic greenhouses in southern China and optimization of PCMs, J. Build. Eng., 57, p. 104882.
11. Huang K., Mi T., Zheng C., Wang L., Yi X. (2024) Development of a novel composite PCM based on paints and brick for energy storage applications in agricultural greenhouses, J. Energy Storage, 89, p. 111734.
12. Bi X., Ma Q., Wang X., Zhang X. (2024) Application of PCM on solar-greenhouse back wall and its effects on indoor thermal environment and cucumber production in winter, J. Build. Eng., 93, p. 109883.
13. Yang A., Xu X., Jia S., Hao W. (2024) Heat storage and release performance of solar greenhouses made of composite PCM comprising methyl palmitate and hexadecanol in cold climate, Therm. Sci. Eng. Progr., 54, p. 102837.
14. Nishad S., Krupa A. (2022) PCMs for thermal energy storage applications in greenhouses: a review, Sustain. Energy Technol. Assess., 52, p. 102241.
15. Zhu J., Zhang X., Hua W., Ji J., Lv X. (2023) Current status and development of research on PCMs in agricultural greenhouses: a review, J. Energy Storage, 66, p. 107104.
16. Nasimi S., Fakhorleslam M., Zarei G., Sadrameli S.M. (2024) Passive energy-efficiency optimization in greenhouses using PCMs: a comprehensive review, J. Energy Storage, 90, p. 111762.

Index

Note: Page numbers in *italics* refer to figures and those in **bold** refer to tables.

D

F

G

H

M

Q

R

S

T

U

V

W

X

Y

Z